Solution Methods
for Integral Equations
Theory and Applications

MATHEMATICAL CONCEPTS AND METHODS IN SCIENCE AND ENGINEERING

Series Editor: **Angelo Miele**
Mechanical Engineering and Mathematical Sciences, Rice University

A Continuation Order Plan is available for this series. A continuation order will bring
delivery of each new volume immediately upon publication. Volumes are billed only upon
actual shipment. For further information please contact the publisher.

Solution Methods
for Integral Equations
Theory and Applications

Edited by
Michael A. Golberg
University of Nevada
Las Vegas, Nevada

PLENUM PRESS · NEW YORK AND LONDON

Library of Congress Cataloging in Publication Data

Main entry under title:

Solution methods for integral equations.
  (Mathematical concepts and methods in science and engineering; v. 18)
  Includes index.
  1. Integral equations–Numerical solutions. I. Golberg, Michael A.
QA431.S59   515'.45   79-17900
ISBN 0-306-40254-8

Acknowledgments

Chapters 2, 4, 6, and 8-13 of this volume have appeared in
essentially similar form in the *Journal of Optimization
Theory and Applications,* Volume 24, No. 1

Printed in the United States of America

Contributors

R. C. Allen, Department of Mathematics, University of New Mexico, Albuquerque, New Mexico

J. M. Bownds, Department of Mathematics, University of Arizona, Tucson, Arizona

J. Casti, Department of Computer Applications and Information Systems, and Department of Quantitative Analysis, New York University, New York, New York

D. Elliott, Department of Applied Mathematics, The University of Tasmania, Hobart, Tasmania, Australia

J. A. Fromme, Department of Mathematics, University of Nevada at Las Vegas, Las Vegas, Nevada

M. A. Golberg, Department of Mathematics, University of Nevada at Las Vegas, Las Vegas, Nevada

A. Goldman, Department of Mathematics, University of Nevada at Las Vegas, Las Vegas, Nevada

H. Kagiwada, President, HFS Associates, Los Angeles, California

R. Kalaba, Departments of Economics and Biomedical Engineering, University of Southern California, Los Angeles, California

L. B. Rall, Mathematics Research Center, University of Wisconsin, Madison, Wisconsin

W. Visscher, Theoretical Division, Los Alamos Scientific Laboratory, Los Alamos, New Mexico

G. Wahba, Department of Statistics, University of Wisconsin, Madison, Wisconsin

G. M. Wing, Department of Mathematics, Southern Methodist University, Dallas, Texas

Preface

In recent years, there has been a growing interest in the formulation of many physical problems in terms of integral equations, and this has fostered a parallel rapid growth of the literature on their numerical solution. This volume, then, is a further contribution to what is becoming a subject of increasing concern to scientists, engineers, and mathematicians.

Although a variety of equations and solution techniques are treated here, two areas of emphasis are to be noted; these are the numerical solution of Cauchy singular integral equations and the use of initial value and related methods, topics which we feel present challenges to both analysts and practitioners alike.

Since this volume appears in the series "Mathematical Concepts and Methods in Science and Engineering," some effort has been made to present papers that deal with subjects of scientific and engineering interest and/or mathematical developments motivated by them. Because we anticipate a varied readership, a survey on "the state of the art" is presented in Chapter 1 which serves as a guide to the field and to the remaining chapters in this book.

Needless to say, a multiple-authored book could not have been assembled without the cooperation of many people, and I would like to take this opportunity to thank all those colleagues, students, and secretaries whose efforts made it possible. In particular, I would like to thank Professor Miele for his encouragement, and my wife Gloria for her patience and understanding.

Las Vegas, Nevada   Michael A. Golberg

Contents

1

A Survey of Numerical Methods for Integral Equations

M. A. GOLBERG[1]

Abstract. A brief survey of the existing literature on numerical methods for integral equations is given. Emphasis is placed on equations in one unknown, although it is noted that many methods can be carried over to multidimensional equations as well. Some discussion is presented on the relation of numerical analysis to applications, and areas are delineated for future research.

1. Introduction

Although the basic theory for many classes of integral equations was well known by the end of the first two decades of this century, the development and analysis of numerical methods have been largely accomplished in the last 20 years (Refs. 1–4). In particular the book literature on this topic is essentially a product of the last 5 years (Refs. 1–6). Due to a variety of technical and mathematical developments we feel that this subject should experience rapid growth and take its place as a major branch of numerical analysis.

As with the related area of numerical methods for differential equations, work seems to have advanced along two distinct lines. First, mathematicians have made extensive analyses of many general algorithms, often using sophisticated functional analytic techniques (Refs. 1, 4, and 6). In contrast, the scientific and engineering communities often depend on special methods for specific technical problems (Refs. 7–8). Mathematical work tends to emphasize convergence proofs and the achievement of high accuracy on model problems (Refs. 2–3), whereas papers on applications tend to ignore such things entirely (Refs. 7–8). Consequently we feel that significant opportunities exist for the applied mathematician and scientist to

---

[1] Professor of Mathematics, University of Nevada at Las Vegas, Las Vegas, Nevada.

cooperate on the numerical solution of integral equations, and one of the purposes of this book is to indicate some of the possibilities for this interaction. As Noble's bibliography shows (Ref. 9), integral equations occur in almost all branches of science and technology, and obtaining solutions, whether analytical or numerical, has been a major occupation of applied mathematicians for most of this century.

In this volume we shall present a sampling of some of the current research being done on the numerical solution of integral equations. Some effort has been made to include topics and methods whose development has been closely related to technical problems in science and engineering. However, because of the scope and complexity of the field, we feel it useful to present a brief overview of the field, both as a guide to the chapters in this book and to the literature in general.

2. Classification of Integral Equations

In some respects the theory of integral equations closely resembles that for partial differential equations, in the sense that there are a large number of types (Refs. 4 and 9), each possessing its own theory and therefore requiring its own form of numerical analysis. For this reason the problem of solving an integral equation must be considered as a large number of distinct topics: A general-purpose algorithm analogous to those for initial value problems for ordinary differential equations cannot be produced.

To begin, we shall give a classification of some of the major types of integral equations that have appeared in the literature as a prelude to a discussion of numerical techniques available for their solution. For the sake of clarity some of the ensuing "zoology" is summarized in diagrammatic form on the following pages.

The first major division is that between one-dimensional and multi-dimensional equations. Since most mathematical work is usually done in the context of equations in one variable (Refs. 1–6), our attention is restricted to them. Much of the terminology, however, is also applicable to the multi-variate case, the major departure apparently being in the nomenclature for Volterra equations (Ref. 10).

Following this it is typical to subdivide equations as linear or nonlinear, linear equations being those in which the unknown function appears only to the first power and nonlinear ones encompassing the remainder. Figure 1 and Table 1 illustrate the variety of possibilities for linear equations, and Fig. 2 and Table 2 are similar charts for nonlinear ones. For the most part the numerical analysis of integral equations has dwelled on linear equations in a single variable, although in principle many of the techniques are applicable to multidimensional and nonlinear equations as well.

Here again we would like to point out some of the disparities between the mathematical and technical literature. Although many practical problems such as airplane design (Refs. 7–8) require the solution of multidimensional equations, mathematical work has tended to focus on one-dimensional ones. Similarly, with respect to type, mathematicians have given detailed treatments of Fredholm and Volterra equations, whereas problems in radiative transfer, aerodynamics, and hydrodynamics often lead to Cauchy singular (or worse) equations (Refs. 5, 11–15), which until recently, in the west; have all but been ignored numerically (Refs. 5, 16–17). Notwithstanding, the engineering community has been routinely solving such equations for many years (Refs. 12, 14, 18).

3. Numerical Methods

Based on our classification in the previous section we shall present a survey of existing methods for numerically solving integral equations. Since many specific areas are considered in depth in later chapters, we shall concentrate mainly on topics not treated there.

3.1. Fredholm Equations of the Second Kind.
Because of their importance, we begin with a discussion of Fredholm equations of the second kind. Until recently most of the mathematical work seems to have been devoted to them. This seems reasonable since such equations occur frequently directly (Refs. 2–4) and because many other types of equations such as Cauchy singular equations and dual integral equations can often be shown to be equivalent to them and so can be solved by methods developed for this particular case.

In general, numerical analysis seems to break down into two distinct areas: solution of the inhomogeneous equation $g(t) \neq 0$ and solution of the eigenvalue problem $g(t) = 0$. (See Table 1.) As for the mechanics of deriving most algorithms, one can usually begin with an inhomogeneous equation and set $g(t) = 0$ to obtain an appropriate method for solving the eigenvalue problem. Mathematically, the implementation and analysis of algorithms for the eigenvalue problem are usually more complex than for the case $g(t) \neq 0$. However, as a practical matter, it appears that eigenvalue problems occur much less frequently, and so we shall devote our attention to the case $g(t) \neq 0$.

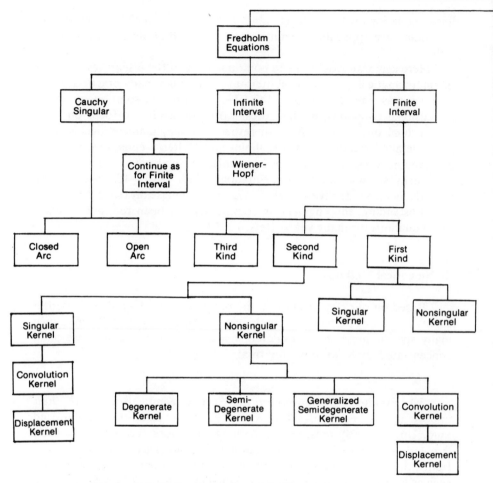

Fig. 1.   Classification of one-dimensional linear integral equations.

A reading of the literature (Refs. (1–6, 9) indicates that techniques for solving Fredholm equations of the second kind can be classified into five broad categories: (1) analytical and semianalytical methods, (2) kernel approximation methods, (3) projection methods, (4) quadrature methods, and (5) Volterra and initial value methods.

3.1.1. Analytical and Semianalytical Methods.   Here we include any methods that rely in whole or in part on a specific analytic representation of the solution. Examples include the use of iteration (Refs. 3, 19), numerical inversion of transforms (Refs. 11, 20), and the numerical implementation of the Wiener–Hopf factorization given by Stenger in Ref. 21.

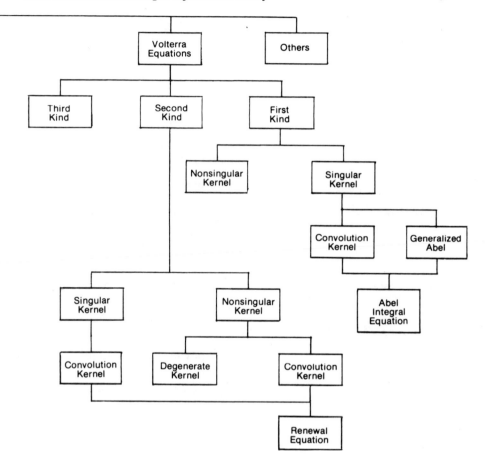

Fig. 1 (continued).

3.1.2. Kernel Approximation Methods. Here, and in the remainder of this section, we assume that we wish to solve

$$u(t) = g(t) + \int_a^b K(t, s)u(s)\, ds, \tag{1}$$

where $-\infty < a < b < \infty$ and Eq. (1) is assumed to have a unique solution.

Kernel approximation methods, as the name implies, work by finding a sequence of kernels $\{K_n(t, s)\}_{n=1}^{\infty}$ such that $K_n(t, s)$ converges to $K(t, s)$ in a suitable topology. Approximations $\{u_n(t)\}_{n=1}^{\infty}$ to $u(t)$ are generated by

Table 1. Forms of linear integral equations.

Name of equation	Form
Fredholm—second kind	$u(t) = g(t) + \int_a^b K(t, s) u(s)\, ds$
Fredholm—first kind	$\int_a^b K(t, s) u(s)\, ds = g(t)$
Fredholm—third kind	$a(t) u(t) = g(t) + \int_a^b K(t, s) u(s)\, ds$ $a(t) = 0$ for at least one $t \in [a, b]$
Wiener–Hopf	$u(t) = g(t) + \int_0^\infty K(t - s) u(s)\, ds$
Volterra—second kind	$u(t) = g(t) + \int_a^t K(t, s) u(s)\, ds$
Volterra—first kind	$\int_a^t K(t, s) u(s)\, ds = g(t)$
Renewal equation	$u(t) = g(t) + \int_a^t K(t - s) u(s)\, ds$
Abel equation	$\int_a^t [u(s)/(t - s)^\alpha]\, ds = g(t),\ 0 < \alpha < 1$
Cauchy singular	$a(t) u(t) + b(t) \int_\Gamma [u(s)/(s - t)]\, ds$ $+ \int_\Gamma K(t, s) u(s)\, ds = g(t)$ Γ = open or closed arc in $\mathcal{R}^2$

solving the equations

$$u_n(t) = g(t) + \int_a^b K_n(t, s) u_n(s)\, ds, \qquad n \geq 1. \qquad (2)$$

Under suitable conditions, to be discussed later, one can easily show that $u_n(t)$ converges in some norm to $u(t)$.

For practical purposes it is necessary to define $\{K_n(t, s)\}_{n=1}^\infty$ so that Eq. (2) can be reduced to a tractable arithmetic problem. Classically only degenerate kernel approximations have been considered (Refs. 3–4). That is, $K_n(t, s)$ is taken to be of the form

$$K_n(t, s) = \sum_{k=1}^n a_{n,k}(t) b_{n,k}(s). \qquad (3)$$

In this case

$$u_n(t) = g(t) + \sum_{k=1}^{n} c_{n,k} a_{n,k}(t), \qquad (4)$$

where

$$c_{n,k} = \int_a^b b_{n,k}(s) u_n(s) \, ds, \qquad (5)$$

and $\{c_{n,k}\}_{k=1}^{n}$ are obtained by solving the linear equations

$$c_{n,k} = g_{n,k} + \sum_{j=1}^{n} \alpha_{n,kj} c_{n,j}, \qquad k = 1, 2, \ldots, n, \qquad (6)$$

with

$$g_{n,k} = \int_a^b g(s) b_{n,k}(s) \, ds, \qquad k = 1, 2, \ldots, n, \qquad (7)$$

and

$$\alpha_{n,kj} = \int_a^b b_{n,k}(s) a_{n,j}(s) \, ds, \qquad k, j = 1, 2, \ldots, n. \qquad (8)$$

More recently various writers (Refs. 2, 4, 22–23) have considered semidegenerate kernel approximations which are defined by

$$K_{m,p}(t, s) = \begin{cases} \sum_{k=1}^{m} a_{n,k}(t) b_{n,k}(s), & a \le s < t, \\ \\ \sum_{k=1}^{p} c_{n,k}(t) d_{n,k}(s), & t \le s \le b. \end{cases} \qquad (9)$$

Such kernels are useful in modeling various mild forms of discontinuity along the line $t = s$. Again $u_n(t)$ may be obtained by solving a finite system of linear equations (Refs. 22–24). A general theory for these kernels is developed in Chapters 11 and 13.

Although degenerate kernel methods play an important role in the theory of Fredholm equations and give a natural type of approximation method, their use in practical problems seems to have taken a back seat to other methods such as collocation and quadrature (Refs. 3–4). In this regard we should point out that many other methods require one to solve a sequence of approximating equations of the form

$$u_n(t) = g_n(t) + \int_a^b K_n(t, s) u_n(s) \, ds, \qquad (10)$$

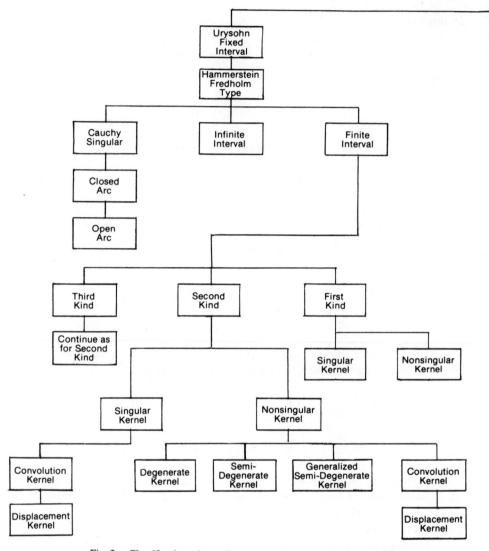

Fig. 2.   Classification of one-dimensional nonlinear integral equations.

where $K_n(t, s)$ is a degenerate kernel. The concept of generating "optimal" degenerate kernels has recently been taken up by Sloan *et al.* in Ref. 25 and will be discussed later in its relation to projection methods.

3.1.3. Projection Methods.   Projection methods include such techniques as collocation, the method of moments, Galerkin's method and least-squares procedures. The most economical way of describing them is via a functional analytic formalism.

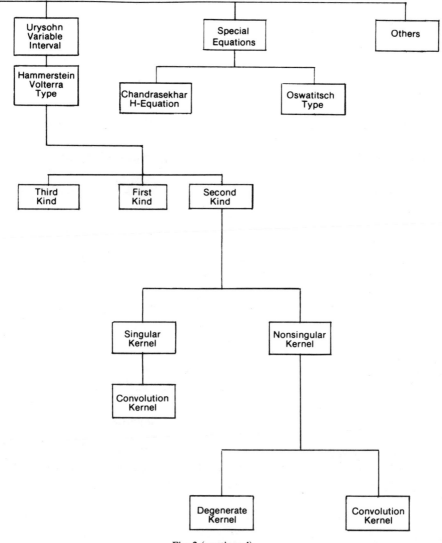

Fig. 2 (continued).

To do this let X be a Banach space and consider the operator $K : X \to X$ defined by

$$Ku = \int_a^b K(t, s)u(s)\, ds. \tag{11}$$

If $K(t, s)$ is continuous or more generally has certain types of integrable singularities, then K will induce a compact operator on X (Refs. 1, 3, 4). The integral equation (1) can then be written in operator form as

$$u = g + Ku. \tag{12}$$

Table 2. Forms of nonlinear integral equations.

Name of equation	Form
Urysohn—second kind	$u(t) = g(t) + \int_a^b F(t, s, u(s))\, ds$
Hammerstein	$u(t) = g(t) + \int_a^b K(t, s)f(s, u(s))\, ds$
Urysohn—first kind	$\int_a^b F(t, s, u(s))\, ds = g(t)$
Urysohn–Volterra	$u(t) = g(t) + \int_a^t F(t, s, u(s))\, ds$
Hammerstein–Volterra—second kind	$u(t) = g(t) + \int_a^t K(t, s)f(s, u(s))\, ds$
Hammerstein–Volterra—first kind	$\int_a^t K(t, s)f(s, u(s))\, ds = g(t)$
Cauchy singular	$a(t)u(t) + b(t)\int_\Gamma [u(s)/(s-t)]\, ds$
	$+ \int_\Gamma F(t, s, u(s))\, ds = g(t)$
Chandrasekhar—H equation	$1 + \lambda \int_a^b [tu(t)u(s)/(t+s)]\, ds = g(t)$

Now let $\{X_n\}_{n=1}^\infty$ be a sequence of finite-dimensional subspaces of X such that $\bigcup_{n=1}^\infty X_n$ is dense in X. Let Y_n be another sequence of finite-dimensional subspaces of X and let $P_n: X \to Y_n$ define a sequence of projection operators; i.e., $P_n^2 = P_n$. To solve (1) we attempt to approximate u by a sequence $\{u_n\}_{n=1}^\infty$ such that $u_n \in X_n$. Let

$$R_n(u_n) = u_n - Ku_n - g \qquad (13)$$

denote the residual. If $u_n = u$, then $R_n(u_n) = 0$. However, in general $u_n \neq u$, and one tries to select u_n so that $R_n(u_n)$ is small. This is accomplished by making the projection of $R_n(u_n)$ onto Y_n equal to zero. That is, u_n is picked to solve

$$P_n R_n(u_n) = 0, \qquad (14)$$

or equivalently

$$P_n u_n - P_n K u_n = P_n g. \qquad (15)$$

The most common implementations of this technique have $X_n = Y_n$, $n = 1, 2, \ldots$, so that $P_n u_n = u_n$ and u_n solves the equation of the second kind

$$u_n - P_n K u_n = P_n g. \tag{16}$$

To do numerical calculations we take $\{\phi_k\}_{k=1}^n$ as a basis for X_n and write

$$u_n = \sum_{k=1}^{n} a_k \phi_k. \tag{17}$$

letting $\{l_k\}_{k=1}^n$ be a basis for the dual space X_n^* of X_n, the coefficients $\{a_k\}_{k=1}^n$ can be obtained by solving the set of equations

$$\sum_{k=1}^{n} l_j(\phi_k) a_k - \sum_{k=1}^{n} l_j(P_n K \phi_k) a_k = l_j(g_n), \qquad j = 1, 2, \ldots, n. \tag{18}$$

To illustrate these ideas in a more concrete fashion, three standard approximation methods are examined via this framework.

3.1.4. Polynomial Collocation. For polynomial collocation the customary setup is to take $X = (C[a, b], \|\cdot\|_\infty)$ where $C[a, b]$ is the space of continuous functions on $[a, b]$ and $\|\cdot\|_\infty$ is the sup norm. X_n is the subspace of polynomials of degree $n - 1$. Here $Y_n = X_n$, and P_n is the operator which maps a function u in X onto the polynomial which interpolates to it on the set of points $\{t_k\}_{k=1}^n$. That is,

$$P_n(u(t)) = \sum_{k=1}^{n} u(t_k) l_k(t), \tag{19}$$

where $\{l_k(t)\}_{k=1}^n$ are the fundamental polynomials of Lagrange interpolation. This gives

$$(P_n K u)(t) = \sum_{k=1}^{n} l_k(t) \int_a^b K(t_k, s) u(s) \, ds. \tag{20}$$

Let $\{p_k(t)\}_{k=1}^n$ be a basis for X_n; then the projection equations (18) become

$$u_n(t) - \sum_{k=1}^{n} l_k(t) \int_a^b K(t_k, s) u_n(s) \, ds = \sum_{k=1}^{n} l_k(t) g(t_k). \tag{21}$$

By evaluating both sides of (21) at $t = t_j$, $j = 1, 2, \ldots, n$, and using the fact that $l_k(t_j) = \delta_{kj}$, we obtain

$$u_n(t_j) - \int_a^b K(t_j, s) u_n(s) \, ds = g(t_j), \qquad j = 1, 2, \ldots, n. \tag{22}$$

Writing

$$u_n(t) = \sum_{k=1}^{n} a_k p_k(t) \tag{23}$$

and substituting into Eq. (22), we obtain the usual collocation equations (Refs. 3, 26),

$$\sum_{k=1}^{n} a_k p_k(t_j) - \sum_{k=1}^{n} a_k \int_a^b K(t_j, s) p_k(s) \, ds = g(t_j), \qquad j = 1, 2, \ldots, n. \quad (24)$$

Other collocation methods, using, for example, splines to represent $u(t)$, can also be cast into the above framework. Further details can be found in Refs. 3–4.

With respect to our previous remarks note that (22) can be written in the form

$$u_n(t) - \int_a^b K_n(t, s) u_n(s) \, ds = g_n(t), \qquad (25)$$

where $K_n(t, s)$ is the degenerate kernel given by

$$K_n(t, s) = \sum_{k=1}^{n} l_k(t) m_k(s), \qquad (26)$$

and

$$m_k(s) = K(t_k, s), \qquad k = 1, 2, \ldots, n. \quad (27)$$

3.1.5. Galerkin's Method. For Galerkin's method we let $X = (L_2[a, b], \|\cdot\|_2)$, where $L_2[a, b]$ is the space of Lebesgue square integrable functions on $[a, b]$ and

$$\|f\|_2 = \left(\int_a^b |f|^2 \, dt \right)^{1/2}.$$

Let $\{\phi_k\}_{k=1}^{\infty}$ be a basis for X. Pick $X_n = \text{Span}(\{\phi_k\}_{k=1}^{n})$, $Y_n = X_n$, and P_n as orthogonal projection onto X_n. Specifically, if $\{\phi_k\}_{k=1}^{\infty}$ are orthonormal, then

$$P_n u = \sum_{k=1}^{n} \langle u, \phi_k \rangle \phi_k,$$

where

$$\langle u, \phi_k \rangle = \int_a^b u(t) \bar{\phi}_k(t) \, dt.$$

By letting

$$u_n(t) = \sum_{k=1}^{n} a_k \phi_k(t),$$

analysis similar to that above shows that $\{a_k\}_{k=1}^n$ can be obtained from

$$\sum_{k=1}^n a_k\langle\phi_k, \phi_j\rangle - \sum_{k=1}^n a_k\langle\phi_k, K\phi_j\rangle = \langle g, \phi_j\rangle, \qquad j = 1, 2, \ldots, n. \qquad (28)$$

In the special case where $\{\phi_k\}_{k=1}^n$ are orthonormal Eq. (28) simplifies to

$$a_j - \sum_{k=1}^n a_k\langle\phi_j, K\phi_k\rangle = \langle g, \phi_j\rangle, \qquad j = 1, 2, \ldots, n. \qquad (29)$$

We observe in this case that $u_n(t)$ satisfies

$$u_n(t) - \int_a^b K_n(t, s)u_n(s)\,ds = g_n(t), \qquad (30)$$

where $K_n(t, s)$ is the degenerate kernel given by

$$K_n(t, s) = \sum_{k=1}^n \phi_k(t)\psi_k(s),$$

with

$$\psi_k(s) = \int_a^b K(t, s)\bar{\phi}_n(t)\,dt.$$

3.1.6. Least Squares. For simplicity we assume that g, K, and thus u are real and write

$$T = I - K.$$

Again we take $X = L_2[a, b]$ and consider approximating u by

$$u_n(t) = \sum_{k=1}^n a_k\phi_k(t).$$

By forming the residual $R_n(u_n)$, $\{a_k\}_{k=1}^n$ are chosen by minimizing $\|R_n(u_n)\|_2^2$. Carrying out the usual operations leads to the equations

$$\sum_{j=1}^n a_j\langle T\phi_k, T\phi_j\rangle = \langle T\phi_k, g\rangle, \qquad k = 1, 2, \ldots, n. \qquad (31)$$

To see that this procedure may be regarded as a projection method take $Y_n = \mathrm{Span}(\{T\phi_k\}_{k=1}^n)$ and let P_n be orthogonal projection onto Y_n. From Eq. (31) it is easily seen that these equations are equivalent to

$$\langle Tu_n - g, T\phi_k\rangle = 0, \qquad k = 1, 2, \ldots, n. \qquad (32)$$

That is, the residual is orthogonal to Y_n, so that Eq. (32) can be written as

$$P_n(Tu_n - g) = 0.$$

Another useful way of regarding least squares as a projection method is to write Eq. (31) as

$$\sum_{j=1}^{n} a_j \langle T^*T\phi_j, \phi_k \rangle = \langle T^*g, \phi_k \rangle, \qquad k = 1, 2, \ldots, n,$$

where T^* is the adjoint of T. In this form one sees that least squares is equivalent to using Galerkin's method on the integral equation

$$T^*Tu = T^*g.$$

Although other projection methods have been discussed in the literature, the above three appear to be the most popular, and where such methods are used in practice collocation appears to be the most widespread (Refs. 7–8).

In fact there are interesting relations among all three and relations between projection and degenerate kernel methods. In this regard we make the following comments.

Collocation is generally the most efficient of the projection methods since the arithmetic needed to form the matrices

$$A = [T\phi_k(t_j)]_{k,j=1}^{n},$$

$$B = [\langle T\phi_k, \phi_j \rangle]_{k,j=1}^{n},$$

$$C = [\langle T\phi_k, T\phi_j \rangle]_{k,j=1}^{n}$$

is minimized for the collocation matrix A, where only single rather than multiple integrals need be calculated. In this light we note that if the inner products $\langle T\phi_k, \phi_j \rangle$ and $\langle g, \phi_j \rangle$ are obtained by quadrature as

$$\langle T\phi_k, \phi_j \rangle \simeq \sum_{p=1}^{n} w_p T\phi_k(t_p)\overline{\phi_j(t_p)}$$

and

$$\langle g, \phi_j \rangle \simeq \sum_{p=1}^{n} w_p g(t_p)\overline{\phi_j(t_p)}$$

and if $\det[\phi_j(t_p)] \neq 0$, then Galerkin's method is numerically equivalent to the collocation method using $\{\phi_k\}_{k=1}^{n}$ as a basis and $\{t_p\}_{p=1}^{n}$ as collocation points. A similar statement is true for the method of least squares provided that A is nonsingular (Ref. 12). Thus, under approximate conditions all three of the standard projection methods are numerically equivalent.

3.1.7. Quadrature Methods. Perhaps the most natural way of approximating the solution of (1) is to replace the integral by a numerical quadrature. Thus

$$u(t) \simeq \sum_{k=1}^{n} w_k K(t, s_k) u(s_k) + g(t), \tag{33a}$$

where $\{w_k\}_{k=1}^{n}$ and $\{s_k\}_{k=1}^{n}$ are the weights and nodes of a quadrature rule Q_n. If $u_n(t)$ is defined as the solution to

$$u_n(t) = g(t) + \sum_{k=1}^{n} w_k K(t, s_k) u_n(s_k), \tag{33b}$$

then evaluating $u_n(t)$ at $t = t_j$, $j = 1, 2, \ldots, n$, gives the following set of algebraic equations for $\{u_n(t_j)\}_{j=1}^{n}$:

$$u_n(t_j) = g(t_j) + \sum_{k=1}^{n} w_k K(t_j, s_k) u_n(s_k), \qquad j = 1, 2, \ldots, n. \tag{34}$$

If Eq. (34) has a unique solution for large enough n, then Eq. (33b) provides a natural interpolation formula for obtaining $u_n(t)$ at all $t \in [a, b]$. Although one can start from Eq. (34) and use some other form of interpolation, the relationship between Eqs. (33b) and (34) is useful, both computationally and theoretically. This observation was apparently first made by Nyström (Ref. 27) and is at the core of the modern analyses given by Anselone, Moore, and Atkinson (Refs. 1, 3).

The Nyström method is generally efficient and flexible provided that the kernel $K(t, s)$ is well behaved. For kernels with singularities special modifications such as singularity subtraction and product integration methods generally need to be used.

Singularity subtraction is helpful if the discontinuities are confined to the line $t = s$. In this case we write

$$\int_a^b K(t, s) u(s) \, ds = \int_a^b K(t, s)[u(s) - u(t)] \, ds + u(t) \int_a^b K(t, s) \, ds. \tag{35}$$

If

$$\lim_{t \to s} K(t, s)[u(s) - u(t)] = 0,$$

then the integral in (35) can be replaced by a quadrature. Proceeding as before leads to the system of linear equations

$$\hat{u}(t_j) = g(t_j) + \sum_{k=1}^{n}{}' w_k K(t_j, t_k)[\hat{u}(t_k) - \hat{u}(t_j)]$$

$$+ \hat{u}(t_j) \int_a^b K(t_j, s) \, ds, \qquad j = 1, 2, \ldots, n, \tag{36}$$

that must be solved to approximate $u(t)$. (Here $\sum'$ indicates that the term for $k = j$ is omitted.)

To illustrate product integration, assume that $K(t, s)$ can be factored as

$$K(t, s) = L(t, s)M(t, s),$$

where $M(t, s)$ is continuous and $L(t, s)$ carries the singularities of $K(t, s)$. Let

$$v_t(s) = M(t, s)u(s), \tag{37}$$

and write the integral in Eq. (1) as

$$\int_a^b L(t, s)v_t(s) \, ds. \tag{38}$$

One now proceeds to approximate this integral along the lines of a general Gaussian quadrature. For example, if we let $L_n(v_t(s))$ denote the Lagrange interpolation polynomial of $v_t(s)$ using $\{t_k\}_{k=1}^n$ as collocation points, then

$$\int_a^b L(t, s)v_t(s) \, ds \simeq \int_a^b L(t, s)L_n(v_t(s)) \, ds$$

$$= \sum_{k=1}^n \int_a^b L(t, s)M(t, s_k)u(s_k)l_k(s) \, ds$$

$$= \sum_{k=1}^n M(t, s_k)u(s_k) \int_a^b L(t, s)l_k(s) \, ds$$

$$= \sum_{k=1}^n w_k(t)M(t, s_k)u(s_k), \tag{39}$$

where

$$w_k(t) = \int_a^b L(t, s)l_k(s) \, ds.$$

Using Eq. (39) in the right-hand side of Eq. (1) gives the following linear equations for approximating $u(t)$:

$$u_n(t_j) = g(t_j) + \sum_{k=1}^n w(t_j)M(t_j, t_k)u_n(t_k), \qquad j = 1, 2, \ldots, n. \tag{40}$$

Again we have the natural Nyström-type interpolation

$$u_n(t) = g(t) + \sum_{k=1}^n w_k(t)M(t, s_k)u_n(s_k). \tag{41}$$

The effectiveness of the method depends on one's ability to obtain a good splitting so that accurate formulas for the quadrature weights may be obtained.

More generally, one can consider kernel decompositions of the form (Ref. 3)

$$K(t, s) = \sum_{k=1}^{n} L_k(t, s) M_k(t, s) \tag{42}$$

and proceed as above for each individual term in Eq. (42).

Last we note that it is not difficult to show that collocation may be considered as a form of product quadrature (Ref. 3), a topic which is discussed further in Section 3.2.3.

3.1.8. Volterra and Initial-Value Methods. In the methods discussed above the general goal was to replace the integral equation by a sequence of algebraic equations which could then be solved numerically to obtain an approximation to $u(t)$. A somewhat different class of procedures is based on trying to show that the solution $u(t)$ satisfies other functional equations and then solving these instead. Procedures of this kind again seem to fall into four distinct categories:

1. Volterra methods (Refs. 24, 28),
2. Interval imbedding (Ref. 2),
3. Parameter imbedding (Ref. 2),
4. Boundary-value methods (Refs. 22–23).

For Volterra methods the kernel must have the form

$$K(t, s) = \begin{cases} \sum_{k=1}^{n} a_k(t) b_k(s), & a \leq s < t, \\ \hat{K}(t, s), & t \leq s \leq b. \end{cases}$$

As shown originally by Aalto (Ref. 24), such equations can be solved in terms of related Volterra equations and thus are amenable to solution by iteration or step-by-step methods. More details of this method are presented in Chapter 11.

Interval imbedding appears to have its origins in the work of Chandrasekhar and Sobolev on the equations of radiative transfer (Ref. 29). Here one considers the solution to be a function of the upper limit of integration b. Under appropriate conditions on the kernel, the solution $u(t, b)$ and the resolvent kernel are shown to satisfy a certain set of integrodifferential equations which are then discretized to obtain ordinary differential equations. These may then be used to calculate an approximation to the solution to $u(t)$. Further elaboration of this approach is given in Chapter 8.

Parameter imbedding occurred as an outgrowth of interval imbedding (Ref. 2) but can be shown to be related to the classical integrodifferential equation of Hilbert for the resolvent kernel $R(t, s, \lambda)$ of $\lambda K(t, s)$. Briefly,

$$u(t) = g(t) + \lambda \int_a^b R(t, s, \lambda) g(s) \, ds, \qquad (43)$$

where

$$R(t, s, \lambda) = K(t, s) + \lambda \int_a^b K(t, \tau) R(\tau, s, \lambda) \, d\tau. \qquad (44)$$

In general

$$\frac{\partial R(t, s, \lambda)}{\partial \lambda} = \int_a^b R(t, \tau, \lambda) R(\tau, s, \lambda) \, d\tau, \qquad (45)$$

and

$$R(t, s, 0) = K(t, s). \qquad (46)$$

Also

$$\frac{\partial u(t, \lambda)}{\partial \lambda} = \int_a^b R(t, s, \lambda) u(s) \, ds, \qquad (47)$$

with

$$u(t, 0) = g(t). \qquad (48)$$

When the integrals in Eqs. (45) and (47) are replaced by quadratures one obtains an initial value problem for calculating $u(t, \lambda)$. A characteristic of such methods is that they can be considered as initial value implementations of the Nyström method (Refs. 30–31). Further details may be found in Refs. 2 and 30–31.

Boundary-value methods are related both to Volterra methods and interval imbedding and require that the kernel be either degenerate, semi-degenerate, or generalized semidegenerate. Here the desire is to obtain an exact representation of the solution in terms of the solution to a related two-point boundary-value problem for a system of ordinary differential equations (Refs. 22–23). The technique is an extension of the work of Bownds and Wood on Volterra equations (Refs. 32–33), which in turn was motivated by a formula given by Goursat for the solution of a Volterra equation with a degenerate kernel (Ref. 34). This work is discussed in greater detail in Chapters 9 and 10.

Last, we remark that relations have been found by Shumitzky and Stenger (Refs. 35–36) between imbedding methods and the classical method of Wiener and Hopf for convolution equations (Ref. 11).

An interesting offshoot of some of this work has been the recent development of "fast" algorithms for the determination of Kalman filters and the solution of optimal control problems. Such procedures are presently called Chandrasekhar algorithms as the quantities of importance are the analogues of the X and Y functions first discovered in problems on radiative transfer. (See Refs. 29 and 37 for more detail.)

3.2. Fredholm Equations: Convergence.
In the previous section we have discussed, in an informal way, a variety of numerical methods for solving Eq. (1). For completeness we now present some of the basic theory necessary to establish the convergence and stability of those algorithms. Since initial value methods are treated more fully later on, we shall confine our attention to the more classical procedures discussed in Sections 3.1.1–3.1.7.

3.2.1. Kernel Approximations.
Let X be a Banach space and denote by $[X]$ the space of bounded linear operators on X. Assume that $K \in [X]$ and that $K_n(t, s)$ induces an operator $K_n \in [X]$, $n \geq 1$. Then Eq. (2) can be written as

$$u_n - K_n u_n = g. \tag{49}$$

Theorem 3.1. Assume that $\|K - K_n\| \to 0$. Then $\exists n_0$ such that for all $n \geq n_0$ $(I - K_n)^{-1} \in [X]$, $\|(I - K_n)^{-1}\|$ is uniformly bounded and $\|u - u_n\| \to 0$. In addition the error estimate

$$\|u - u_n\| \leq \|(I - K_n)^{-1}\| \|K_n - K\| \|u\| \tag{50}$$

holds.

A proof of Theorem 3.1 may be found in Ref. 3. We shall consider the following important special cases.

1. $X = C[a, b]$, $K(t, s)$ is continuous, and $\{K_n(t, s)\}_{n=1}^{\infty}$ is a sequence of uniformly convergent polynomial approximations to $K(t, s)$. In particular, if $K(t, s)$ is analytic, then $K_n(t, s)$ may be taken to be a truncated Taylor series (Ref. 3).

Polynomial approximations can also be generated via interpolation, as in Eq. (26), or by various optimal approximation schemes. For some useful results we recommend the paper of Bownds and Wood (Ref. 32).

2. $X = L_2[a, b]$ and $\{K_n(t, s)\}_{n=1}^{\infty}$ are taken as generalized Fourier series approximations, such as those generated by Galerkin's method.

The latter methods are often costly to apply because of the arithmetic necessary to form the matrix in Eq. (6). However, the rate of convergence is independent of the smoothness of $g(t)$ and may be superior to related projection methods when $g(t)$ is not well behaved (Ref. 3).

3.2.2. Projection Methods. Here Eq. (16) is written as

$$u_n - K_n u_n = g_n, \tag{51}$$

where

$$K_n = P_n K \quad \text{and} \quad g_n = P_n g. \tag{52}$$

To establish convergence the usual assumptions made are

(i) $\|K - K_n\| \to 0$,   (ii) $P_n g \to g$.

If K is compact, then it is sufficient (but not necessary) that $P_n x \to x$ for all $x \in X$ and $\|P_n\| \le M$, $n \ge 1$ (Refs. 3–4). In this case we have the following convergence theorem.

Theorem 3.2. Assume that (i) and (ii) hold. Then for all $n \ge n_0$, $(I - K_n)^{-1} \in [X]$, $\|(I - K_n)^{-1}\|$ is uniformly bounded, and $\|u - u_n\| \to 0$. The error estimate

$$\|u - u_n\| \le \|(I - K_n)^{-1}\| \, \|u - P_n u\| \tag{53}$$

holds.

Proof. A proof may be found in Ref. 3. We shall only comment that $P_n u \to u$ since $P_n g \to g$ and $K_n \to K$, and so Eq. (53) clearly exhibits convergence.   □

We shall now make several observations concerning Theorem 3.2. First, in one of the more important applications, that of polynomial collocation where $X = (C[a, b], \|\cdot\|_\infty)$, one cannot always guarantee that the method converges without careful choice of the collocation points and suitable smoothness of $K(t, s)$ and $g(t)$. For example, it is well known that polynomial interpolation using equally spaced nodes is generally divergent (Ref. 3). However, an approach due to Vainikko (Ref. 38) may be used to establish L_2 convergence of orthogonal collocation. This result was first established for kernels which were Green's functions for ordinary differential operators and was developed to prove the convergence for a class of collocation methods for solving two-point boundary-value problems (Refs. 38–39). Since the idea seems not to be well known, a short exposition is presented here.

For simplicity, assume that $K(t, s)$ and $g(t)$ are continuous. Let $\rho(t)$ be a positive integrable weight function on $[a, b]$ and let $\{\phi_n(t)\}_{n=1}^\infty$ denote the set of polynomials orthonormal with respect to $\rho(t)$. That is,

$$\int_a^b \rho(t)\phi_n(t)\phi_m(t)\, dt = \delta_{nm}. \tag{54}$$

[We use the convention that $\deg(\phi_n(t)) = n - 1$.] Then it is well known that $\phi_n(t)$ has $n - 1$ distinct real zeros $\{t_k\}_{k=1}^n \subseteqq [a, b]$.

Consider the collocation method using $\{\phi_k(t)\}_{k=1}^n$ as a basis and the zeros of $\phi_{n+1}(t)$ as the collocation points. To establish convergence, let $X = L_\rho^2[a, b]$ be the space of functions which are square integrable with respect to $\rho(t)$. Consider K as a compact operator from $L_\rho^2[a, b] \to (C[a, b], \|\cdot\|_\infty)$. P_n is then assumed to act from $C[a, b] \to L_\rho^2[a, b]$. Using properties of Gaussian quadrature, one can show that $P_n \in [C[a, b], L_\rho^2[a, b]]$ (Refs. 38 and 40). To apply Theorem 3.2 it must be shown that $\|K - K_n\| \to 0$ and that $P_n g \to g$. This is established using the following theorem of Erdös and Turan (Ref. 40).

Theorem 3.3. Let f be a continuous function and let $L_n(f)$ be the polynomial interpolating to f at $\{t_k\}_{k=1}^n$, the zeros of $\phi_{n+1}(t)$. Then

$$\lim_{n \to \infty} \int_a^b \rho(t)|f - L_n(f)|^2 \, dt = 0.$$

By using this it follows that $P_n f \to f$ for all $f \in C[a, b]$. Since K is compact, the Banach–Steinhaus theorem implies that $\|K - K_n\| \to 0$ (Ref. 3). It now follows from Eq. (53) that

$$\lim_{n \to \infty} \int_a^b \rho(t)|u - u_n|^2 \, dt = 0. \tag{55}$$

Thus $u_n(t)$ converges in mean square to $u(t)$.

To obtain a uniformly convergent sequence, we use $v_n(t)$ defined by

$$v_n(t) = g(t) + \int_a^b K(t, s) u_n(s) \, ds. \tag{56}$$

Theorem 3.4. $v_n(t)$ converges uniformly to $v(t)$.

Proof. Subtracting Eq. (56) from Eq. (1) gives

$$u(t) - v_n(t) = \int_a^b K(t, s)[u(s) - u_n(s)] \, ds.$$

Using the Cauchy–Schwarz inequality, we get

$$|u(t) - v_n(t)| \leq \left(\int_a^b [|K(t, s)|^2 / \rho(s)] \, ds \right)^{1/2} \left(\int_a^b \rho(s)|u(s) - u_n(s)|^2 \, ds \right)^{1/2}.$$

The result now follows from Eq. (55). □

We note that Theorems 3.3 and 3.4 can be generalized to some weakly singular kernels, and a similar proof has recently been used to establish L_2 convergence of a collocation method for a class of Cauchy singular integral equations (Refs. 41–42).

The most important case of this method occurs when $a = -1$, $b = 1$, $\rho(t) = 1/\sqrt{1-t^2}$ so that $\phi_{n+1}(t) = T_n(t)$, the nth Chebyshev polynomial. In this case, in addition to L_2 convergence, one often obtains uniform convergence if K and g are smooth enough (Refs. 3–4). Further discussion may be found in Refs. 38 and 41.

To establish the convergence of Galerkin's method, take $X = L_2[a, b]$ and assume that

$$\int_a^b \int_a^b |K(t, s)|^2 \, ds \, dt < \infty.$$

We shall only consider the case where $\{\phi_n\}_{n=1}^\infty$ is a complete orthonormal basis so that $P_n x \to x$ for all $x \in X$ and $\|P_n\| = 1$, $n \geq 1$. Thus $\|K - K_n\| \to 0$, and Theorem 3.2 gives

$$\lim_{n \to \infty} \|u - u_n\|_2 = 0.$$

By using the fact that least squares can be considered as Galerkin's method applied to

$$u - K^*u - Ku + K^*Ku = g - K^*g,$$

the above result establishes convergence here as well. [*Note*: Since u satisfies $T^*Tu = T^*g$, the Fredholm alternative shows that $N(T^*) = N(T) = 0$, giving the existence of $(T^*)^{-1}$. Thus one arrives at $Tu = g$.][2]

As in our discussion of collocation, iterating once on $u_n(t)$ determined by Galerkin's method gives

$$v_n(t) = g(t) + \int_a^b K(t, s)u_n(s) \, ds,$$

which is easily shown to be uniformly convergent if

$$\max_{t \in [a,b]} \int_a^b |K(t, s)|^2 \, ds < \infty.$$

It is of some interest to examine other important convergence properties of the sequence $\{v_n(t)\}_{n=1}^\infty$ apparently noted first by Sloan *et al.* in Ref. 25. To see what happens, consider iterating as above on u_n obtained from

---

[2] $N(T)$ denotes the null-space of T.

Eq. (51). This gives the sequence v_n, $n \geq 1$, defined by

$$v_n = g + Ku_n. \tag{57}$$

Applying P_n to both sides of Eq. (57), we get

$$P_n v_n = P_n g + K_n u_n = u_n. \tag{58}$$

Thus the projection of v_n onto X_n is u_n. Intuitively one expects that v_n will be a better approximation to u than u_n. For u_n given by Galerkin's method this has been numerically demonstrated in Ref. 25 and for collocation by Phillips in Ref. 26. (See also Section 3.2.3.)

To continue, we substitute $u_n = P_n v_n$ into Eq. (57), giving v_n as the solution to

$$v_n = g + KP_n v_n. \tag{59}$$

By letting $L_n = KP_n$ it is seen that

$$v_n - L_n v_n = g, \tag{60}$$

which is in the form of a kernel approximation method. In the special case where u_n is generated by Galerkin's method we have the following remarkable estimate (Ref. 25):

$$\|u - u_n\|_2 \leq \|(I - L_n)^{-1}\| \, \|K - L_n\| \, \|u - P_n u\|_2. \tag{61}$$

To see this, observe that $\|K - L_n\| \to 0$ (Refs. 3, 25) so that for n sufficiently large

$$v_n = (I - L_n)^{-1} g. \tag{62}$$

Subtracting this from Eq. (1) gives

$$u - v_n = (I - K)^{-1} g - (I - L_n)^{-1} g$$
$$= (I - L_n)^{-1}[(K - L_n)u]. \tag{63}$$

Decomposing u as

$$u = P_n u + (I - P_n)u \tag{64}$$

and substituting into Eq. (63), we obtain

$$v - v_n = (I - L_n)^{-1}[(K - L_n)(u - P_n u)]. \tag{65}$$

Taking norms on both sides of Eq. (65) yields Eq. (61).

This shows that v_n converges to u at a rate which is the product of that for a degenerate kernel method and a projection method. Unfortunately one cannot guarantee improved convergence for collocation methods via this estimate since in general one does not have $\|K - L_n\| \to 0$. However, as pointed out above, numerically one does see improved convergence, as illustrated by the results of Ref. 26. The theoretical justification for this appears to come from analyzing collocation as a form of product integration and will be taken up in Section 3.2.3.

Further work by Sloan (Ref. 43) has shown for Galerkin's method that for all n sufficiently large v_n is always a better approximation to u than any element $u_n \in X_n$, in the sense that $\|u - v_n\|_2 \leq \|u - u_n\|_2$, and has extended this result to higher-order iterates (Ref. 43). Since forming v_n requires little additional arithmetic, Atkinson (Ref. 3) strongly suggests implementing Galerkin's method using Eq. (57).

As a last observation we note that if $x_n \in X_n$ then

$$Kx_n = L_n x_n, \tag{66}$$

and this was the starting point for the work in Ref. 25. The motivation there was to produce good degenerate kernel approximations L_n to K: good in the sense that Eq. (66) holds.

More recently Sloan has extended these ideas to the eigenvalue–eigenvector problem for K (Ref. 44), and further developments have appeared in the work of Chatelin (Ref. 45).

3.2.3. Quadrature Methods. To discuss the convergence of quadrature methods we use the Nyström interpolation given by Eq. (33b). This makes it possible to compare $u_n(t)$ and $u(t)$ as elements in the same space. Formally, let $X = C[a, b]$ and assume that $K(t, s)$ is continuous. Define $K_n: X \to X$ by

$$K_n u = \sum_{k=1}^{n} w_k K(t, s_k) u(s_k). \tag{67}$$

K_n is a bounded finite rank operator and thus compact.

By using Eq. (67), Eq. (33b) can be written as

$$u_n = K_n u_n + g. \tag{68}$$

Assume that n is sufficiently large so that the operator $(I - K_n)^{-1}$ exists and is an element of $[X]$. Then

$$u_n = (I - K_n)^{-1} g, \tag{69}$$

and subtracting Eq. (69) from Eq. (1) gives

$$u - u_n = (I - K_n)^{-1}(Ku - K_n u). \tag{70}$$

If it can be shown that

$$\|(I - K_n)^{-1}\| \leq \tau, \qquad n \geq n_0, \tag{71}$$

we arrive at the estimate

$$\|u - u_n\|_\infty \le \tau \|Ku - K_nu\|_\infty. \tag{72}$$

If the quadrature rule Q_n has the property that

$$\lim_{n \to \infty} Q_n(f) = \int_a^b f(t) \, dt \tag{73}$$

for all $f \in C[a, b]$, then it is easily shown (Ref. 3) that $\|Ku - K_nu\|_\infty \to 0$ so that u_n converges uniformly to u. Now

$$\|Ku - K_nu\|_\infty = \max_{t \in [a,b]} \left| \int_a^b K(t, s)u(s) \, ds - \sum_{k=1}^n w_k K(t, s_k)u(s_k) \right|, \tag{74}$$

showing that the approximation error can be calculated in terms of the quadrature error for $K(t, s)u(s)$. It is important to note that the quadrature error is determined by the product $K(t, s)u(s)$ and not by $K(t, s)$ alone. Thus, even if the kernel is poorly behaved, it may be possible to achieve high accuracy if $u(t)$ is smooth enough. For example, if $K(t, s)u(s)$ is C^4 in s and Simpson's rule is used, then

$$\|u - u_n\|_\infty = O(h^4),$$

where h is the step size. Furthermore, known asymptotic error expansions for quadrature lead to similar expansions for $u - u_n$ and thus provide a convenient basis for the applications of extrapolation techniques (Ref. 4).

To complete the analysis it is necessary to establish the validity of Eq. (72). It is at this point that the convergence theory becomes technically more involved than for kernel approximation or projection methods. As pointed out above, if $\{Q_n\}_{n=1}^\infty$ is a sequence of convergent quadrature rules, then $K_n x \to Kx$ uniformly for all $x \in X$. However, in general $\|K - K_n\| \not\to 0$ so that proving the existence and boundedness of $(I - K_n)^{-1}$ cannot be carried out by a simple application of Banach's lemma (Ref. 3).

In 1960 Brakhage (Ref. 46) proved that

$$\|(K - K_n)K\| \to 0 \qquad \text{and} \qquad \|(K - K_n)K_n\| \to 0 \tag{75}$$

and showed that these relations were sufficient to establish the validity of Eq. (73). In 1964 Anselone and Moore (Ref. 47) observed that the sequence of operators $\{K_n\}_{n=1}^\infty$ had the important property of being collectively compact, i.e.,

$$S = \{K_n x : \|x\|_\infty \le 1, n \ge 1\}$$

had compact closure, and that Eq. (75) was an immediate consequence of that fact. Throughout the 1960's they and colleagues made an extensive

development of the theory of collectively compact operators and its application to quadrature methods for integral equations (Ref. 1). Of particular
value is its use in establishing the convergence of product quadrature
methods (Refs. 1, 3). Here one has to deal with operators of the form

$$K_n u = \sum_{k=1}^{n} w_k(t) M(t, s_k) u(s_k), \tag{76}$$

which are the analogues of those in Eq. (67). The general theory allows one
to obtain error estimates of the form

$$\|u - u_n\|_\infty \le \sigma \|Ku - K_n u\|_\infty$$

for u_n defined by Eq. (41). Further details may be found in the books
by Anselone and Atkinson and in the paper of de Hoog and Weiss (Ref.
48).

In closing we want to point out that the theory of collectively compact
operators enables one to obtain improved error estimates for collocation by
using the previously mentioned relation between it and product quadrature.
To see this, consider Eq. (22), where we use the representation

$$u_n(t) = \sum_{k=1}^{n} l_k(t) u_n(t_k). \tag{77}$$

Then

$$u_n(t_j) = \sum_{k=1}^{n} w_k(t_j) u_n(t_k) + g(t_j), \qquad j = 1, 2, \ldots, n, \tag{78}$$

where

$$w_k(t_j) = \int_a^b K(t_j, s) l_k(s) \, ds. \tag{79}$$

Thus the equations now have the form of the equations for a product
quadrature method, and if we interpolate via Eq. (41), we get

$$v_n(t) = g(t) + \sum_{k=1}^{n} w_k(t) v_n(t_k), \tag{80}$$

where $v_n(t_j) = u_n(t_j)$, provided that n is large enough. Using this, we arrive at
the improved error estimate

$$\|u - v_n\|_\infty \le \sigma \|Ku - K_n u\|_\infty. \tag{81}$$

For particular cases we again refer the reader to Atkinson's book or the
paper of de Hoog and Weiss.

3.3. Convergence Acceleration. If both the kernel $K(t, s)$ and $g(t)$ are smooth functions, then all of the methods discussed previously will generally perform satisfactorily. Collocation methods may be preferred if one has to keep the size of matrices small (Ref. 3), with quadrature methods usually being more efficient if the cost of evaluating the kernel is high.

However, if $K(t, s)$ and/or $g(t)$ are nonsmooth, or possibly discontinuous, then primary methods may have to be modified in order to give an efficient algorithm. In general one probably has to examine each case on its own merits. Consequently we shall present only a brief description of some of the more important methods that have been proposed for accelerating convergence.

The case of badly behaved kernels is considered first. For quadrature methods we have seen that the solution error depends on the error in integrating $K(t, s)u(s)$. If $u(s)$ is smooth but $K(t, s)$ is not, then collocation or product integration should be the preferred method.

In this situation it may also pay to look at special methods for treating the kernel. This is one of the motivations for imbedding and related methods, which can often handle kernels with Green's-function-type singularities exactly (Refs. 2, 22–23). From another point of view the classical approximation methods generally replace the kernel by a smooth kernel of finite rank, and discontinuities in the kernel are smoothed out in the process of approximation. Kernel approximations modeling the singularities could be considered in this situation (Refs. 4, 22–23).

Another approach to this problem is to regularize the kernel in some way. To do this, operate with K on both sides of Eq. (12), giving

$$Ku = Kg + K^2u. \tag{82}$$

But $Ku = u - g$ so that

$$u = g + Kg + K^2u. \tag{83}$$

For example, if

$$K(t, s) = H(t, s)/|t - s|^\alpha, \qquad 0 < \alpha < 1, \tag{84}$$

where $H(t, s)$ is continuous, K^2 will be smoother than K, and furthermore for some p, K^p will be continuous. The above process can then be repeated to produce an equation with a continuous kernel. For practical reasons one can, at best, consider $p = 2$. Then if -1 is not an eigenvalue of K, Eq. (83) will have a unique solution, and thus u can be obtained from this somewhat better behaved equation. The method seems to be restricted to situations where both K^2 and Kg can be obtained analytically.

If a projection method is used, it has been shown that iteration has convergence-accelerating properties. Although there is limited reported numerical evidence, the results of Sloan and Phillips (Refs. 25–26) seem to indicate that one can obtain an order of magnitude decrease in error using a single iteration.

Still yet another device is to use some form of extrapolation to accelerate convergence. For quadrature methods one has to have sufficient smoothness of $K(t, s)u(s)$ so that the necessary asymptotic error expansions can be derived (Refs. 3–4). In Ref. 4, Baker provides a detailed analysis of this procedure. It would also be useful to have similar results available for projection methods. Some *ad hoc* procedures for polynomial collocation are given in Chapter 5.

Atkinson, in Refs. 3 and 49, extending an idea of Brakhage, develops a somewhat different approach to iteration. Here again the desire is to keep the size of the system of equations that need to be solved small. To illustrate this approach we shall examine a simple version of the algorithms discussed in Ref. 3.

Suppose one has obtained an approximation to u satisfying the equation

$$u_m = K_m u_m + g, \tag{85}$$

where m is large enough so that $(I - K_m)^{-1}$ exists. Now let $n > m$ and consider solving

$$u_n = K_n u_n + g. \tag{86}$$

Rewriting Eq. (25) in the form

$$u_n - K_m u_n = (K_n - K_m)u_n + g \tag{87}$$

enables one to construct the recursion scheme

$$u_n^{(k+1)} - K_m u_n^{(k+1)} = (K_n - K_m)u_n^{(k)} + g \tag{88}$$

for solving Eq. (86).

The effect of this procedure is to require only the inversion of the lower-order operator $I - K_m$ rather than $I - K_n$. If $\|(I - K_m)^{-1}(K_n - K_m)\| < 1$, then $u^{(k)} \to u_n$, and operation counts show that this procedure is often more economical than solving Eq. (86) directly. The proper choice of m and n requires some care, and Atkinson has developed several programs which do this automatically (Ref. 3) when quadrature methods are employed. These programs appear to be some of the few available for solving integral equations with automatic error control.

Although the programs have been written to apply to smooth kernels for one-dimensional problems, the real advantage of this approach appears to be for solving multidimensional equations and/or equations with singular kernels.

If the kernel is smooth but $g(t)$ is not, iterative variants may also be applied to improve the behavior of primary algorithms. However, other strategies may also be considered. The simplest of these is the regularization method of Kantorovich (Refs. 3, 50). For this we let $v = u - g$ and observe that v satisfies

$$v = Kv + Kg. \tag{89}$$

If Kg can be calculated conveniently and accurately, then Eq. (89) can be solved by any of the methods given in Section 3.2. This will give rise to an approximation u_0 to u of the form

$$u_0 = v_0 + g, \tag{90}$$

where v_0 is the approximate solution of Eq. (89).

This technique was apparently first used as a method for guaranteeing convergence of a projection method when $P_n g$ did not converge to g or more generally converged more slowly than $P_n K$ did to K. To see the nature of the approximation in this case, let

$$v_n = P_n K v_n + P_n K g \tag{91}$$

and put

$$u_n = v_n + g. \tag{92}$$

From Eqs. (91) and (92) it follows that

$$u_n = P_n K u_n + g, \tag{93}$$

so that applying a projection method to the regularized equation (89) produces a degenerate kernel method. From Theorem 3.1 we get

$$\|u - u_n\| \le \tau \|K - P_n K\| \|u\|,$$

so that u_n converges to u at a rate independent of the smoothness of g.

As with kernel regularization this method may be extended by defining

$$v_p = u - \sum_{s=0}^{p-1} K^s g$$

and solving

$$v_p = K v_p + K^p g.$$

The practicality of this method usually ceases for $p > 1$.

In closing we want to mention two other devices for treating poorly behaved g's. Observe from Eq. (43) that one has

$$u(t) = g(t) + \int_a^b R(t, s, 1) g(s) \, ds, \tag{94}$$

where the resolvent kernel $R(t, s, \lambda)$ is given by Eq. (44). From Eq. (94) it is seen that u generally inherits the behavior of $g(t)$ if K is smooth, so that it may be advantageous to calculate the resolvent kernel and obtain $u(t)$ from Eq. (94). Several computational schemes have been put forward for doing this (Refs. 2, 23). This approach has also been the subject of several recent papers by Williams (Refs. 51–53), who was motivated by some particularly difficult problems in aerodynamics.

Another way of avoiding the calculation of u is based on the following observation. Often in practice various linear functionals of the solution $u(t)$ are desired rather than u itself (Refs. 51 and 54). Suppose, for example, that we wish to calculate

$$l(u) = \int_a^b l(t)u(t)\, dt, \tag{95}$$

where $l(t)$ is a smooth function. Let u_* be the solution to

$$u_* = K^*u_* + l(t); \tag{96}$$

then (Ref. 55)

$$l(u) = \int_a^b u_*(t)g(t)\, dt. \tag{97}$$

Thus Eq. (96) can be solved and $l(u)$ approximated using Eq. (97). We note that this idea is quite general and has been applied to Abel equations (Ref. 54) and to various one- and multidimensional singular equations as well (Refs. 14 and 18).

3.4. Stability.

In the previous section we discussed the convergence of many of the schemes which have been proposed for the solution of equations of the second kind. Although convergence is an obvious property that a numerical method should possess, its presence does not necessarily tell the whole story. Since at each stage one must solve a finite-dimensional problem, it is important to know how errors due to truncation and roundoff are propagated. Thus it is necessary to study the stability of both the equation and its finite-dimensional approximations.

Consider the problem of solving the linear equation

$$Ax = b, \qquad A \in [X], x \in X, b \in X, \tag{98}$$

where $A^{-1} \in [X]$ so that Eq. (98) is uniquely solvable. For practical purposes, due to various errors in the representations of A and b, one actually solves the perturbed equation

$$(A + \Delta A)(x + \Delta x) = b + \Delta b, \tag{99}$$

where ΔA, Δb, and Δx represent the errors in A, b, and x. In discussing stability one is concerned with either the absolute error $\|\Delta x\|$ or the relative error $\|\Delta x\|/\|x\|$. It is well known (Refs. 3–4) that the absolute error is determined by $\|A^{-1}\|$ and the relative error by the condition number

$$C(A) = \|A^{-1}\| \|A\|.$$

In both cases the analysis generally rests on the estimation of $\|A^{-1}\|$.

For an integral equation of the second kind there are two types of stability questions that one wishes to answer. First, is the original equation stable? Second, is the solution method stable? For the first we have observed that $A = I - K$ has a bounded inverse provided that K is compact. Thus $\|A^{-1}\| = \|(I-K)^{-1}\|$, and the equation is stable if this quantity is not too large. This will be true if 1 is not close to an eigenvalue of K (Ref. 3).

For the second, we observe that for most classical solution methods one solves a sequence of approximating problems

$$(I - K_n)u_n = g_n, \tag{100}$$

where numerically one has to consider the related sequence of matrix problems

$$A_n \tilde{u}_n = \tilde{g}_n, \qquad (\tilde{u}_n, \tilde{g}_n) \in \mathbb{C}^n \quad \text{or} \quad R^n, \tag{101}$$

where A_n is an n-dimensional representation of the operator $I - K_n$. The stability of the numerical algorithm is then determined by a knowledge of $\|A_n^{-1}\|_n$, where for each $n \| \cdot \|_n$ is a matrix norm on R^n or $\mathbb{C}^n$ induced by a vector norm compatible with the norm on X. If the approximation method is such that $(I - K_n)^{-1}$ converges in norm to $(I - K)^{-1}$, then $\|(I - K_n)^{-1}\|$ converges to $\|(I - K)^{-1}\|$, so that one has for some n_0

$$\sup_{n \geq n_0} \|(I - K_n)^{-1}\| \leq \tau,$$

and of course $\|(I - K_n)^{-1}\| \approx \|(I - K)^{-1}\|$ for large enough n. Thus one might conclude that if the original problem is stable then the sequence $\{A_n\}_{n=n_0}^{\infty}$ will be also, and one should then have a bound

$$\|A_n^{-1}\|_n \leq \tau. \tag{102}$$

Unfortunately Eq. (102) is not true in general, so one must be careful not to draw naive conclusions.

For some algorithms such as Galerkin's method with a complete orthonormal basis or the Nyström method Eq. (102) will hold, but for others, such as polynomial collocation, it does not.

To illustrate these ideas we shall develop the analysis of $\|A_n^{-1}\|_n$ for collocation (Ref. 3). Take $x = C[a, b]$ and consider the set of collocation equations

$$\sum_{k=1}^{n} a_k \left\{ \phi_k(t_j) - \int_a^b K(t_j, s)\phi_k(s)\, ds \right\} = g(t_j), \qquad j = 1, 2, \cdots, n.$$

Assume all quantities are real and let the norm on R^n be given by

$$\|x\|_\infty^n = \max |x_i|, \qquad x = (x_1, x_2, \ldots, x_n).$$

$\|A\|_n$ is then the induced matrix norm. Let $b = (b_i)_{i=1}^n \in R^n$, and define $g(t) \in C[a, b]$ such that

$$g(t_i) = b_i, \qquad i = 1, 2, \ldots, n,$$

and in addition having the property that

$$\|g\|_\infty = \|b\|_\infty^n.$$

Also the solution $u_n(t)$ satisfies

$$u_n(t) = \sum_{k=1}^{n} a_k \phi_k(t), \tag{103}$$

where

$$A_n \mathbf{a} = b, \qquad \mathbf{a} = (a_1, a_2, \ldots, a_n).$$

Now

$$(I - P_n K) u_n = P_n g,$$

so that

$$\|u_n\|_\infty \le \|(I - P_n K)^{-1}\|_\infty \|P_n\| \|g\|_\infty \le \tau \|P_n\| \|g\|_\infty. \tag{104}$$

Let

$$\Phi_n = [\phi_k(t_j)]_{j,k=1}^n, \qquad \Phi_n^{-1} = [C_{kj}]_{j,k=1}^n;$$

then Eq. (103) yields

$$u_n(t_j) = \sum_{k=1}^{n} a_k \phi_k(t_j), \qquad i = 1, 2, \ldots, n,$$

giving

$$\mathbf{a} = \Phi_n^{-1} \mathbf{u}_n, \qquad \mathbf{u}_n = (u_n(t_j))_{j=1}^n.$$

Thus

$$\|\mathbf{a}_n\|_\infty^n \le \|\Phi_n^{-1}\|_n \|\mathbf{u}_n\|_\infty^n. \tag{105}$$

Using Eq. (104) and (105) and the fact that $\|\mathbf{u}_n\|_\infty^n \leq \|u_n\|_\infty$, we obtain

$$\|\mathbf{a}\|_\infty^n \leq \|\Phi_n^{-1}\|_n \|\mathbf{u}_n\|_\infty^n \leq \|\Phi_n^{-1}\|_n \|u_n\|_\infty$$

$$\leq \tau \|\Phi_n^{-1}\|_n \|P_n\| \|g\|_\infty = \tau \|\Phi_n^{-1}\|_n \|P_n\| \|b\|_\infty^n. \tag{106}$$

By the definition of $\|A_n^{-1}\|_n$, Eq. (106) gives

$$\|A_n^{-1}\|_n \leq \tau \|\Phi_n^{-1}\|_n \|P_n\|. \tag{107}$$

As a particular case we shall examine Eq. (107) for polynomial collocation. From the estimate it is seen that stability depends both on the choice of basis and collocation points (convergence depends only on the latter). If we use $\phi_k(t) = l_k(t)$, then $\phi_k(t_j) = \delta_{kj}$ and $\|\Phi_n^{-1}\| = 1$. To minimize $\|P_n\|$ one should choose $\{t_j\}$ as the zeros of the shifted Chebyshev polynomial (Ref. 3)

$$T_n(t) = \cos\{n \cos^{-1}[(2t - a - b)/(b - a)]\}, \qquad a \leq t \leq b.$$

In this case (Ref. 3)

$$\|P_n\| = O(\log n),$$

so A_n will generally be well conditioned if τ is not large.

As indicated above, similar results exist for the Galerkin and Nyström methods so that it appears that the stability of projection and quadrature methods is generally well understood. The situation is less satisfactory for kernel approximation methods and initial value methods. For degenerate kernels of the type generated by a projection method the stability of the related projection methods yields a stability result for the degenerate kernel method. For other types of kernel approximations no general theory seems to exist (Ref. 3).

For initial value methods various authors have indicated that the methods are stable; however, no general proof appears to exist. Some preliminary results may be found in Refs. 2 and 56. (See also Chapter 13.)

3.5. Cauchy Singular Equations.

Closely related to Fredholm equations, and until recently seemingly neglected by numerical analysts, are the Cauchy singular equations

$$a(t)u(t) + b(t) \oint_\Gamma u(s)/(t - s) \, ds + \oint_\Gamma K(t, s)u(s) \, ds = g(t), \tag{108}$$

where Γ is either an open or closed contour in the plane. In the particular case where $a^2(t) + b^2(t) \neq 0$, $t \in \Gamma$, it is known that such equations are equivalent to Fredholm equations of the second kind (Ref. 57) so that one expects their numerical analysis to be somewhat similar.

Until recently, most of the theoretical numerical work has been confined to the Soviet Union, with the book by Ivanov summarizing the state of the art circa 1968 (Ref. 5). Since there is an almost total absence of numerical results in his book, it is not easy to assess the usefulness of many of the algorithms proposed there. However, there is one important subclass of problems which has been successfully treated in the technical literature for many years. It corresponds to taking $\Gamma = [a, b]$ (usually $a = -1$, $b = 1$ for convenience) with $a(t) = 0$, $b(t) = $ constant. Such equations occur frequently in aerodynamics (Refs. 12, 42, 51–53), hydrodynamics (Ref. 15), elasticity (Ref. 58), and fracture mechanics (Ref. 59). With respect to the above restrictions Eq. (108) becomes

$$\int_{-1}^{1} [u(s)/(t-s)] \, ds + \int_{-1}^{1} K(t, s)u(s) \, ds = g(t). \tag{109}$$

In practice one often has (Refs. 12, 51–53)

$$K(t, s) = a \, \log|t - s| + K_c(t, s), \tag{110}$$

where $K_c(t, s)$ is continuous. If $a = K_c = 0$, then Eq. (109) reduces to the classical airfoil equation studied by Söhngen (Ref. 60), Tricomi (Refs. 61–62), and others (Ref. 57). The principal characteristic of Eq. (109) is that in general a nonzero solution exists to the homogeneous equation

$$\int_{-1}^{1} [u(s)/(t-s)] \, ds + \int_{-1}^{1} K(t, s)u(s) \, ds = 0, \tag{111}$$

and numerical methods must account for this (Refs. 12 and 51–53). Although the recently developed methods of Elliott and Dow (Ref. 16) apply to this case, it is somewhat more convenient to use the physics of the problem in order to calculate a unique solution directly. In aerodynamics and hydrodynamics one usually imposes the auxiliary condition $u(t_0) = 0$, $t_0 \in [-1, 1]$ in order to establish uniqueness. Using the known fact that $u(t)$ behaves like $1/\sqrt{1 - t^2}$ as $t \to \pm 1$, one is led to seek solutions of the form

$$u(t) = [(t_0 - t)/\sqrt{1 - t^2}]\psi(t). \tag{112}$$

Using Eq. (112) in Eq. (111) allows the reformulation

$$\int_{-1}^{1} \{[(t_0 - s)/\sqrt{1 - s^2}][\psi(s)/(t - s)]\} \, ds$$

$$+ \int_{-1}^{1} [(t_0 - s)/\sqrt{1 - s^2}]K(t, s)\psi(s) \, ds = g(t). \tag{113}$$

In the most common situations $t = \pm 1$, so that Eq. (113) becomes

$$\int_{-1}^{1} [\sqrt{1-s/(1+s)}][\psi(s)/(t-s)]\, ds + \int_{-1}^{1} [\sqrt{1-s/(1+s)}]K(t,s)\psi(s)\, ds = g(t)$$

$$(114)$$

or

$$\int_{-1}^{1} \sqrt{(1+s)/(1-s)}[\psi(s)/(t-s)]\, ds$$

$$+ \int_{-1}^{1} \sqrt{(1+s)/(1-s)}K(t,s)\psi(s)\, ds = g(t). \qquad (115)$$

Since the operator

$$H\psi = \int_{-1}^{1} \sqrt{(1-s)/(1+s)}[\psi(s)/(t-s)]\, ds \qquad (116)$$

is now invertible (Refs. 5, 12) with

$$H^{-1}\psi = \int_{-1}^{1} \sqrt{(1+s)/(1-s)}[\psi(s)/(s-t)]\, ds, \qquad (117)$$

Eqs. (114) and (115) can be written in operator form as

$$H\psi + K\psi = g \qquad (118)$$

or

$$-H^{-1}\psi + K\psi = g. \qquad (119)$$

Multiplying through by H^{-1} or H gives the equivalent equation of the second kind,

$$\psi + H^{-1}K\psi = H^{-1}g, \qquad (120)$$

or

$$-\psi + HK\psi = Hg. \qquad (121)$$

Thus Eq. (109) is now amenable to solution by any of the methods given in Section 3.1 for Fredholm equations of the second kind. This approach, although feasible, is inefficient for complicated kernels, a case of frequent occurrence. It turns out that it is more convenient to deal with Eqs. (118) and (119) directly. The development of this approach is presented in Chapters 4 and 5 for Eq. (107), and the general singular equation (108) is treated in Chapter 3.

3.6. Linear Integral Equations of the First Kind.   Integral equations of the first kind are characterized by having the unknown function only under the integral sign, whereas in equations of the second kind the unknown function appears both under and outside the integral sign. This small structural difference changes both the theory and the numerical analysis drastically.

For ease of exposition we shall consider equations of the form

$$\int_{A(t)} K(t, s)u(s)\, ds = g(t), \tag{122}$$

where $A(t) \subseteq R$ is a set generally dependent on t. For example, if $A(t) = [a, b]$, then Eq. (122) becomes the Fredholm equation

$$\int_a^b K(t, s)u(s)\, ds = g(t), \tag{123}$$

while for $A(t) = [a, t]$ or $[t, b]$ we obtain Volterra equations. Since this framework would, in principle, include some Cauchy singular equations, we shall make the assumption throughout this section that the kernel $K(t, s)$ is integrable. We shall begin with a discussion of the Fredholm case.

As we have seen in Sections 3.2 and 3.4, solving an equation of the second kind is a well-posed problem. That is, small perturbations in the data usually result in small perturbations of the solution. Formally this derives from the fact that if K is compact, then $(I - K)^{-1}$, if it exists, is bounded. For equations of the first kind with an integrable kernel the compactness of K is the source of the difficulty. If K^{-1}: Range $K \to X$ exists, it is generally unbounded, so no estimate of the form

$$\|u\| \le C\|g\|$$

exists. Thus solving Eq. (123) is an inherently unstable problem, the instability manifesting itself in the ill-conditioning of matrix approximations to K (Ref. 4).

Notwithstanding, it is possible at times to solve Eq. (123) by methods which are entirely analogous to those used for equations of the second kind (Ref. 4). Thus one can consider the use of such techniques as collocation, Galerkin and quadrature methods, and least squares as possibilities, provided that the system of linear equations that need to be solved is not too ill-conditioned (Refs. 4, 58). If only low-order accuracy is desired, then depending on the smoothness of the kernel, such a straightforward approach may prove successful (Ref. 4).

To get a somewhat better perspective on the problem, it is useful to consider the source of the instability in solving Eq. (122) in a somewhat more concrete manner. Many points of view are possible (Refs. 4, 54), and we shall examine two. First, suppose the kernel in Eq. (123) is the Green's function of a differential operator L. That is, we assume that

$$L_t(K(t, s)) = \delta(t - s), \tag{124}$$

where $\delta(t - s)$ is the Dirac delta function. Using Eq. (124) in Eq. (123) yields

$$L_t \int_a^b K(t, s)u(s) \, ds = \int_a^b \delta(t - s)u(s) \, ds = u(t).$$

This gives the solution as

$$u = L(g(t)),$$

provided that g is in the domain of L. For such problems it is seen that solving Eq. (123) numerically is a generalization of the problem of numerical differentiation. It is well known that this problem is ill-posed and that there is a lower bound on the discretization step that can be used; eventually roundoff error dominates even if one has arbitrarily small truncation error (Refs. 4, 63).

This point of view is also useful when dealing with Volterra equations. Consider the generalized Abel equation

$$\int_a^t [K_1(t)K_2(s)/(t^p - s^p)^\alpha]u(s) \, ds = g(t), \qquad 0 < \alpha < 1, \tag{125}$$

an equation of current interest in its relation to problems in particle size statistics (Refs. 54, 64) and other inverse problems (Refs. 54, 65). By suitable changes of the independent and dependent variables one can reduce Eq. (125) to the classical Abel equation

$$\int_0^t [\tilde{u}(s)/(t - s)^\alpha] \, ds = \tilde{g}(t). \tag{126}$$

The inversion formula

$$\tilde{u}(t) = \frac{\sin \pi\alpha}{\pi} \frac{d}{dt} \int_0^t \frac{\tilde{g}(s)}{(t - s)^{1-\alpha}} \, ds \tag{127}$$

shows that solving Eq. (125) may be considered as a problem of fractional differentiation (Ref. 25). Because the order of differentiation is less than 1, Anderssen refers to such problems as being "mildly ill-posed" (Ref. 54). As $g(t)$ is often obtained from inexact measurements, the effect of a straightforward application of Eq. (127) will in many cases merely serve to amplify the noise (Ref. 54).

Returning to Fredholm equations, another, more classical viewpoint can be adopted. Assume, for simplicity, that $K(t, s)$ is symmetric, positive definite, and nondegenerate. Then there exists a complete set of orthonormal eigenfunctions $\{\phi_n\}_{n=1}^{\infty}$ and associated eigenvalues $\{\lambda_n\}_{n=1}^{\infty}$ for $K(t, s)$ (Ref. 4). That is,

$$\int_a^b K(t, s)\phi_n(s) \, ds = \lambda_n \phi_n(s), \qquad n \geq 1,$$

where $\lambda_1 > \lambda_2 > \lambda_3 > \cdots > \lambda_n \cdots$. If $u \in L_2[a, b]$, then

$$u = \sum_{n=1}^{\infty} \langle u, \phi_n \rangle \phi_n. \tag{128}$$

Expanding g as in Eq. (128) and substituting into Eq. (123), we obtain

$$K\left(\sum_{n=1}^{\infty} a_n \phi_n\right) = \sum_{n=1}^{\infty} \langle g, \phi_n \rangle \phi_n. \tag{129}$$

Using the orthonormality of $\{\phi_n\}_{n=1}^{\infty}$ gives

$$a_n = \langle g, \phi_n \rangle / \lambda_n, \tag{130}$$

so that

$$u = \sum_{n=1}^{\infty} \{\langle g, \phi_n \rangle / \lambda_n\} \phi_n, \tag{131}$$

provided that

$$\sum_{n=1}^{\infty} |\langle g, \phi_n \rangle|^2 / \lambda_n^2 < \infty. \tag{132}$$

Since $\lambda_n \to 0$ as $n \to \infty$, errors in g are magnified by arbitrary large factors, so again the problem is shown to be ill-posed. For nonsymmetric kernels, a similar analysis may be given in terms of the singular values and functions of K (Ref. 4).

The representation given in Eq. (131) is also important in seeing how the smoothness of $K(t, s)$ affects the stability. Generally the smoother $K(t, s)$ is, the more rapidly the eigenvalues λ_n decrease. This leads to the rule of thumb that the more derivatives that $K(t, s)$ has, the more difficult Eq. (123) will be to solve. This, of course, is in sharp contrast to the situation for equations of the second kind.

Picturing the solution process as a "noise amplifier," one can take the filtering point of view of trying to evaluate the series in Eq. (131) by damping out the high-frequency components. A simple device for doing this is to add a component λ to λ_n and then define $u_\lambda(t)$ by

$$u_\lambda(t) = \sum_{n=1}^{\infty} \{\langle g, \phi_n \rangle / (\lambda + \lambda_n)\} \phi_n(t).$$ (133)

For example, if K is positive definite, then choosing $\lambda > 0$, one can show that

$$\lim_{\lambda \to 0} \|u_\lambda - u\|_2 = 0.$$

Thus for large λ the solution is smooth, whereas for $\lambda \to 0$, $u_\lambda \to u$, while the problem becomes more ill-posed. One tends to feel that there should exist an "optimal" λ which gives the best compromise between accuracy and stability. How to calculate it is a question of some difficulty.

Viewing this procedure in another way, one sees that $u_\lambda(t)$ satisfies

$$\lambda u_\lambda(t) + \int_a^b K(t, s) u_\lambda(s)\, ds = g(t).$$ (134)

Thus Eq. (123) can be stabilized by the addition of the operator λI to K. More generally one can consider regularization by addition of an operator λG to K and then solving

$$\lambda G u_\lambda + K u_\lambda = g.$$ (135)

For general kernels we can appeal to the Picard theory (Ref. 4) and start with the normal equations

$$K^* K u = K^* g$$

obtained by minimizing

$$\|Ku - g\|_2^2.$$ (136)

Generalizing Eq. (135), one seeks to regularize it by solving

$$\lambda R u_\lambda + K^* K u_\lambda = K^* g, \qquad \lambda > 0.$$ (137)

The choice of both λ and R is again crucial. Since R is added to smooth the solution, a logical choice is a differential operator (Ref. 4). Equation (137) is now an integrodifferential equation so that solutions will be differentiable by definition. To see how one might pick R the least-squares approach again comes into play. Here we take u_λ to minimize

$$\|Ku_\lambda - g\|^2 + \lambda \|u_\lambda\|_s^2,$$ (138)

where $\|u\|_s$ is a seminorm introduced to control the differentiability of u. A popular choice [sometimes disguised (Refs. 4, 66)] is

$$\|u\|_s = \|u''\|_2.$$

If we use this, then minimizing the expression in Eq. (139) leads to $Ru = u''''$. This approach has been considered in detail by the Russian school led by Tikhonov (Refs. 4, 67) and is also discussed by Baker in Ref. 4. Again the method of choosing λ is left partly open. Baker suggests solving the variational problem

$$\min_{u_\lambda} \{\|Ku_\lambda - f\|_2^2 + \lambda \|u_\lambda\|_s^2\}$$

for "large" values of λ and then using these to extrapolate back to $\lambda = 0$. No numerical examples are presented in Ref. 4 to test this procedure.

More recently the problem of determining λ has been taken up by Wahba, who considers solving Eq. (123) as a generalization of the classical regression problem and uses the method of weighted cross-validation (Ref. 68). Considerable numerical success has been achieved and is reported in Chapter 7.

Other approaches to solving Eq. (122) have been the use of regularization applied to discretized versions of Eq. (122) (Ref. 66) and the use of iteration (Ref. 69), while for Abel equations the adjoint method of Section 3.4 has been proposed (Refs. 54–55).

The field is still undergoing development, and much work remains to be done.

3.7. Linear Volterra Equations.

The two principal forms of linear Volterra equations are

$$u(t) = g(t) + \int_a^t K(t, s)u(s)\, ds \tag{139}$$

and

$$g(t) = \int_a^t K(t, s)u(s)\, ds. \tag{140}$$

Equation (139) is an equation of the second kind, whereas Eq. (140) is an equation of the first kind. Since numerical methods for these equations are similar to those that are used for their nonlinear counterparts, we shall defer our discussion of this topic to the following section.

3.8. Nonlinear Equations.

Since nonlinear equations can come in an infinite variety of types, we shall restrict our attention to those whose development parallels that of related linear equations. In contrast to the linear case the theory of nonlinear integral equations is by no means as complete, and in general is considerably more intricate. In fact it is probably fair to say that each specific situation will require its own special investigation. (See Refs. 4 and 70–71 for specific examples.)

However, for some important types of equations, many of the basic numerical methods developed for linear equations can be carried over with little difficulty. The cases of most importance are Urysohn and Hammerstein equations and their Volterra analogues, so we shall concentrate on them.

3.8.1. Urysohn and Hammerstein Equations. Consider the integral equation

$$u(t) = g(t) + \int_a^b K(t, s, u(s))\, ds, \qquad -\infty < a < b < \infty. \qquad (141)$$

Equation (141) is usually called an Urysohn equation. The particular case where

$$F(t, s, u) = K(t, s)f(s, u(s)) \qquad (142)$$

gives rise to the Hammerstein equation

$$u(t) = g(t) + \int_a^b K(t, s)f(s, u(s))\, ds, \qquad (143)$$

which can be considered the nonlinear analogue of a Fredholm equation of the second kind. For simplicity we shall restrict our attention to Eq. (143) and leave it to the reader to supply the necessary generalizations for Eq. (141).

As for Fredholm equations numerical methods generally fall into the categories of (1) analytic methods, (2) kernel approximation and projection methods, (3) quadrature methods, and (4) initial value techniques. Since the mechanics closely resemble those given in Section 3.1, we shall be somewhat more brief. The reader is referred to Refs. 1, 2, 4, and 70 and to Chapters 8 and 13 for further information.

Chief among the analytic methods is iteration. However, as is shown below, the implementation of iteration often requires some form of discretization, and we shall defer our analysis to later in this section.

To define projection methods Eq. (143) is written in operator form as

$$u = g + K(u), \qquad (144)$$

where $K(u)\colon U \subset X \to X$ is defined by

$$K(u) = \int_a^b K(t, s)f(s, u(s))\, ds, \qquad (145)$$

and for ease of discussion we shall take $U = X$. With reference to Section 3.1.3 a sequence of approximations u_n to u is given by

$$P_n(u_n - K(u_n) - g) = 0. \tag{146}$$

In the special case where $P_n u = u_n$, Eq. (146) becomes

$$u_n - P_n K(u_n) = P_n g. \tag{147}$$

To illustrate the use of Eq. (147) we shall consider polynomial collocation. This yields

$$u_n(t) - \sum_{k=1}^{n} l_k(t) \int_a^b K(t_k, s) f(s, u_n(s)) \, ds = \sum_{k=1}^{n} l_k(t) g(t_k). \tag{148}$$

Evaluating Eq. (148) at $t = t_j$, $j = 1, 2, \ldots, n$, and using $u_n = \sum_{k=1}^{n} a_k \phi_k(t)$, we obtain the nonlinear equations

$$\sum_{k=1}^{n} a_k \phi_k(t_j) - \int_a^b K(t_j, s) f\left(s, \sum_{k=1}^{n} a_k \phi_k(s)\right) ds = g(t_j), \qquad j = 1, 2, \ldots, n, \tag{149}$$

for the determination of $\{a_k\}_{k=1}^{n}$. In general Eq. (149) will have to be solved by iteration. If f is sufficiently smooth, then Newton's method seems to be preferred (Refs. 4 and 70). For nonsmooth f's very little work seems to have been reported; however, the recent development of quasi-Newton methods (Refs. 4, 7) should be of value for such problems.

Galerkin's method and least-squares procedures can also be developed along the lines of those for linear equations.

Quadrature methods are also straightforward. Replacing the integral in Eq. (143) by a numerical integration gives

$$u(t) \simeq g(t) + \sum_{k=1}^{n} w_k K(t, s_k) f(s_k, u(s_k)). \tag{150}$$

Defining $u_n(t)$ by

$$u_n(t) = g(t) + \sum_{k=1}^{n} w_k K(t, s_k) f(s_k, u_n(s_k)), \tag{151}$$

gives

$$u_n(t_j) = g(t_j) + \sum_{k=1}^{n} w_k K(t_j, s_k) f(s_k, u_n(s_k)), \qquad j = 1, 2, \ldots, n, \tag{152}$$

which is a set of nonlinear equations to determine approximations to $u(t)$ at $t = t_j$. Equation (151) then provides the analogue of the Nyström interpolant for linear equations and can be written in operator form as

$$u_n = g(t) + K_n(u_n), \tag{153}$$

where

$$K_n(u) = \sum_{k=1}^{n} w_k K(t, s_k) f(s_k, u(s_k)). \tag{154}$$

Again Eq. (153) generally has to be solved by iteration, with Newton's method the most popular (Refs. 1, 4, 70).

For nonsmooth kernels one can use singularity subtraction based on the formula

$$u(t) = g(t) + \int_a^b K(t, s)[f(s, u(s)) - f(t, u(t))]\, ds + f(t, u(t)) \int_a^b K(t, s)\, ds. \tag{155}$$

Likewise, by proceeding as in Section 3.1.7, product quadrature methods can be defined by

$$u_n(t_j) = g(t_j) + \sum_{k=1}^{n} w_k(t_j) M(t_j, s_k) f(s_k, u(s_k)), \qquad j = 1, 2, \ldots, n, \tag{156}$$

with the interpolation

$$u_n(t) = g(t) + \sum_{k=1}^{n} w_k(t) M(t, s_k) f(s_k, u(s_k)). \tag{157}$$

Volterra, imbedding, and boundary value methods have also been developed for the solution of nonlinear equations. A discussion of the Volterra method may be found in the paper of Rall (Ref. 28; see also Chapter 11). Imbedding methods are treated in the papers and book of Kagiwada and Kalaba (Ref. 4), and boundary value methods are briefly examined in Chapter 13. As with linear equations some of these procedures need restrictions on the type of kernel and/or the smoothness of K and g (Refs. 2, 28, 56). Some of these techniques have the attraction of being able to be implemented in a noniterative fashion and can therefore make use of well-developed software for nonlinear initial and boundary value problems. Boundary value methods may also be viewed as a convenient way of implementing kernel approximation methods (Ref. 56).

A somewhat different approach to solving Eq. (142) or (143) is via some sort of linearization procedure, usually Newton's method [or quasi-linearization, as it is often called (Ref. 72)]. If one assumes that $K(u)$ is Fréchet differentiable, then we can write

$$K(u) = K(v) + K'(v)(u - v) + \theta(u - v), \tag{158}$$

where $K'(v)$ is the Fréchet derivative of K at v (Ref. 70) and

$$\lim_{\|u-v\|\to 0} \|\theta(u-v)\|/\|u-v\| = 0.$$

Thus the right-hand side of Eq. (144) is approximately equal to

$$g + K'(v) + K'(v)(u-v),$$

so that

$$v \simeq g + K(v) + K'(v)(u-v),$$

where v is taken as some initial approximation to u. Iterating leads to the Newton–Kantorovich scheme

$$u_{n+1} = g + K(u_n) + K'(u_{n+1})(u_{n+1} - u_n), \tag{159}$$

or equivalently

$$u_{n+1} = g + K'(u_n)u_{n+1} + \Delta_n. \tag{160}$$

Writing Eq. (159) out explicitly shows that if $\lim_{n\to\infty} u_n(t) = u(t)$, then $u(t)$ may be obtained by solving the sequence of linear Fredholm equations of the second kind

$$u_{n+1}(t) = g(t) + \Delta_n(t) + \int_a^b K(t,s)f_u(s, u_n(s))u_{n+1}(s)\, ds. \tag{161}$$

The solution to Eq. (161) may then be approximated by any of the methods discussed in Section 3.2. An important point is that this generally does not provide an algorithm different from those discussed previously. This commutativity of linearization and discretization is illustrated for the Nyström method given by Eqs. (151) and (152).

Consider the operator form of Eq. (151) and linearize as above, giving

$$u_{n,m+1} = g + K_n(u_{n,m}) - K'(u_{n,m})u_{n,m} + K'(u_{n,m+1})u_{n,m+1}. \tag{162}$$

Discretizing Eq. (159) using the same quadrature rule as for Eq. (151) gives

$$u_{m+1,n}(t) = g(t) - \sum_{k=0}^n w_k K(t, s_k)f_j(s_k, u_m(s_k))u_{m,n}(t)$$

$$+ \sum_{k=1}^n w_k K(t, s_k)f(s_k, u_{n,m}(s_k))$$

$$+ \sum_{k=1}^n K(t, s_k)f_u(s_k, u_{n,m}(s_k))u_{m+1,n}(s_k). \tag{163}$$

To compare these, choose $u_{n,0}(t) = u_{0,m}(t)$. Assuming that each of Eqs. (162) and (163) has a unique solution shows that

$$u_{n,m} = u_{m,n}, \qquad (m, n) \ge 0. \tag{164}$$

A general discussion of the commutativity of Newton linearization and discretization for general functional equations is given by Ortega and Rheinboldt in Ref. 73. Although such results hold for other processes as well, we know of no other paper that discusses this point in a comprehensive fashion. We suggest that in general one make an independent check for a result of this type when other solution methods are used.

The convergence of quadrature and projection algorithms has been discussed by several authors, principally Vainikko in the Soviet Union (Ref. 74) and Anselone in the United States. Since the proofs are relatively technical, we shall content ourselves with an informal presentation and refer the reader to Refs. 1 and 4 for a more complete account.

For projection methods satisfying Eq. (147) we obtain, subtracting Eq. (147) from Eq. (144),

$$u - u_n = g - P_n g + K(u) - P_n K(u_n). \tag{165}$$

Adding and subtracting $P_n K(u)$, we obtain

$$
\begin{aligned}
u - u_n &= g - P_n g + K(u) - P_n K(u_n) + P_n K(u) - P_n K(u) \\
&= g - P_n g + (K(u) - P_n K(u)) + P_n K(u) - P_n K(u_n).
\end{aligned} \tag{166}
$$

Assuming that $K(u)$ is differentiable gives

$$K(u) - K(u_n) = K'(u_n)(u - u_n) + \theta(u - u_n). \tag{167}$$

Using Eq. (167) in Eq. (166) and rearranging, we obtain

$$
\begin{aligned}
(I - P_n K'(u_n))(u - u_n) &= g - P_n g + K(u) - P_n K(u) + \theta(u - u_n) \\
&= u - P_n u + \theta(u - u_n).
\end{aligned} \tag{168}
$$

Dropping the last term on the right-hand side and making the assumption that $(I - P_n K'(u_n))^{-1} \in [X]$, we find that

$$u - u_n \simeq (I - P_n K'_n(u_n))(u - P_n u). \tag{169}$$

From Eq. (169) we expect for n sufficiently large that

$$\|u - u_n\| \le C\|u - P_n u\|. \tag{170}$$

A more rigorous argument (Ref. 4) shows that Eq. (170) gives a correct bound.

For the Nyström method we get

$$u - u_n = K(u) - K_n(u_n).$$ (171)

Proceeding as above, we find that

$$u - u_n = K'_n(u_n)(u - u_n) + (K(u) - K_n(u_n)) + \theta(u - u_n).$$ (172)

From this it follows that

$$u - u_n \simeq (I - K'_n(u_n))^{-1}(K(u) - K_n(u_n)),$$ (173)

so that the error in u_n, as in the linear case, can be estimated in terms of the quadrature error for $K(u)$.

One can also carry our iteration on projection approximations, and formally similar properties to those in the linear case can be derived. It would be interesting to show that improved error estimates, such as those obtained for Galerkin's method and collocation for linear equations, hold.

  3.8.2. *Volterra Equations.*   The nonlinear analogues of Eqs. (139) and (140) are

$$u(t) = g(t) + \int_a^t F(t, s, u(s)) \, ds$$ (174)

and

$$g(t) = \int_a^t F(t, s, u(s)) \, ds.$$ (175)

We shall consider the solution of Eq. (174) first. One approach is to make the observation that if

$$\tilde{F}(t, s, u) = \begin{cases} F(t, s, u), & a \leq s < t, \\ 0, & t \leq s \leq b, \end{cases}$$ (176)

then Eq. (174) can be written as a Fredholm equation,

$$u(t) = g(t) + \int_a^b \tilde{F}(t, s, u(s)) \, ds.$$ (177)

In this form any of the methods in the previous section may be applied. However, if we consider the particular case where

$$F(t, s, u) = K(t, s)f(s, u(s))$$ (178)

so that

$$\tilde{F}(t, s, u) = \tilde{K}(t, s)f(s, u(s))$$ (179)

with

$$\tilde{K}(t, s) = \begin{cases} K(t, s), & a \le s \le t, \\ 0, & t < s \le b, \end{cases} \tag{180}$$

then Eq. (177) has a discontinuous kernel unless $K(t, t) = 0$. Consequently most "global" methods will tend to be inefficient.

A more fruitful and generally recommended approach is to view Eq. (174) as a generalization of the initial value problem for ordinary differential equations and to make use of methods which exploit this analogy (Refs. 4, 75). Here there seems to be three distinct approaches: (1) step-by-step and related block-by-block methods (Refs. 4, 75), (2) Runge–Kutta methods (Refs. 4, 76), and (3) degenerate kernel methods (Refs. 32–33).

In the simplest step-by-step methods we solve Eq. (174) over the interval $[a, b]$. Let $h = (b - a)/n$ and define $t_k = a + kh$, $k = 0, 1, 2, \ldots, n$. Then

$$u(t_k) = g(t_k) + \int_a^{t_k} F(t_k, s, u(s)) \, ds. \tag{181}$$

If $F(t, s, u)$ is smooth and the integral in Eq. (181) is approximated by a quadrature rule using values of the integrand at the points t_j, $j = 0, 1, 2, \ldots, k$, then

$$u(t_k) \simeq g(t_k) + \sum_{j=0}^{k} w_j F(t_k, t_j, u(t_j)). \tag{182}$$

Using Eq. (182), we seek approximate values of $u(t_k)$ by solving the set of nonlinear equations

$$\mathbf{u}(t_k) = g(t_k) + \sum_{j=0}^{k} w_{kj} F(t_k, t_j, \mathbf{u}(t_j)), \qquad k = 0, 1, 2, \ldots, n. \tag{183}$$

Examination of Eq. (183) shows that it is triangular and that it is possible to solve these equations in the forward direction to obtain $\mathbf{u}(t_0), \mathbf{u}(t_1), \ldots, \mathbf{u}(t_n)$ in a sequential fashion. For example, taking $\mathbf{u}(t_0) = g(t_0) = g(a)$, we get

$$u(t_1) = g(t_1) + w_{10} F(t_1, t_0, \mathbf{u}(t_0)) + w_{11} F(t_1, t_1, \mathbf{u}(t_1)). \tag{184}$$

By using $\mathbf{u}(t_0) = g(a)$, Eq. (184) is seen to give an equation involving only $\mathbf{u}(t_1)$. Simple induction now shows that the kth equation can be solved for $\mathbf{u}(t_k)$ alone once $\mathbf{u}(t_j)$, $j = 0, 1, 2, \ldots, k - 1$, have been determined. It is this feature of being able to solve for the values $\mathbf{u}(t_k)$ one at a time, as for ordinary differential equations, that makes these methods attractive.

By making different choices of the weights $\{w_{kj}\}$ one arrives at a large variety of potential methods for solving Eq. (174). Proper choices should give high-order truncation error and, as for differential equations, should be stable. Baker (Ref. 4) presents a comprehensive account of many such procedures. Noble (Ref. 58) suggests using "something simple," such as the repeated trapezoid rule for smooth F's and product quadrature based on a piecewise linear representation of $u(t)$ otherwise. Equation (174) may then be solved for step sizes h, $2h$, $4h$, ... and the results extrapolated to obtain higher-order accuracy.

If $F(t, s, u)$ has the form given in Eq. (180), then the analogy between Eq. (174) and the initial value problem may be exploited in a somewhat different fashion. Formally expand the kernel $K(t, s)$ as

$$K(t, s) = \sum_{n=1}^{\infty} a_n(t) b_n(s) \qquad (185)$$

and substitute into Eq. (174), giving

$$u(t) = g(t) + \sum_{n=1}^{\infty} a_n(t) \int_a^t b_n(s) f(s, u(s))\, ds$$

$$= g(t) + \sum_{n=1}^{\infty} a_n(t) z_n(t), \qquad (186)$$

where

$$z_n(t) = \int_a^t b_n(s) f(s, u(s))\, ds.$$

Under appropriate continuity conditions we have

$$z_n'(t) = b_n(t) f(t, u(t)),$$
$$z_n(a) = 0, \qquad n \geq 1. \qquad (187)$$

Equations (185) and (186) have been used in two ways. Bownds and Wood in Refs. 32–33 consider truncating the series for some finite N, leading to the system of ordinary differential equations

$$\mathbf{z}_n'(t) = f(t, g(t) + \sum_{k=1}^{N} a_n(t) \mathbf{z}_n(t)),$$
$$\mathbf{z}_n(a) = 0, \qquad n = 1, 2, \ldots, N, \qquad (188)$$

which may then be integrated using standard techniques. For smooth kernels the method is fast and produces accurate solutions with readily available software. This approach is discussed further in Chapters 9 and 10. If the solution $u(t)$ is required at m points, then this method requires $O(m)$ kernel evaluations in contrast to the $O(m^2)$ required by most step-by-step methods (Refs. 4 and 75–76).

In another direction Pouzet has applied Runge–Kutta methods for ordinary differential equations to the system in Eq. (187) and thus derived a class of Runge–Kutta methods for Eq. (174). This approach also leads to methods that require $O(m^2)$ kernel evaluations, which is interesting since the direct integration of Eq. (188) requires only $O(m)$ evaluations.

Our discussions of Eq. (175) will be limited to the cases where $F(t, s, u)$ is linear in u and so has the form

$$F(t, s, u) = K(t, s)u. \tag{189}$$

In practice one usually has $K(t, s)$ continuous or

$$K(t, s) = H(t, s)/(t - s)^\alpha, \qquad 0 < \alpha < 1, \tag{190}$$

with $H(t, s)$ continuous. As for Eq. (174) one can write Eq. (175) as

$$\int_a^b \tilde{K}(t, s)u(s)\, ds = g(t), \tag{191}$$

where $\tilde{K}(t, s)$ is given by Eq. (180). Given our assumptions on the kernel, the operator defined by the right-hand side of Eq. (194) will be compact on $C[a, b]$, and so it represents an ill-posed, or mildly ill-posed, problem. For this reason care must be taken in devising adequate numerical methods.

When $K(t, s)$ is continuous an approximate solution may be obtained by the use of quadrature methods analogous to those discussed previously for equations of the second kind. Discretizing, as in Eq. (183), leads to the set of equations

$$\sum_{j=0}^k w_{kj} K(t_k, t_j)\mathbf{u}(t_j) = g(t_k), \qquad k = 0, 1, 2, \ldots, n, \tag{192}$$

for approximating $u(t)$ at $t = t_k$, $k = 0, 1, 2, \ldots, n$. We note that the system in Eq. (192) consists of n equations in $n + 1$ unknowns, and so a starting value for $\mathbf{u}(t_0)$ is required. If $g'(a)$ exists, $K(a, a) \neq 0$, and $\partial K(t, s)/\partial t$ is continuous, then $u(a)$ may be obtained by differentiation of Eq. (174), giving

$$u(a) = g'(a)/K(a, a). \tag{193}$$

By setting

$$\mathbf{u}(t_0) = \mathbf{u}(a) = g'(a)/K(a, a),$$

Eq. (192) may now be solved in a step-by-step fashion. Usually low-order quadrature rules such as the repeated trapezoidal are used, and some form of extrapolation and smoothing are needed to provide good accuracy and to control oscillations in the computed values (Ref. 4).

In view of the ill-posed nature of Eq. (174) regularization methods may also be employed. This has been proposed by Schmaedeke and studied extensively by Anderssen and his colleagues (Refs. 54, 77).

A third method is to use the indirect approach of converting Eq. (175) to an equation of the second kind. If $K(t, t) \neq 0$ and $g'(t)$ and $\partial K(t, s)/\partial t$ exist and are continuous, then differentiation of Eq. (175) gives

$$K(t, t)u(t) + \int_a^t [\partial K(t, s)/\partial t]u(s)\, ds = g'(t). \qquad (194)$$

Provided that $g'(t)$ and $\partial K(t, s)/\partial t$ are readily available, $u(t)$ may be computed by methods of the type discussed above for equations of the second kind.

For kernels of the form given in Eq. (190) quadrature methods may be replaced by product quadrature techniques, or regularization may be used (Ref. 54). In some circumstances it may be feasible to convert Eq. (174) to one of the second kind (Refs. 4, 33) and apply the methods discussed for these equations.

As pointed out in Section 3.6, equations of this type often arise in situations where $g(t)$ is obtained from noisy data. In practical applications, such as particle statistics (Refs. 54, 64), linear functionals of the solution are important. Here Anderssen advocates a procedure which we have shown to be equivalent to solving an appropriate adjoint problem (Refs. 54–55).

4. Applications

In the previous sections we have discussed in some detail the existing literature on the numerical solution of integral equations. Needless to say, such work should be considered in its relation to specific applications. As we have indicated in Section 1, and at other points, integral equations occur in many branches of science and technology, and in many circumstances there has been substantial two-way interaction between the mathematical theory and applications. In fact the impetus for much of the classical Fredholm theory came from the work of nineteenth-century mathematicians on the Dirichlet problem for Laplace's equation. (See Ref. 65 for an interesting historical discussion.) This symbiotic relationship has continued with the development of the theory of singular integral equations being closely related to problems in elasticity and potential theory (Ref. 57) through to the more recent work on imbedding methods as an outgrowth of investigations in radiative transfer (Refs. 2, 29).

Several good reference sources are Noble's bibliography (Ref. 9) and the recent articles of Lonseth and Anderssen (Refs. 54, 65). Further extensive bibliographies may be found in many of the chapters in this book.

Possibly the most common source of integral equations is the reformulation of boundary value problems in mathematical physics. Although much of present numerical analysis of such problems appears to focus on direct finite difference and finite element methods, many problems have characteristics which make numerical solution via conversion to an equivalent integral equation attractive (Refs. 8, 12, 15, 57). Some features of this type are (1) complicated, particularly multiply connected domains; (2) noncompact domains; (3) solutions with known singularities; (4) problems for higher-order equations such as the biharmonic; (5) problems where dimensionality reduction is important; and (6) problems with mixed boundary conditions. Many classical problems in potential theory and elasticity have some or all of these features (Refs. 57, 65), and their integral formulations are well known to mathematicians. A somewhat less known area with similar features is that of aerodynamics. Here one is interested in solving Laplace's, Helmholtz's, and other linear and nonlinear equations which arise in the description of flows past airfoils, wings, and aircraft, both in free air and in wind tunnels. These problems exhibit many of the complexities described above, and integral equation techniques (often called kernel function methods) have been the preferred method of solution for many years (Refs. 7, 8, 12). Although at one time mathematicians seem to have been actively involved in this area (Ref. 65), most recent work appears to have been accomplished by engineers. A wealth of literature exists; however, much of it is confined to governmental and industrial technical reports (Refs. 8, 12) and so is not readily accessible to the mathematics community. There are a large number of interesting problems, and we believe that the field can be a fertile one for numerical analysts. For example, after almost 60 years of numerical calculations on equations of the type given by Eqs. (109) and (110) convergence proofs have only recently appeared (Refs. 12, 16, 17, 41). Some current work in this area is discussed in detail in Chapters 4 and 5. One sees there that boundary value problems of some complexity can be conveniently solved numerically in terms of an equivalent integral equation. For practical purposes computational times are 1–2 orders of magnitude smaller than can be achieved by a direct solution of the partial differential equation. In fact the whole area of subsonic and transonic flows past airfoils appears to be an important testing ground of the relative merits of an integral equation approach versus the more traditional (at least to numerical analysts) solution of the differential equation. Since we believe that this area has not received much recent exposure among mathematicians, we include a separate bibliography on this topic following Chapter 5.

Of course integral equations occur in other ways as well. Abel equations arise in problems of stereology, particle statistics, and geophysics (Refs. 55, 64–65); Wiener–Hopf and certain related nonlinear equations occur in problems of radiative transfer (Refs. 2, 14, 29); and Volterra equations of the second kind appear in the description of heat flow and probability distributions (Refs. 55, 78).

One could continue of course, and the remaining chapters, along with the cited reference material, should be consulted for further specific details.

5. Conclusions

During the last 15 years there has been rapid development of numerical techniques for the solution of integral equations along with increased interest in the application of these methods to significant scientific and technological problems. In this chapter we have tried to survey some of the highlights of this work. In the remainder of the book many of the ideas that we have just touched on are developed more fully.

It goes without saying that it would be impossible to give a comprehensive treatment of so vast a subject in a single volume. Where topics have not been covered we have tried to supply adequate references to the literature, and the interested reader is urged to consult them.

Although it is difficult to predict the directions research will take, there seem to be certain problems that should be tackled in the near future. First, it would be useful to reach the level of software development that has been achieved for many problems in differential equations (Refs. 79–80). This is important not only for the purpose of having efficient, readily available programs but also so that one may begin the formal type of comparison work that such people as Hull and Enright (Refs. 81–82) have initiated for ordinary differential equations. This will probably require closer cooperation among the numerical analyst, the scientist, and the computer scientist (Ref. 58). For example, it is not even clear what a good measure of computational efficiency is. Minimum solution times are not the only thing that should be considered. Since computer charging algorithms often use storage as a major component, methods that take longer but use less memory may be preferred to those whose computation time is shorter (Ref. 3). With the exception of the work of Atkinson (Ref. 3), little seems to have been accomplished along these lines.

Second, for problems where an integral equation approach competes with a direct differential equation formulation, it would be helpful to know when the integral equation is to be preferred. One difficulty here is that where comparisons are made (Refs. 71, 83) "first-cut" programs for integral equations are put up against "state-of-the-art" codes for partial differential equations. Even so, the integral equation may win out (Ref. 71).

Of course, further development of new algorithms and more efficient implementation of existing ones should take place, and one would also like to see more attention being paid to multidimensional equations.

We have found the area to be challenging, both mathematically and computationally, and we hope the reader does as well. So read on, and "bon appetit!"

References

1. ANSELONE, P. M., *Collectively Compact Operator Approximation Theory and Applications to Integral Equations*, Prentice-Hall, Englewood Cliffs, New Jersey, 1971.
2. KAGIWADA, H., and KALABA, R. E., *Integral Equations Via Imbedding Methods*, Addison-Wesley Publishing Company, Reading, Massachusetts, 1974.
3. ATKINSON, K. E., *A Survey of Numerical Methods for the Solution of Fredholm Integral Equations of the Second Kind*, Society for Industrial and Applied Mathematics, Philadelphia, Pennsylvania, 1976.
4. BAKER, C. T. H., *The Numerical Treatment of Integral Equations*, Cambridge University Press, Oxford, England, 1977.
5. IVANOV, V. V., *The Theory of Approximate Methods and Their Application to the Numerical Solution of Singular Integral Equations*, Translated by A. Ideh, edited by R. S. Anderssen and D. Elliott, Noordhoff International Publishing Company, Leyden, Holland, 1976 (Russian edition published in 1968).
6. DELVES, L. M., and WALSH, J., *Numerical Solution of Integral Equations*, Clarendon Press, Oxford, England, 1974.
7. LOCK, R. C., *Methods for Elliptic Problems in External Aerodynamics*, Computational Methods and Problems in Aeronautical Fluid Dynamics, Edited by B. L. Hewitt, C. R. Illingworth, R. C. Lock, K. W. Mangler, J. H. McDowell, C. Richards, and I. Walkden, Academic Press, New York, New York, 1976.
8. ROWE, W., REDMAN, M., EHLERS, F., and SEBASTIAN, J., *Prediction of Unsteady Loadings Caused by Leading and Trailing Edge Control Surface Motions in Subsonic Compressible Flow*, National Aeronautics and Space Administration, Contractor Report No. 2543, 1975.
9. NOBLE, B., *A Bibliography on Methods for Solving Integral Equations*, Mathematics Research Center Technical Summary Report No. 1176, Madison, Wisconsin, 1971.
10. DEFRANCO, R. J., *Stability Results for Multiple Volterra Equations*, University of Arizona, PhD Thesis, 1973.

11. DAVIES, B., *Integral Transforms and Their Applications*, Springer-Verlag, New York, New York, 1978.

12. FROMME, J., and GOLBERG, M., *Unsteady Two Dimensional Airloads Acting on Oscillating Thin Airfoils in Subsonic Ventilated Wind Tunnels*, National Aeronautics and Space Administration, Contractor Report No. 2967, 1978.

13. FROMME, J., and GOLBERG, M., *Numerical Solution of a Class of Integral Equations Arising in Two Dimensional Aerodynamics*, Journal of Optimization Theory and Applications, Vol. 24, No. 1, 1978.

14. MILNE, R., *Application of Integral Equations for Fluid Flows in Unbounded Domains*, Finite Elements in Fluids, Vol. 2, Edited by R. H. Gallagher, J. T. Oden, and O. C. Zienkiewicz, John Wiley and Sons, New York, New York, 1975.

15. NISHIYAMA, T., *Lifting Surface Theory of Fully Submerged Hydrofoils*, Journal of Ship Research, Vol. 8, No. 4, 1965.

16. DOW, M. L., and ELLIOTT, D., *The Numerical Solution of Singular Integral Equations over* $[-1, 1]$ Society for Industrial and Applied Mathematics Journal on Numerical Analysis, Vol. 16, No. 1, 1979.

17. LINZ, P., *An Analysis of a Method for Solving Singular Integral Equations*, BIT, Vol. 17, pp. 329–337, 1977.

18. LANDAHL, M. T., and STARK, V. J. F., *Numerical Lifting Surface Theory— Problems and Progress*, Journal of the American Institute of Aeronautics and Astronautics, Vol. 6, No. 11, 1968.

19. ALLEN, R. C., and WING, G. M., *A Method for Accelerating the Iterative Solution of a Class of Fredholm Integral Equations*, Chapter 2, this volume.

20. BELLMAN, R., KALABA, R., and LOCKETT, J. A., *Numerical Inversion of the Laplace Transform*, American Elsevier Publishing Company, New York, New York, 1966.

21. STENGER, F., *Connection Between a Cauchy System Representation of Kalaba and Fourier Transforms*, Applied Mathematics and Computation, Vol. 1, No. 1, 1975.

22. GOLBERG, M. A., *The Conversion of Fredholm Integral Equations to Equivalent Cauchy Problems*, Applied Mathematics and Computation, Vol. 2, pp. 1–18, 1976.

23. GOLBERG, M. A., *The Conversion of Fredholm Integral Equations to Equivalent Cauchy Problems—Computation of Resolvents*, Applied Mathematics and Computation, Vol. 3, No. 1, 1977.

24. AALTO, S. K., *Reduction of Fredholm Integral Equations with Green's Function Kernels of Volterra Equations*, Oregon State University, MS Thesis, 1966.

25. SLOAN, I., BURN, B., and DATYNER, N., *A New Approach to the Numerical Solution of Integral Equations*, Journal of Computational Physics, Vol. 18, No. 1, 1975.

26. PHILLIPS, J., *The Use of Collocation as a Projection Method for Solving Linear Operator Equations*, Society for Industrial and Applied Mathematics Journal on Numerical Analysis, Vol. 9, pp. 14–27, 1972.

27. NYSTRÖM, E. J., *Über die Praktische Auflösung von Integralgleichingen mit Anwerdungen auf Randwertanfgaben*, Acta Mathematica, Vol. 54, pp. 185–204, 1930.

28. RALL, L., *Resolvent Kernals of Green's Function Kernels and Other Finite Rank Modifications of Fredholm and Volterra Kernels*, Journal of Optimization Theory and Applications, Vol. 24, No. 1, 1978.

29. CHANDRASEKHAR, S., *Radiative Transfer*, Clarendon Press, Oxford, England, 1950.

30. GOLBERG, M. A., *Initial Value Methods in the Theory of Fredholm Integral Equations*, Journal of Optimization Theory and Applications, Vol. 9, pp. 112–119, 1972.

31. GOLBERG, M. A., *Convergence of an Initial Value Method for Solving Fredholm Integral Equations*, Journal of Optimization Theory and Applications, Vol. 12, pp. 334–356, 1973.

32. BOWNDS, J. M., and WOOD, B., *On Numerically Solving Non-linear Volterra Integral Equations with Fewer Computations*, Society for Industrial and Applied Mathematics Journal on Numerical Analysis, Vol. 13, pp. 705–719, 1976.

33. BOWNDS, J. M., *On Solving Weakly Singular Volterra Equations of the First Kind with Galerkin Approximations*, Mathematics of Computation, Vol. 30, pp. 747–757, 1976.

34. GOURSAT, E., *Determination de la Resolvante d'une Equation Volterra*, Bulletin des Sciences et Mathematiques, Vol. 57, pp. 144–150, 1933.

35. SHUMITZKY, A., *On the Equivalence Between Matrix Riccati Equations and Fredholm Resolvents*, Journal of Computer and System Science, Vol. 3, pp. 76–87, 1968.

36. STENGER, F., *The Approximate Solution of Wiener–Hopf Integral Equations*, Journal of Mathematical Analysis and Applications, Vol. 3, pp. 687–724, 1972.

37. KAILATH, T., *Some New Algorithms for Recursive Estimation in Constant Linear Systems*, Institute for Electrical and Electronic Engineering Transactions on Information Theory, Vol. 19, pp. 750–760, 2973.

38. VAINIKKO, G. M., *On the Stability and Convergence of the Collocation Method*, Differential Equations, Vol. 1, pp. 186–195, 1965.

39. WALSH, J., *Boundary-Value Problems in Ordinary Differential Equations*, The State of the Art in Numerical Analysis, Edited by D. A. H. Jacobs, Academic Press, New York, New York, 1977.

40. SZËGO, G., *Orthogonal Polynomials*, American Mathematical Society Colloquium Publications, Vol. 23, Fourth Edition, Providence, Rhode Island, 1975.

41. FROMME, J. A., and GOLBERG, M. A., *On the L_2 Convergence of Collocation for the Generalized Airfoil Equation* (to appear).

42. FROMME, J. A., and GOLBERG, M. A., *Computation of Aerodynamic Interference Effects on Oscillating Airfoils with Flaps*, National Aeronautics and Space Administration, Contractor Report No. 2967, May 1978.

43. SLOAN, I. H., *Improvement by Iteration for Compact Operator Equations*, Mathematics of Computation, Vol. 30, pp. 758–764, 1976.

44. SLOAN, I. H., *Iterated Galerkin Method for Eigenvalue Problems*, Society for Industrial and Applied Mathematics Journal on Numerical Analysis, Vol. 13, pp. 753–760, 1976.
45. CHATELIN, F., *Theorie de l'Approximation des Operateurs Lineaires—Application au Calcul del Valeurs Propres D'operateurs Differentiels et Integraux*, Universite Scientifique et Medicale de Grenoble, Institute de Recherche en Mathematiques Avancees, Grenoble, France, 1977.
46. BRAKHAGE, H., *Über die Numerische Behandlung von Integralgleichungen Nach der Quadrature Formelmethode*, Numerische Mathematik, Vol. 2, pp. 183–196, 1960.
47. ANSELONE, P. M., and MOORE, R. H., *Approximate Solutions of Integral and Operator Equations*, Journal of Mathematical Analysis and Applications, Vol. 9, pp. 268–277, 1964.
48. DE HOOG, F., and WEISS, R., *Asymptotic Expansions for Product Integration*, Mathematics of Computation, Vol. 27, pp. 295–306, 1973.
49. ATKINSON, K., *Iterative Variants of the Nyström Method for the Numerical Solution of Integral Equations*. Numerische Mathematik, Vol. 22, pp. 17–31, 1973.
50. KANTOROVICH, L. V., *Functional Analysis and Applied Mathematics*, Uspekhi Mathematika Nauk, Vol. 3, pp. 89–185, 1948.
51. WILLIAMS, M. H., *The Resolvent of Singular Integral Equations*, Quarterly of Applied Mathematics, Vol. 28, pp. 99–110, 1977.
52. WILLIAMS, M. H., *Exact Solutions in Oscillating Airfoil Theory*, Journal of the American Institute of Aeronautics and Astronautics, Vol. 15, pp. 875–877, 1977.
53. WILLIAMS, M. H., *The Solution of Singular Integral Equations by Jacobi Polynomials* (to appear).
54. ANDERSSEN, R. S., *Application and Numerical Solution of Abel Type Integral Equations*, Mathematics Research Center, University of Wisconsin, Madison, Technical Summary Report No. 1787, 1977.
55. GOLBERG, M. A., *A Method of Adjoints for Solving Some Ill-posed Equations of the First Kind*, Journal of Applied Math. and Computation, Vol. 5, No. 2, pp. 123–130, 1979.
56. GOLBERG, M. A., *Boundary and Initial-Value Methods for Solving Fredholm Equations with Semidegenerate Kernels*, Journal of Optimization Theory and Applications, Vol. 24, No. 1, 1978.
57. MUSKHELISHVILI, N. I., *Singular Integral Equations*, Noordhoff International Publishing Company, Amsterdam, Holland, 1953.
58. NOBLE, B., *The Numerical Solution of Integral Equations*, The State of the Art in Numerical Analysis, Edited by D. A. H. Jacobs, Academic Press, New York, New York, 1977.
59. ERDOGAN, F., GUPTA, G. D., and COOK, T. S., *Numerical Solution of Singular Integral Equations*, Mechanics of Fracture, Vol. 1, pp. 368–425, 1973.
60. SÖHNGEN, H., *Zur Theorie der Endlichen Hilbert Transformation*, Mathematische Zeitschift, Vol. 60, pp. 31–51, 1954.
61. TRICOMI, F., *On the Finite Hilbert Transformation*, Quarterly Journal of Mathematics, Vol. 2, pp. 199–211, 1951.

62. TRICOMI, F., *Integral Equations*, Interscience Publishers, New York, New York, 1957.
63. RALSTON, A., and RABINOWITZ, P., *A First Course in Numerical Analysis*, Second Edition, McGraw-Hill Book Company, New York, New York, 1978.
64. GOLDMAN, A., and VISSCHER, W., *Applications of Integral Equations in Particle Size Statistics*, Chapter 6, this volume.
65. LONSETH, A. T., *Sources and Applications of Integral Equations*, Society for Industrial and Applied Mathematics Review, Vol. 19, No. 2, 1977.
66. TWOMEY, S., *The Application of Numerical Filtering to the Solution of Integral Equations Encountered in Indirect Sensing Measurements*, Journal of the Franklin Institute, Vol. 279, pp. 95–109, 1965.
67. TIKHONOV, A. N., *On the Solution of Ill-posed Problems and the Method of Regularization*, Soviet Mathematics, Vol. 4, pp. 1035–1038, 1963.
68. WAHBA, G., *Practical Approximate Solutions to Linear Operator Equations when the Data Are Noisy*, Society for Industrial and Applied Mathematics Journal on Numerical Analysis, Vol. 14, No. 4, 1977.
69. STRAND, O. N., *Theory and Methods Related to the Singular Function Expansion and Landweber's Iteration for Integral Equations of the First Kind*, Society for Industrial and Applied Mathematics Journal on Numerical Analysis, Vol. 11, pp. 798–825, 1974.
70. RALL, L. B., *Computational Solution of Nonlinear Operator Equations*, John Wiley and Sons, New York, New York, 1969.
71. KRAFT, E. M., and LO, C. F., *Analytical Determination of the Blockage Effect in a Perforated Wall Wind Tunnel*, Journal of the American Institute of Aeronautics and Astronautics, Vol. 15, No. 4, 1977.
72. BELLMAN, R., and KALABA, R. E., *Quasilinearization and Boundary Value Problems*, American Elsevier Publishing Company, New York, New York, 1965.
73. ORTEGA, J. M., *On Discretization and Differentiation of Operators with Applications to Newton's Method*, Society for Industrial and Applied Mathematics Journal on Numerical Analysis, Vol. 3, No. 1, 1966.
74. VAINIKKO, G. M., *Galerkin's Perturbation Method and the General Theory of Approximate Methods for Non-linear Equations*, U.S.S.R. Computational Mathematics and Mathematical Physics, Vol. 7, pp. 1–41, 1967.
75. GAREY, L., *Solving Nonlinear Second Kind Volterra Equations by Modified Increment Methods*, Society for Industrial and Applied Mathematics Journal on Numerical Analysis, Vol. 12, pp. 501–508, 1975.
76. POUZET, P., *Methode l'Integration Numerique des Equations Integrales et Integro-differentielles du type de Volterra de Second Espece-Formules de Runge-Kutta*, Symposium on the Numerical Treatment of Ordinary Differential Equations, Integral and Integro-differential Equations, Birkhauser-Verlag, Basel, Switzerland, 1960.
77. SCHMAEDEKE, W. W., *Approximate Solutions of Volterra Integral Equations of the First Kind*, Journal of Mathematical Analysis and Applications, Vol. 24, pp. 604–613, 1968.

78. MILLER, R. K., *Nonlinear Volterra Integral Equations*, W. A. Benjamin Inc., Menlo Park, California, 1971.
79. SHAMPINE, L. F., and WATTS, H. A., *Solving NonStiff Ordinary Differential Equations—The State of the Art*, Society for Industrial and Applied Mathematics Review, Vol. 18, pp. 376–411, 1976.
80. SCOTT, M. R., and WATTS, H. A., *Computational Solution of Linear Two Point Boundary Value Problems via Orthonormalization*, Society for Industrial and Applied Mathematics Journal on Numerical Analysis, Vol. 14, pp. 40–70, 1977.
81. HULL, T. E., ENRIGHT, W. H., FELLEN, B. M., and SEDGWICK, A. F., *Comparing Numerical Methods for Ordinary Differential Equations*, Society for Industrial and Applied Mathematics Journal on Numerical Analysis, Vol. 9, pp. 603–637, 1972.
82. HULL, T. E., and ENRIGHT, W. H., *Test Results on Initial Value Methods for Non-stiff Ordinary Differential Equations*, Society for Industrial and Applied Mathematics Journal on Numerical Analysis, Vol. 13, pp. 944–961, 1976.
83. NIXON, D., *An Extended Integral Equation Method for the Unsteady Transonic Flow Past a Two-dimensional Airfoil*, Computational Methods and Problems in Aeronautical Fluid Dynamics, Edited by B. L. Hewitt, C. R. Illingsworth, R. C. Lock, K. W. Mangler, J. H. McDonnel, C. Richards, and F. Walkden, Academic Press, New York, New York, 1976.

2

A Method for Accelerating the Iterative Solution of a Class of Fredholm Integral Equations[1,2]

R. C. ALLEN[3] AND G. M. WING[4]

Abstract. The present paper extends the synthetic method of transport theory to a large class of integral equations. Convergence and divergence properties of the algorithm are studied analytically, and numerical examples are presented which demonstrate the expected theoretical behavior. It is shown that, in some instances, the computational advantage over the familiar Neumann approach is substantial.

1. Introduction

In 1962, Kopp (Ref. 1) introduced the so-called *synthetic method* as a device for performing certain nuclear reactor calculations. The basic idea can be traced to early work of Cesari (Ref. 2). The synthetic algorithm is often very successful in speeding up transport calculations. However, in some instances, its behavior is anomalous; indeed, on occasion one finds that the iterative scheme apparently diverges even though the physics of the problem suggests no reason for this behavior.

In order to try to get some basic idea of the phenomenon involved, the authors analyzed the method in a very simple model: the classical constant-parameter slab geometry (Ref. 3). That investigation revealed that the synthetic method may indeed diverge, and a criterion for this phenomenon was found. Various numerical calculations supplemented and complemented these investigations.

---

[1] The authors acknowledge with pleasure conversations with Paul Nelson. Thanks are due also to Janet E. Wing, whose computer program was used in making the calculations reported in Section 8.

[2] This work was performed in part under the auspices of USERDA at the Los Alamos Scientific Laboratory of the University of California, Los Alamos, New Mexico.

[3] Associate Professor, Department of Mathematics, University of New Mexico, Albuquerque, New Mexico.

[4] Professor of Mathematics, Southern Methodist University, Dallas, Texas.

It was also noted in the course of the study that the synthetic method need by no means be confined to problems arising from transport or reactor theory. Rather, it provides a very effective algorithm for numerically analyzing a large and important class of integral equations, namely Fredholm equations of the second kind with displacement kernels. It is this generalization of the method that we shall discuss in this chapter, both from analytical and numerical viewpoints.

In Section 2, we state the problem in more specific terms and derive the synthetic algorithm. Actually, we shall find that the algorithm is really a special case of the splitting method of matrix theory applied to more general operators. In the following section, we construct an approximate operator which plays a key role in the synthetic scheme. The principal eigenfunction and eigenvalue of this approximate operator are also exhibited.

The basis of our analytical arguments is knowledge of the highest eigenvalue of the original Fredholm operator. To ask that this eigenvalue be known exactly is far too great a requirement. Rather, we exhibit classes of displacement kernels whose highest eigenvalue can be estimated from above and below quite accurately. Such classes are by no means small or uninteresting. This entire matter is discussed in Section 4, and the necessary estimates are obtained.

The next two sections are devoted to divergence and convergence results. These are based upon the estimates given in Section 4.

In Section 7, we discuss briefly the choice of *best* approximate operators, where best is defined in terms of the effectiveness of the algorithm. It must be admitted that our results here are not entirely satisfactory. Only a relatively crude numerical device has been found to expedite the making of this choice. One suspects that a more sophisticated analytical criterion must be available, but it eludes us. However, the method does give some reasonable information on this optimal choice, as Section 8 verifies. There, several numerical examples of the synthetic method applied to a variety of displacement kernels are discussed in some detail. The results illustrate quite dramatically the efficacy of the method. We include, of course, the transport problem which originally occupied our attention.

We conclude the chapter with a discussion of some problems which are left open for investigation. In many of the cases mentioned, it is quite possible to simply apply the synthetic algorithm numerically in the hope that it will remain effective. However, the dramatic success of the method demonstrated in Section 8 is accompanied by equally dramatic failure in some cases. If one is to apply the device in new situations, the need for careful analysis and greater understanding is very clear.

2. Statement of the Problem and Derivation of the Algorithm

Throughout this chapter, we shall be studying integral equations of the form

$$\varphi(x) = \int_{-a}^{a} K(|x - y|)\{\gamma\varphi(y) + S(y)\}\, dy. \tag{1}$$

Unless explicit mention is made to the contrary, we shall always be working in an L_2-context, so that the classical Hilbert–Schmidt theory applies. All functions are real, a is finite, and γ is a real positive parameter.

It is well known that (1) is soluble in L_2 provided γ is less than the reciprocal of the spectral radius of the operator K:

$$K \cdot = \int_{-a}^{a} K(|x - y|) \cdot dy. \tag{2}$$

Moreover, this solution may be obtained by formation of the usual Neumann series, equivalent to the iterative scheme,

$$\varphi_{i+1}(x) = \int_{-a}^{a} K(|x - y|)\{\gamma\varphi_i(y) + S(y)\}\, dy,$$
$$\varphi_0(x) \equiv 0. \tag{3}$$

However, it is also the case that this scheme usually converges very slowly when γ^{-1} is near the spectral radius of K, often making this approach numerically useless. Moreover, many other numerical methods for solving (1) also become ill-behaved under these circumstances. It is this difficulty that we propose to investigate.

Let us consider Eq. (1) written in operator form as

$$\varphi = K(\gamma\varphi + S). \tag{4}$$

We introduce a new operator K_L (called the low-order operator in reactor theory literature), which, for the moment, we may think of as simply some sort of an approximation to K. Write

$$(I - \gamma K_L + \gamma K_L - \gamma K)\varphi = KS, \tag{5}$$

or, equivalently,

$$\varphi = (I - \gamma K_L)^{-1}\gamma(K - K_L)\varphi + (I - \gamma K_L)^{-1}KS. \tag{6}$$

Equation (6) suggests the iterative scheme

$$\varphi_{i+1} = (I - \gamma K_L)^{-1}\gamma(K - K_L)\varphi_i + (I - \gamma K_L)^{-1}KS. \tag{7}$$

This is actually equivalent to the synthetic method, although (7) usually appears written in a quite different form. We shall discuss that form when the algorithm is implemented in Section 8. There, we shall also choose a reasonable value for φ_0. For now, we note that, when K_L is the null operator and $\varphi_0 \equiv 0$, (7) reduces to (3).

It is clear that Eq. (7) may be difficult to implement, unless the operator $(I - \gamma K_L)^{-1}$ is accessible and well-behaved. We suppose that K_L has been so chosen that this is the case. The presence of the term $K - K_L$ suggests that, if K_L is a good approximation to K, then the iterative scheme may indeed converge faster than the classical Neumann series. It is this observation which we intend to exploit.

To conclude this section, we impose certain additional conditions on the kernel $K(u)$ which will hold throughout our work:

Assumption (A1): $K(u)$ is defined for all $u \geq 0$, and $K(u) \geq 0$;
Assumption (A2): $\int_0^\infty K(u)\, du = \frac{1}{2}k_0 < \infty$;
Assumption (A3): $\int_0^\infty u^2 K(u)\, du = k_2 < \infty$.

3. Construction of an Operator K_L

There are obviously many possibilities for an approximate operator K_L. In the original transport theory context, it is customary to use one of the well-known approximations to the transport operator. Of these, the diffusion operator is the simplest. To construct K_L, we shall mimic one of the many derivations of diffusion theory. No claim is being made that we are presenting a *rigorous derivation* of a generalized diffusion operator. Rather, we present a plausibility argument. Throughout, the symbol $\sim$ should be interpreted as meaning "is approximately equal to" in the broadest possible sense.

Many kernels $K(u)$ which arise in practice are peaked in the vicinity of $u = 0$ and die off rather rapidly as u gets large. Thus, in (1), $K(|x-y|)$ is somewhat delta-function-like and picks off both φ and S in the vicinity of
$$x = y.$$
Moreover, for a large, the limits in the integral of (1) may be approximated by $\pm\infty$. Thus,

$$\varphi(x) \sim \int_{-\infty}^\infty K(|x-y|)\{\gamma\varphi(y) + S(y)\}\, dy$$

$$\sim \int_{-\infty}^\infty K(|t|)\{\gamma\varphi(x+t) + S(x+t)\}\, dt$$

$$\sim \int_{-\infty}^\infty K(|t|)\{\gamma[\varphi(x) + t\varphi'(x) + \tfrac{1}{2}t^2\varphi''(x)] + [S(x) + tS'(x) + \tfrac{1}{2}t^2 S''(x)]\}\, dt.$$

$$(8)$$

If we now further suppose that S'' is small, we get

$$\varphi(x) \sim \gamma k_0 \varphi(x) + \gamma k_2 \varphi''(x) + k_0 S(x). \tag{9}$$

We shall find in the next section that $\gamma k_0 \sim 1$ is of most interest; therefore, we write

$$(1 - \gamma k_0)/\gamma k_0 \sim 1 - \gamma k_0, \qquad k_0/\gamma k_0 \sim k_0. \tag{10}$$

Thus, it is plausible that the function φ approximately satisfies the differential equation

$$(1/k_0)[-(k_2/k_0)\varphi''(x) + \varphi(x)] = \gamma\varphi(x) + S(x). \tag{11}$$

Equation (11) is, apart from constants, the classical diffusion equation. In transport theory, one of the standard boundary conditions used is

$$\varphi(\pm a) = 0, \tag{12}$$

and we agree to impose this. Another plausibility argument using (1) may be made. For further comment, see Section 9.

Thus, we have found a way of obtaining an operator K_L. For our purposes, we generalize (11) somewhat. Again referring to transport theory, we recall that the diffusion constant that occurs there is frequently modified to better simulate the physics. Analogously, we replace (11) by

$$L\varphi = (1/k_0)[-(1/\alpha^2)\varphi''(x) + \varphi(x)] = \gamma\varphi(x) + S(x), \tag{13}$$

where α is unspecified. While it is to be expected that its value is approximately $\sqrt{(k_0/k_2)}$, we shall try to choose α in any given problem so as to optimize the convergence properties of the synthetic method.

Finally, we recognize that K_L must actually be an integral operator. Clearly, it is the Green's function associated with L and the boundary conditions (12):

$$K_L \cdot = \int_{-a}^{a} G(x, y, \alpha) \cdot dy, \tag{14}$$

$$G(x, y, \alpha) = [\alpha k_0/2 \sinh(2a\alpha)] \cdot \{\cosh[\alpha(2a - |x - y|)] - \cosh[\alpha(x + y)]\}. \tag{15}$$

We make no apologies for this crude construction of K_L. Obviously, there are many other possibilities. However, it should be noted that the diffusion operator is easy to work with and that it often provides very satisfactory results in mathematical physics. Moreover, as we shall see in the latter sections of this paper, K_L does approximate K very well for many problems. A more sophisticated approximation might be found. However, it could well be sufficiently more complicated to actually increase the work of computation; obviously, there is a trade off between elegance and utility.

To complete this section, we note that the eigenvalues and eigen-functions of K_L are most easily found by turning back to the operator L [Eq. (13)]. In particular, the largest eigenvalue of K_L is

$$\Lambda_0 = k_0[1 + (\pi/2a\alpha)^2]^{-1}, \tag{16-1}$$

corresponding to the normalized eigenfunction

$$\Phi_0(x) = (1/\sqrt{a})\cos(\pi x/2a). \tag{16-2}$$

4. Eigenvalues of K

As might be expected, we shall find that an investigation of con-vergence and divergence properties of the synthetic method requires rather detailed information about the spectral radius of K. Note first that Assumption (A1) implies that K has a positive eigenvalue λ_0 which is larger in absolute value than any other eigenvalue of K. The corresponding eigenfunction φ_0 is nonnegative. A crude estimate of λ_0 may be obtained by the following simple argument. Write

$$\lambda_0\varphi_0(x) = \int_{-a}^{a} K(|x-y|)\varphi_0(y)\,dy$$

$$\leq \sup_{t\in[-a,a]} \varphi_0(t) \int_{-a}^{a} K(|x-y|)\,dy$$

$$\leq \sup_{t\in[-a,a]} \varphi_0(t) \int_{-\infty}^{\infty} K(|u|)\,du$$

$$= k_0 \sup_{t\in[-a,a]} \varphi_0(t). \tag{17}$$

Taking the supremum of the left-hand side yields

$$\lambda_0 \leq k_0. \tag{18}$$

While the argument is trivial, the result is important enough for us to state formally the following theorem.

Theorem 4.1. Let K satisfy the assumption imposed in Sections 1 and 2. Then, λ_0, the largest positive eigenvalue of K, satisfies (18).

We now seek a lower bound on λ_0, and turn to the classical Rayleigh–Ritz method.

Theorem 4.2. In addition to (A1) through (A3), suppose that $K(u)$ can be represented by a Fourier cosine transform

$$K(u) = \int_0^\infty k(s)\cos(su)\,ds, \tag{19}$$

where

$$k(s) \in L_1(0, \infty).$$

For $0 \le s \le 1$, suppose that

$$k(s) = k(0) + k''(0)s^2/2 + R(s)s^4, \qquad |R(s)| \text{ bounded.} \tag{20}$$

Then, for a large,

$$k_0 - k_2\pi^2/4a^2 + O(1/a^3) \le \lambda_0. \tag{21}$$

Proof. Using the function Φ_0 of (17), we have

$$(K\Phi_0, \Phi_0) = (1/a)\int_{-a}^{a}\int_{-a}^{a} K(|x-y|)\cos(\pi x/2a)\cos(\pi y/2a)\,dx\,dy$$

$$= (1/a)\int_{-a}^{a}\int_{-a}^{a}\int_0^\infty k(s)\cos[s(x-y)]\cos(\pi x/2a)$$

$$\times \cos(\pi y/2a)\,ds\,dx\,dy$$

$$= (1/a)\int_0^\infty k(s)\,ds\int_{-a}^{a}\int_{-a}^{a}[\cos(sx)\cos(sy)+\sin(sx)\sin(sy)]$$

$$\times \cos(\pi x/2a)\cos(\pi y/2a)\,dx\,dy$$

$$= (\pi^2/a^3)\int_0^\infty \{k(s)\cos^2(sa)/[s^2-(\pi/2a)^2]\}\,ds. \tag{22}$$

We now propose to estimate $(K\Phi_0, \Phi_0)$ for large a. Write

$$(K\Phi_0, \Phi_0) = \pi^2\int_0^\infty \{k(w/a)\cos^2 w/[w^2-(\pi/2)^2]^2\}\,dw = \pi^2\left[\int_0^a + \int_a^\infty\right]$$

$$= \pi^2[J_1 + J_2]. \tag{23}$$

For a large, we have

$$|J_2| \le 2\int_a^\infty [|k(w/a)|/w^4]\,dw = (2/a^3)\int_1^\infty [|k(t)|/t^4]\,dt < M/a^3. \tag{24}$$

Turn now to J_1 and use (20) to yield

$$J_1 = \int_0^a \{k(0)+[k''(0)/2](w/a)^2$$

$$+ R(w/a)(w/a)^4\}\cdot\{\cos^2 w/[w^2-(\pi^2/4)]^2\}\,dw. \tag{25}$$

Next, observe that, since $R(s)$ is bounded, $0 \leq s \leq 1$,

$$\left| \int_0^a \{R(w/a)(w/a)^4/[w^2 - (\pi^2/4)]^2\} \cos^2 w \, dw \right|$$

$$\leq (M/a^4) \int_0^a \{w^4 \cos^2 w/[w^2 - (\pi^2/4)]^2\} \, dw. \quad (26)$$

But

$$w^4 \cos^2 w/[w^2 - (\pi^2/4)]^2$$

is itself bounded on $0 \leq w \leq a$, independently of a. Thus,

$$\left| \int_0^a \{R(w/a)(w/a)^4/[w^2 - (\pi^2/4)]^2\} \cos^2 w \, dw \right| < M/a^3. \quad (27)$$

Returning to (25), we write, for a large,

$$\int_0^a \{k(0) + [k''(0)/2](w/a)^2\}\{\cos^2 w/[w^2 - (\pi^2/4)]^2\} \, dw$$

$$= \int_0^\infty \cdots dw - \int_a^\infty \cdots dw = K_1 + K_2,$$

$$|K_2| \leq \int_a^\infty \{|k(0)| + [|k''(0)|/2](w^2/a^2)\}\{1/[w^2 - (\pi^2/4)]^2\} \, dw$$

$$\leq M \int_a^\infty (1/w^4) \, dw + (M/a^2) \int_a^\infty (1/w^2) \, dw < M/a^3. \quad (28)$$

Summarizing, we have

$$(K\Phi_0, \Phi_0) = k(0) \int_0^\infty \{\cos^2 w/[w^2 - (\pi^2/4)]^2\} \, dw$$

$$+ [k''(0)/2a^2] \int_0^\infty \{w^2 \cos^2 w/[w^2 - (\pi^2/4)]^2\} \, dw + O(1/a^3). \quad (29)$$

Straightforward contour integration yields

$$\int_0^\infty \{w^{2n} \cos^2 w/[w^2 - (\pi^2/4)]^2\} \, dw = (1/\pi)(\pi/2)^{2n}, \qquad n = 0, 1.$$

$$(30)$$

Finally, noting that

$$(K\Phi_0, \Phi_0) \leq \lambda_0,$$

and using (23), we have

$$\pi k(0) + [k''(0)/8a^2]\pi^3 + O(1/a^3) \leq \lambda_0. \quad (31)$$

Now, under the conditions imposed on K and k, the Fourier inversion formula is valid:

$$k(s) = (2/\pi) \int_0^\infty K(u) \cos su \, du. \tag{32}$$

Thus,

$$k(0) = (2/\pi) \int_0^\infty K(u) \, du = k_0/\pi, \tag{33}$$

$$k''(0) = (2/\pi) \int_0^\infty u^2 K(u) \, du = -2k_2/\pi. \tag{34}$$

Using these values in (31), we obtain (21), and the theorem is proved. $\square$

Theorem 4.1 uses only the basic properties of K. To obtain Theorem 4.2, further assumptions were needed. It should be noted that these conditions are only sufficient for the validity of (21); they are by no means necessary. Indeed, the kernel

$$K(u) = E_1(u)$$

of interest in transport theory cannot be represented as in (19), with

$$k(s) \in L_1.$$

Nevertheless, direct calculation reveals that (21) holds. For this reason, future results will be stated in terms of (21) holding, rather than in terms of specific representations of K, etc.

5. Conditions for Divergence of the Algorithm

Let us write (7) in the form

$$(\varphi + \epsilon_{i+1}) = T(\varphi + \epsilon_i) + (I - \gamma K_L)^{-1} KS, \tag{35}$$

where

$$T = (I - \gamma K_L)^{-1} \gamma (K - K_L). \tag{36}$$

It is clear that the algorithm will diverge if the *error sequence* $\{\epsilon_i\}$ fails to approach zero. We shall show that this can happen by demonstrating that the operator T can have an eigenvalue μ_0 whose absolute value exceeds unity.

The analysis of T is complicated by the fact that T is not Hermitian. However, T does belong to the class of symmetrizable operators (see Refs. 4–5). Specifically, T is a symmetric operator when considered in the

Hilbert space H^*, where H^* is equipped with the inner product

$$(f, g)^* = (Bf, g), \tag{37}$$

with

$$B = (I - \gamma K_L). \tag{38}$$

In (37), $(\cdot, \cdot)$ denotes the usual inner product in the space

$$H = L_2.$$

It is readily shown that H^* is a valid Hilbert space.

Next, we note that

$$T = B^{-1}(K - K_L)$$

is compact in H. For the eigenvalues of B are bounded away from zero (see Section 3); therefore, B^{-1} is bounded. Further, the operator $K - K_L$ is a Hilbert–Schmidt operator in H.

In view of these facts, T may be treated as a compact, symmetric operator provided one confines the study to H^*. There, the Rayleigh–Ritz principle applies, ensuring that T does have an eigenvalue. Let μ_0 be the eigenvalue of largest absolute value. Then,

$$|\mu_0| = \sup_{\|g\|^* = 1} |(Tg, g)^*|, \tag{39}$$

where

$$\|g\|^* = \sqrt{(g, g)^*}. \tag{40}$$

Let us choose

$$f = m\Phi_0 \tag{41}$$

where Φ_0 is given by (17) and m is chosen so that $\|f\|^* = 1$. Thus, with Λ_0 as in (16),

$$(\|f\|^*)^2 = ((I - \gamma K_L)f, f) = m^2(1 - \gamma\Lambda_0) = 1. \tag{42}$$

From (39),

$$|\mu_0| \geq |(Tf, f)^*| = |\gamma((K - K_L)f, f)|$$
$$= [\gamma/(1 - \gamma\Lambda_0)]|(K\Phi_0, \Phi_0) - \Lambda_0|. \tag{43}$$

We now assume that (21) holds. If we expand for large a, we obtain

$$|(K\Phi_0, \Phi_0) - \Lambda_0| = (\pi^2/4a^2)|(k_0/\alpha^2 - k_2) + O(1/a)|. \tag{44}$$

As yet, γ is unspecified. We wish to select it in such a way that

$$|\mu_0| > 1,$$

so that the synthetic algorithm is not convergent. However, we also wish γk_0 to be less than unity, so that the ordinary Neumann series converges. To this end, we require

$$1 - \gamma \Lambda_0 = 1 - \gamma k_0 \{1 - \pi^2/4a^2\alpha^2 + O(1/a^4)\}$$
$$\geq (1 + 1/a)(\pi^2/4a^2\alpha^2). \tag{45}$$

Clearly, for a sufficiently large,

$$\gamma k_0 = [1 - (1 + 1/a)(\pi^2/4a^2\alpha^2)]/[1 - \pi^2/4a^2\alpha^2 + O(1/a^4)]$$
$$= 1 - \pi^2/4a^3\alpha^2 + O(1/a^4) < 1. \tag{46}$$

However, using (44), we find that

$$|\mu_0| \geq [\gamma/(1 + 1/a)](4a^2\alpha^2/\pi^2)(\pi^2/4a^2)|(k_0/\alpha^2 - k_2) + O(1/a)|$$
$$= [\gamma k_0/(1 + 1/a)]/[1 - (k_2/k_0)\alpha^2] + O(1/a). \tag{47}$$

Finally we note that, if

$$\alpha > \surd(2k_0/k_2),$$

then

$$|\mu_0| > 1$$

for large a, and the divergence result is established.

Theorem 5.1.   If (21) holds and

$$\alpha > \surd(2k_0/k_2),$$

then the synthetic algorithm can diverge, even though the ordinary Neumann series converges.

It may be appropriate to remark that, analytically, divergence is assured only for certain error sequences $\{\epsilon_i\}$. Clearly, if $\epsilon_0 = 0$, then the method converges. From a computational viewpoint, however, the presence of roundoff error will almost certainly introduce into the computation a component of the eigenfunction of T belonging to μ_0. For

$$|\mu_0| > 1$$

and a sufficiently large then, divergence will occur.

6. Conditions for Convergence of the Algorithm

To establish convergence results, we shall prove that, under certain conditions, T is a contracting operator. Actually, we shall establish this in

H^*, and then prove that convergence in H^* implies convergence in

$$H = L_2.$$

We first consider the norm of T in H^*:

$$\|T\|^* = \sup_{\|f\|^*=1} |(Tf, f)^*| = \sup_{\|f\|^*=1} |\gamma((K - K_L)f, f)|. \qquad (48)$$

The restriction

$$\|f\|^* = 1$$

is equivalent to

$$((I - \gamma K_L)f, f) = 1. \qquad (49)$$

Thus, (48) becomes

$$\|T\|^* = \sup_{\|f\|^*=1} |\gamma(Kf, f) - (f, f) + 1|, \qquad (50)$$

or, after a bit of manipulation,

$$\|T\|^* = \sup_{\|g\|=1} |\{[1 - \gamma(Kg, g)]/[1 - \gamma(K_Lg, g)]\} - 1| \qquad (51)$$

Recall that

$$\gamma(Kg, g) \le \gamma\lambda_0 < \gamma k_0.$$

As in the work of Section 5, we require that

$$\gamma k_0 < 1,$$

so that convergence of the ordinary Neumann series is assured. Also, since K_L is positive definite,

$$0 \le \gamma(K_Lg, g) \le \gamma\Lambda_0 < \gamma k_0 < 1. \qquad (52)$$

Thus, the quantity in braces in (51) is positive. Hence, to show that $\|T\|^* < 1$ (which will provide convergence in H^*), we need only establish that

$$[1 - \gamma(Kg, g)]/[1 - \gamma(K_Lg, g)] < 2. \qquad (53)$$

Since K is positive, the Perron–Frobenius theorem implies that

$$|(Kg, g)| \le \lambda_0. \qquad (54)$$

Hence (53) is true if

$$(1 + \gamma\lambda_0)/(1 - \gamma\Lambda_0) < 2, \qquad (55)$$

or if

$$1 + \gamma k_0 < 2 - 2\gamma k_0 / [1 + (\pi/2a\alpha)^2]. \tag{56}$$

Now, the left-hand side of (56) is fixed and is less than 2. For given a, however large, we may select α sufficiently small so that (56) holds.

Thus, provided α is so chosen,

$$\|T\|^* < 1,$$

and the synthetic algorithm converges in H^*. We must still establish convergence in the original Hilbert space,

$$H = L_2.$$

To do so, let

$$\epsilon_{i+1} = T\epsilon_i, \tag{57}$$

so that

$$\lim_{i \to \infty} \|\epsilon_i\|^* = 0.$$

From the definition of H^*, this implies that

$$\lim_{i \to \infty} ((I - \gamma K_L)\epsilon_i, \epsilon_i) = 0. \tag{58}$$

But

$$((I - \gamma K_L)\epsilon_i, \epsilon_i) = \|\epsilon_i\|^2 - \gamma(K_L\epsilon_i, \epsilon_i) \ge \|\epsilon_i\|^2(1 - \gamma\Lambda_0). \tag{59}$$

Since

$$1 - \gamma\Lambda_0 > 0,$$

it follows that

$$\epsilon_i \to 0$$

in H.

Theorem 6.1. Suppose that $\gamma k_0 < 1$, so that the Neumann series generated by (3) converges in

$$H = L_2.$$

Then, for given a, the parameter α may always be chosen so small that the synthetic algorithm also converges in L_2.

We might remark that the above theorem is, in some ways, disappointing. While it assures us that the synthetic method does not *destroy* convergence, provided α is properly chosen, it provides no hint of an optimal α-value (see Section 7), and it also leaves a large *gap* between α-values which surely produce convergence and those for which divergence can occur.

There are two *consolation* results which we wish to discuss. The first of these is rather easy to prove, and we only state it.

Theorem 6.2. Suppose that, for given γ_0, a_0, α_0,

$$\|T\|^* < 1.$$

Then, the synthetic method converges in H for (γ, a, α) in a neighborhood of $(\gamma_0, a_0, \alpha_0)$.

This follows from the readily verifiable fact that $\|T\|^*$ is continuous in the various parameters.

The other result is much more interesting, from both the mathematical and computational viewpoints.

Theorem 6.3. Let γ, a, α be such that

$$\|T\|^* < 1, \qquad \gamma k_0 < 1.$$

Suppose that

$$a' \le a, \qquad \gamma' \le \gamma.$$

Then, provided α remains fixed, the synthetic method converges in L_2 for γ' and a'.

Proof. Writing

$$T = T(\gamma, a),$$

we shall show that

$$\|T(\gamma', a')\|^* \le \|T(\gamma, a)\|^* < 1$$

for

$$\gamma' \le \gamma, \qquad a' \le a.$$

We note that

$$\|T\|^* = \max\{-\mu_0^-, \mu_0^+\}, \tag{60}$$

where $\mu_0^\pm$ are the largest positive and smallest negative eigenvalues of T, respectively. Now [see (51)],

$$\mu_0^+ = \sup_{\|g\|=1} \{[\gamma(Kg, g) - 1]/[1 - \gamma(K_L g, g)] + 1\}. \tag{61}$$

Since the quantity in brackets is nonpositive, μ_0^+ (if it exists) is less than unity. Thus, we may concentrate our attention on μ_0^-. In an obvious notation,

$$-\mu_0^-(\gamma, a) = \sup_{\|g\|=1} \{\gamma[(K_L(a)g, g) - (K(a)g, g)]/[1 - \gamma(\bar{K}_L(a)g, g)]\}$$

$$= \gamma[(K_L(a)\psi_0^-, \psi_0^-) - (K(a)\psi_0^-, \psi_0^-)]/$$

$$[1 - \gamma(K_L(a)\psi_0^-, \psi_0^-)], \quad (62)$$

where ψ_0^- is the eigenfunction belonging to $\mu_0^-(\gamma, a)$,

$$\|\psi_0^-\| = 1.$$

Choose

$$\gamma' \le \gamma, \qquad a' \le a.$$

Assume that $\mu_0^-(\gamma', a')$ exists, and let $\tilde{\psi}_0^-$ be the corresponding normalized eigenfunction. Define

$$g = \tilde{\psi}_0^-, \qquad -a' \le x \le a', \qquad g = 0, \qquad \text{elsewhere.}$$

By the basic integral definition of K [Eq. (2)]

$$(K(a')\tilde{\psi}_0^-, \tilde{\psi}_0^-) = (K(a)g, g). \quad (63)$$

The operator K_L is a bit more difficult to handle. Explicit use of the Green's function [Eq. (15)] reveals that, for any $a'' \ge a'$,

$$(2\alpha/k_0)(K_L(a'')g, g) = [1/\sinh(2\alpha a'')] \int_{-a''}^{a''} \int_{-a''}^{a''} \{\cosh(2\alpha a'')$$

$$\times [\cosh(\alpha x)\cosh(\alpha y)$$

$$- \sinh(\alpha x)\sinh(\alpha y)] - \sinh(2\alpha a'')\sinh(\alpha|x-y|)$$

$$- \cosh(\alpha x)\cosh(\alpha y)$$

$$- \sinh(\alpha x)\sinh(\alpha y)\}g(x)g(y)\,dx\,dy$$

$$= \coth(2\alpha a'')(I_1 - I_2) - \operatorname{csch}(2\alpha a'')(I_1 + I_2)$$

$$- \int_{-a'}^{a'} \int_{-a'}^{a'} \sinh(\alpha|x-y|)g(x)g(y)\,dx\,dy, \quad (64)$$

where

$$I_1 = \left(\int_{-a'}^{a'} g(x)\cosh(\alpha x)\,dx \right)^2, \qquad I_2 = \left(\int_{-a'}^{a'} g(x)\sinh(\alpha x)\,dx \right)^2.$$

$$(65)$$

We now wish to determine how $(K_L(a'')g, g)$ changes as a'' increases to a. Recall the definition of g, and observe that I_1, I_2 and the double integral over $\sinh(\alpha|x-y|)$ are independent of a''. Hence,

$$(2\alpha/k_0)(d/da'')(K_L(a'')g, g) = -2\alpha \, \mathrm{csch}^2(2\alpha a'')(I_1 - I_2)$$
$$+ 2\alpha \, \mathrm{csch}(2\alpha a'') \coth(2\alpha a'')(I_1 + I_2)$$
$$= [2\alpha/\sinh^2(2\alpha a'')]$$
$$\times [-(I_1 - I_2) + \cosh(2\alpha a'')(I_1 + I_2)]$$
$$\geq [4\alpha/\sinh^2(2\alpha a'')]I_2 \geq 0, \tag{66}$$

since

$$I_1 + I_2 \geq 0.$$

Therefore, $(K_L(a'')g, g)$ is a nondecreasing function of a'',

$$a'' \geq a'.$$

Hence,

$$(K_L(a')g, g) \leq (K_L(a)g, g). \tag{67}$$

Rewriting (63) gives

$$-\mu_0^-(\gamma', a') = \gamma'[(K_L(a')\tilde\psi_0^-, \tilde\psi_0^-) - (K(a')\tilde\psi_0^-, \tilde\psi_0^-)]/[1 - \gamma'(K_L(a')\tilde\psi_0^-, \tilde\psi_0^-)]. \tag{68}$$

Since $-\mu_0^- > 0$, the numerator in (68) must be positive. Therefore, using (63) and (67), we have

$$\gamma'[(K_L(a')\tilde\psi_0^-, \tilde\psi_0^-) - (K(a')\tilde\psi_0^-, \tilde\psi_0^-)] \leq \gamma[(K_L(a)g, g) - (K(a)g, g)]. \tag{69}$$

But K_L is positive definite, so that

$$\gamma'(K_L(a')g, g) \leq \gamma(K_L(a)g, g), \tag{70}$$

or

$$1 - \gamma'(K_L(a')\tilde\psi_0^-, \tilde\psi_0^-) \geq 1 - \gamma(K_L(a)g, g). \tag{71}$$

Putting (69) and (71) in (68) and employing (62) finally gives

$$-\mu_0^-(\gamma', a') = \gamma[(K_L(a)g, g) - (K(a)g, g)]/[1 - \gamma(K_L(a)g, g)] \leq -\mu_0^-(\gamma, a). \tag{72}$$

Hence,

$$\mu_0^-(\gamma', a') \geq \mu_0^-(\gamma, a) > -1.$$

Therefore,

$$\|T(\gamma', a')\|^* < 1,$$

completing the proof of the theorem. $\qquad\square$

7. Some Remarks on the Optimization of α

It is clear that it is highly desirable to choose α in such a fashion as to make the synthetic algorithm converge as rapidly as possible. This indicates that one should try to minimize the norm of T in the space H. However, the previous section suggests that this may be very difficult. If a truly good estimate of this norm were available, our convergence result would have been much sharper.

As a compromise, we have chosen to study the norm of $K - K_L$, since that is the *principal ingredient* in T. We have also found that, for computational reasons, it is much easier to deal with a sup norm of this quantity rather than with the L_2-norm. Thus, for any given kernel K, we compute [see (14)]

$$\max_{-a \le x \le a} \int_{-a}^{a} |K(x, y) - G(x, y, \alpha)| \, dy \qquad (73)$$

as a function of α. The results obtained are crude, and it cannot be anticipated that this approach will lead to precise estimates concerning the operator T.

We have, however, usually found that the minimum of the expression in (73) occurs at *about* the value of α which is found to be most effective in the actual implementation of the synthetic algorithm to be described in the next section. Thus, as a practical device, one may compute this value and then confine one's actual investigations to α-values nearby. We fully recognize that this procedure is inelegant and theoretically unsatisfactory. A device for obtaining sharper estimates on optimal α-values is much needed.

8. Implementation of the Synthetic Algorithm and Some Numerical Results

As noted earlier, Eq. (7) is equivalent to the synthetic scheme, although it may not be at once recognizable as such. The more standard form is actually more convenient in computations. To derive it, rewrite (7) as

$$\varphi_{i+1} = (I - \gamma K_L)^{-1}\gamma(K - K_L)\varphi_i + (I - \gamma K_L)^{-1}KS$$
$$= (I - \gamma K_L)^{-1}[K(\gamma\varphi_i + S) - \gamma K_L\varphi_i]$$
$$= (I - \gamma K_L)^{-1}[K(\gamma\varphi_i + S) - \varphi_i + \varphi_i - \gamma K_L\varphi_i]$$
$$= (I - \gamma K_L)^{-1}[K(\gamma\varphi_i + S) - \varphi_i] + \varphi_i$$
$$= (I - \gamma K_L)^{-1}R_i + \varphi_i, \tag{74}$$
$$R_i = K(\gamma\varphi_i + S) - \varphi_i. \tag{75}$$

Adding and subtracting R_i on the right-hand side of (74) and manipulating a bit eventually yields

$$\varphi_{i+1} = \varphi_i + R_i + (I - \gamma K_L)^{-1}K_L(\gamma R_i). \tag{76}$$

It is this form of the algorithm which we shall implement. For the initial function, we choose

$$\varphi_0 = (I - \gamma K_L)^{-1}K_L S. \tag{77}$$

This choice is the usual one in transport calculations, and it is suggested by the fact that the solution to (4) is

$$\varphi = (I - \gamma K)^{-1}KS.$$

Obviously, we need an analytical representation of the operator

$$(I - \gamma K_L)^{-1}K_L.$$

Consider

$$\mathscr{L}\psi = (1/k_0)[-(1/\alpha^2)\psi'' + (1 - \gamma k_0)\psi] = f, \qquad \psi(\pm a) = 0. \tag{78}$$

This may be written as [see (13)]

$$\mathscr{L}\psi = L\psi - \gamma\psi = f, \tag{79}$$

or

$$\psi = \gamma K_L\psi + K_L f. \tag{80}$$

Thus,

$$\psi = (I - \gamma K_L)^{-1}K_L f. \tag{81}$$

But ψ may also be found by calculating the Green's function for $\mathscr{L}$:

$$\hat{G}(x, y, \gamma, \alpha) = \alpha k_0/2\sqrt{(1 - \gamma k_0)}\sinh[2a\alpha\sqrt{(1 - \gamma k_0)}]$$
$$\cdot \{\cosh[\alpha\sqrt{(1 - \gamma k_0)}(2a - |x - y|)] - \cosh[\alpha\sqrt{(1 - \gamma k_0)}(x + y)]\}. \tag{82}$$

Hence,

$$(I - \gamma K_L)^{-1} K_L \cdot = \int_{-a}^{a} \hat{G}(x, y, \gamma, \alpha) \cdot dy. \tag{83}$$

All operators in (76), and (76) itself, thus have integral representations, facilitating calculations.

We shall now describe some of the results obtained when the synthetic method was implemented for a variety of kernels. The numerical scheme employed was rather crude, in that the trapezoidal rule was used to approximate all integrals. The step was varied to ensure consistent accuracy in the results. To economize on storage, the function S [see (1)] was chosen to be even; this had the effect of making all functions which occur in the problem even. All calculations were done on a CDC 7600 computer in single-precision arithmetic. Storage considerations constrained the range of a-values which we could consider. For most problems, however, we found that we could study a sufficiently large set of a, γ, and α-values to adequately demonstrate the behavior of the algorithm.

Our convergence criterion in all cases was

$$\max_{x} |\varphi_{i+1}(x) - \varphi_i(x)| < 10^{-3}. \tag{84}$$

Problems were so devised that the solution generally varied between values of about one and ten. Therefore, use of the absolute criterion (84), rather than a relative criterion, seems justified. Because of the transport origin of the problem, we first studied the kernel

$$K(u) = E_1(u),$$

with

$$S(u) = 1.$$

For this case,

$$k_0 = 1, \qquad k_2 = 3,$$

so that the divergence criterion was

$$\alpha > \sqrt{6}.$$

Tables 1–3 give some typical results in parameter ranges of interest. Entries indicate the number of iterations required. The symbol (*) indicates that the scheme had failed to converge after 50 iterations, but seemed to be converging, while the symbol (**) indicates that the algorithm actually diverged.

Table 1.   Numerical results, $a = 3.0$.

α \ γ	0.3	0.6	0.9	0.99	0.999
1.2	5	7	13	23	26
1.4	5	7	12	19	21
1.6	4	6	11	16	18
$\sqrt{3}$ (***)	4	6	10	16	16
2.0	4	6	9	14	14
2.2	4	6	8	12	12
2.4	4	5	7	11	13
2.6	4	5	7	14	26
2.8	4	5	7	23	(*)
3.0	4	5	8	(*)	(**)

(***) This value corresponds to classical diffusion theory.

Table 2.   Numerical results, $a = 6.0$.

α \ γ	0.3	0.6	0.9	0.99	0.999
1.2	5	7	13	25	35
1.4	5	7	12	21	27
1.6	4	6	11	18	22
$\sqrt{3}$ (***)	4	6	10	16	20
2.0	4	6	9	14	16
2.2	4	6	9	13	15
2.4	4	5	8	12	15
2.6	4	5	8	18	34
2.8	4	5	8	42	(*)
3.0	4	5	10	(**)	(**)

(***) This value corresponds to classical diffusion theory.

Table 3.   Numerical results, $a = 8.0$.

α \ γ	0.9	0.99	0.999
1.2	13	26	(*)
1.4	12	22	41
1.6	11	19	33
$\sqrt{3}$ (***)	10	17	29
2.0	9	15	23
2.2	9	14	20
2.4	8	13	18
2.6	8	20	25
2.8	8	(*)	(**)
3.0	9	(**)	(**)

(***) This value corresponds to classical diffusion theory.

The improvement over the Neumann iteration is quite remarkable. For instance, for the case

$$a = 6.0, \qquad \gamma = 0.999,$$

the classical method required 693 iterations as compared with the 18 which suffices when α was chosen as 2.4. It might be noted that the CDC 7600 computer has a word length of about 16 decimal digits. It seems quite likely that convergence in the Neumann case might not have been obtained at all, had a machine with a shorter word length been used.

The optimal α-value is also of interest. The quantity α seems to flirt dangerously close to the divergence value of

$$\sqrt{6} \sim 2.45$$

in cases in which the algorithm was most effective, namely, those with large a and γ nearly unity. A similar phenomenon was observed in many of the other examples which we shall discuss.

Suspecting that the method might be somewhat sensitive to the function S [through the choice of φ_0, Eq. (77)], we tried a variety of such functions, always choosing them even to make use of the symmetry mentioned earlier. The synthetic algorithm was affected very little by changes in S.

Our next effort involved the kernels

$$K(u) = 1/(1+u)^4$$

and

$$K(u) = \exp(-u) \sin^2 u.$$

The first kernel again demonstrated the rather spectacular advantage of the method. In one instance, an iteration count of 17 was recorded versus 493 by Neumann iteration. The second kernel gave good results, but the improvement was not as marked.

In an effort to understand this, we reexamined the construction of the *diffusion operator* (Section 3). There, the assumption was made that the kernel is peaked in the vicinity of the origin. The E_1-function is, in fact, mildly singular there. The function $1/(1+u)^4$ is nicely peaked, although continuous, while $\exp(-u) \sin^2 u$ obviously vanishes at the origin, although it reaches its maximum value nearby.

To further analyze this apparent behavior, we studied kernels of the form

$$K(u) = A_n/(1+u)^{2n}$$

for several values of $n \geq 2$. The constant A_n was chosen so as to make $k_0 = 1$ in each case. Results significantly improved as n increased.

Finally we investigated

$$K(u) = \exp[-(u-a)^2].$$

This kernel has exactly the wrong shape. It was found that the algorithm gave satisfactory results only for the most uninteresting cases. While convergence was observed where expected, it was usually so slow as to make the method totally uninteresting.

While it is always unwise to draw conclusions from relatively few numerical examples, we feel that the synthetic algorithm with K_L as described seems to have the following general properties:

(i)   It is most useful when $K(u)$ is peaked near $u = 0$.

(ii)   It provides little or no advantage over the Neumann series, unless a is relatively large and γk_0 is near unity.

(iii)   The optimal value of α is often quite close to values of α for which the method may diverge.

(iv)   Quite remarkable improvements in convergence over the Neumann series are frequently achieved.

9. Remarks and Conclusions

We have shown how the synthetic method of transport theory may be generalized in such a way as to be applicable to a large class of integral equations. Convergence and divergence properties of the algorithm have been studied analytically, and numerous numerical examples have demonstrated the expected behavior. In some instances, the computational advantage over the familiar Neumann approach is quite remarkable.

Numerous questions are left open. We have confined our study to *low-order operators* K_L which are diffusion-like. A single parameter α has been left free to adjust the degree of approximation of the operator to the given one and thus to optimize the convergence properties of the scheme. No really satisfactory analytical or semi-analytical methods of making this optimal choice have as yet been devised.

While further investigation of the problem of choosing α seems called for, one should not lose sight of the fact that possibly a quite different operator K_L should be introduced. The fact that the integral representation of our K_L does not have a difference kernel suggests the desirability of such a change. Indeed, the kernel

$$\mathcal{G} = C \exp[-\alpha|x-y|]$$

is the Green's function for the diffusion operator L, but with the boundary conditions of (12) replaced by *radiation-type* conditions. The possibility of using $\mathscr{G}$ or some similar function to generate K_L is tempting. The structure suggests that one might hope to obtain a better convergence result than that given by Theorem 6.1. This and allied matters are currently being studied both analytically and numerically.

References

1. KOPP, H. J., *Synthetic Method Solution of the Transport Equation*, Nuclear Science and Engineering, Vol. 17, pp. 65–74, 1963.
2. CESARI, L. *Sulla Risoluzione dei Sistemi di Equazioni Lineari par Approsimaz-ioni Successive*, Atti della Accademia Nationale dei Lincei, Rendiconti della Classe di Scienze, Fisiche, Matematiche, e Naturali, Vol. 25, pp. 422–428, 1937.
3. WING, G. M., *An Introduction to Transport Theory*, John Wiley and Sons, New York, New York, 1962.
4. COCHRAN, J. A., *Analysis of Linear Integral Equations*, McGraw-Hill Publishing Company, New York, New York, 1972.
5. ZAANEN, A. C., *Linear Analysis*, John Wiley and Sons (Interscience Publishers), New York, New York, 1953.

3

The Approximate Solution
of Singular Integral Equations

D. Elliott[1]

Abstract. We present a survey of numerical methods for solving Cauchy singular integral equations on both open and closed arcs in the plane. For completeness, necessary theory is reviewed, particularly the method of regularization. For closed arcs we discuss collocation methods based on piecewise polynomial or rational representations of the solution. Emphasis here, as for the open arc case, is on regularizable equations. For open arcs a detailed discussion is given of a degenerate kernel method developed recently by Dow and Elliott. In addition to this, a generalization of a Galerkin method due to Karpenko is presented. Attention is drawn to the relation of Cauchy singular equations and solving rectangular systems of linear equations. The possibility of exploiting this for the direct solution of such equations is discussed, and some direction for future research is given.

1. Introduction

In the context of integral equations, the word "singular" has many different meanings, so it is essential that we make clear from the outset what is to be discussed in this chapter. We shall be considering equations of the form

$$a(t)\phi(t) + \int_{\mathscr{C}} [M(t, \tau)/(\tau - t)]\phi(\tau)\, d\tau = f(t), \qquad t \in \mathscr{C}, \qquad (1)$$

where $\mathscr{C}$ is some arc or contour in the complex plane; $a, f,$ and M are given functions; and the integral is taken in the sense of the Cauchy principal value. Suppose that all circles with center at the point t and radii $\leq \epsilon$ intersect $\mathscr{C}$ at exactly two points. Let $\mathscr{C}_\epsilon$ denote that part of $\mathscr{C}$ cut out by the circle of

[1] Professor of Applied Mathematics, The University of Tasmania, Hobart, Tasmania, Australia.

radius ϵ; then the Cauchy principal value integral of a function g is denoted and defined by

$$\oint_{\mathscr{C}} g(\tau)\, d\tau = \lim_{\epsilon \to 0} \int_{\mathscr{C}-\mathscr{C}_\epsilon} g(\tau)\, d\tau. \qquad (2)$$

It is customary to rearrange (1) so that it is written in the form

$$a(t)\phi(t) + [b(t)/\pi i] \oint_{\mathscr{C}} [\phi(\tau)/(\tau - t)]\, d\tau + \lambda \int_{\mathscr{C}} k(t, \tau)\phi(\tau)\, d\tau = f(t), \qquad (3)$$

where $b(t) = \pi i M(t, t)$ and $\lambda k(t, \tau) = (M(t, \tau) - M(t, t))/(\tau - t)$, λ being a scalar. It will be convenient to write (3) symbolically in terms of operators. First, we introduce the operator S, which is defined by

$$(S\phi)(t) = (1/\pi i) \oint_{\mathscr{C}} [\phi(\tau)/(\tau - t)]\, d\tau, \qquad t \in \mathscr{C}. \qquad (4)$$

Next, we introduce operators L, A, and K so that (3) can be written as

$$L\phi \equiv A\phi + \lambda K\phi = f. \qquad (5)$$

Here $A\phi = a\phi + bS\phi$ is known as the *dominant part* of $L\phi$, and $(K\phi)(t) = \int_{\mathscr{C}} k(t, \tau)\phi(\tau)\, d\tau$ is known as its *regular part*. As we shall see later, these names are particularly apt since the dominant part of L has a profound influence on the behavior of the solutions of (3).

Equations of the form (3) are of considerable importance in solving boundary-value problems in many branches of applied mathematics. Traditionally, they have been widely used in the theories of elasticity and hydrodynamics. [A brief summary of these applications is to be found in Chapter 7 of Zabreyko *et al.* (Ref. 1).] More recently, they have appeared in such topics as the theory of radiative transfer, neutron transport theory, dispersion theory, particle physics, quantum field theory, and the theory of automatic control. With the importance of such equations in applied mathematics having been established, one might expect that algorithms for their approximate solution would have been extensively developed by now, but this appears not to be the case. Although, for example, the last two decades have seen considerable development of algorithms for the approximate solution of Fredholm integral equations [see the survey by Atkinson (Ref. 2)], there has been an almost total neglect by numerical analysts of singular integral equations. Since the production and analysis of algorithms requires a knowledge of the theory we shall, in Section 2, give an outline of the theory of singular integral equations. But before so doing, we shall briefly sketch its historical development.

The foundations of the theory were laid down almost simultaneously, but independently, by Hilbert (Refs. 3–4) (who was working on analytic function theory) and Poincaré (Ref. 5) (who was studying the general theory of tides) in the first decade of this century. This was at about the same time as Fredholm published his papers (Refs. 6–7) on the solution of the integral equations which now bear his name and which were to have such a profound influence on the development of what we now call functional analysis. However, the theory of singular integral equations made less rapid progress. In 1921, Noether (Ref. 8) published his theorems, which have formed the basis for all subsequent theoretical development of singular integral equations, and a year later Carleman (Ref. 9) gave his explicit solution of the dominant equation $A\phi = f$, in the case when $\mathscr{C}$ is the arc $(-1, 1)$ and b is a constant. For the next two or three decades, progress in the theory was due almost entirely to Russian mathematicians. The prime movers were Muskhelishvili and Privalov, each of whom wrote a standard treatise on the topic (see Refs. 10 and 11). Muskhelishvili's interest in the theory of singular equations followed from his work in the theory of elasticity, in which field he has also written a well-known treatise (Ref. 12). The work of Muskhelishvili and Privalov has been continued by other Russian mathematicians, and one might mention in particular the books written by Gakhov (Ref. 13), Vekua (Ref. 14), and Mikhlin (Ref. 15). Each has developed the basic theory in different ways. In particular, Vekua has considered *systems* of singular integral equations, whereas Mikhlin has considered *multidimensional* singular integral equations. During the first 25 years or so of this vigorous Russian development, which dates from about 1930, the contribution to the theory of singular integral equations by mathematicians not living in Russia appears to have been comparatively small.

When it comes to the problem of constructing algorithms for the approximate solution of singular integral equations, it is perhaps not surprising that once again we must look to the work of Russian mathematicians. The only book so far published that has a large part devoted to approximate solutions of singular integral equations is that by Ivanov (Ref. 16). This was published in 1968, and in it Ivanov has given a summary of all developments to that time. The first half of his book is devoted to an abstract approach to the approximate solution of operator equations, and in the second half he applies these methods to the particular problem of singular integral equations. The book is well conceived, and Ivanov has attempted to include everything known on the topic up to the time of writing. To keep the book within reasonable bounds, the author has had to omit a lot of detail, so much so that Hyers (Ref. 17) has recently remarked, "For a book on numerical analysis, there are remarkably few numbers included." Thus although Ivanov has pointed the way, a considerable amount of developmental work remains to be done.

In the next four sections we shall consider some aspects of the approximate solution of Eq. (3). Inevitably, the author has had to be highly selective in his material, and the survey is not in any way complete. Since the theory of singular integral equations is not widely known, we have given an outline of it in Section 2, concentrating on those results that seem particularly appropriate in this context. In Section 3, we have considered two different methods of approximate solution for the particular case when $\mathscr{C}$ is the unit circle γ, with center at the origin, and the functions a, b, f, and k are Hölder continuous on γ. In Section 4, we have considered the special problems that arise when $\mathscr{C}$ is an arc, which we have chosen to be $(-1, 1)$. Within this section we have described, in some detail, the use of regularization to give an approximate solution of (3) and in addition have outlined a Galerkin method which may be used directly without regularization. Finally, in Section 5, we have tried to indicate areas in which further work on approximate solutions of singular integral equations remains to be done.

2. An Outline of the Theory of Singular Integral Equations

In this section, we shall do little more than quote results, and the reader should consult either Gakhov's book (Ref. 13) or Muskhelishvili's book (Ref. 10) for proofs. It will also simplify our analysis if we assume that $\mathscr{C}$ is a simple, smooth, closed contour which is defined positively in the anticlockwise direction and which divides the complex plane into two domains, D^+ (bounded by $\mathscr{C}$) and D^- (containing the point at infinity). Again, we shall assume that a, b, and f are Hölder continuous on $\mathscr{C}$, with k being Hölder continuous on $\mathscr{C} \times \mathscr{C}$. [A function g is said to be *Hölder continuous* on $\mathscr{C}$ if there exist constants $A \geq 0$ and α, $0 < \alpha \leq 1$, such that for any two points t_1, $t_2 \in \mathscr{C}$, $|g(t_1) - g(t_2)| \leq A|t_1 - t_2|^\alpha$. We shall write $g \in H$ or $g \in H_\alpha$ if we wish to stress the order α.]

Fundamental to the development of the theory is the *Cauchy integral*, where for a given function ϕ defined on $\mathscr{C}$ its Cauchy integral is denoted and defined by

$$\Phi(z) = (1/2\pi i) \int_{\mathscr{C}} [\phi(\tau)/(\tau - z)]\, d\tau, \qquad z \notin \mathscr{C}. \tag{6}$$

If $\Phi^\pm$ denotes the function Φ for $z \in D^\pm$, respectively, then $\Phi^\pm$ are each analytic functions of z in their respective domains of definition. We say that Φ is *sectionally analytic* in the complex plane with $\mathscr{C}$ being the boundary of discontinuity. If we let $\Phi^\pm(t)$ denote the limiting values of $\Phi^\pm$ as z approaches $t \in \mathscr{C}$ through points of $D^\pm$, respectively, then the *Sokhotski–*

Plemelj formulas are given by (see Ref. 10, Chapter 2)

$$\Phi^+(t) - \Phi^-(t) = \phi(t), \qquad t \in \mathscr{C},$$

$$\Phi^+(t) + \Phi^-(t) = (1/\pi i) \int_{\mathscr{C}} [\phi(\tau)/(\tau - t)] \, d\tau, \qquad t \in \mathscr{C}. \tag{7}$$

By means of these formulae, (3) may be transformed into the so-called *Riemann boundary value problem* (note that Muskhelishvili calls it the Hilbert problem, and other authors, the Riemann–Hilbert problem). In particular on applying (7) to the dominant equation $A\phi = f$, we are now required to find that sectionally analytic function Φ, with $\Phi(\infty) = 0$, such that

$$\Phi^+(t) = G(t)\Phi^-(t) + g(t), \qquad t \in \mathscr{C}, \quad \left.\begin{array}{r}\\\\\end{array}\right.$$

where $\tag{8}$

$$G = (a - b)/(a + b), \qquad g = f/(a + b). \quad \left.\begin{array}{r}\\\end{array}\right.$$

We shall assume throughout that the function r, defined by $r = \sqrt{a^2 - b^2}$, is such that $r(t) \neq 0$, $t \in \mathscr{C}$, so that G is neither zero nor infinite on $\mathscr{C}$. The theory has to be modified if $r(t) = 0$ for some t on $\mathscr{C}$, and we shall not consider such modification here (see Ref. 13).

At this point we can introduce an integer, known as the *index* of (3), which plays an important role in the theory of singular integral equations. If $\Delta_{\mathscr{C}} g(t)$ denotes the change in the value of the function g as $\mathscr{C}$ is described once in the positive direction, then the index of the equation $L\phi = f$ (or of the operator L) is denoted and defined by

$$\kappa(L) = (1/2\pi)\Delta_{\mathscr{C}}[\arg G(t)] = (1/2\pi i)\Delta_{\mathscr{C}}[\log G(t)]. \tag{9}$$

We observe that $\kappa(L)$ depends only on a and b; that is, it depends only on the dominant part A and is independent of the regular part K. The significance of the index is given by Noether's theorems (Ref. 8), but to give this we must introduce the *adjoint* operator L^*. This is an operator such that for all functions ϕ, ψ that are Hölder continuous on $\mathscr{C}$, we have

$$\int_{\mathscr{C}} \phi(t)(L\psi)(t) \, dt = \int_{\mathscr{C}} \psi(t)(L^*\phi)(t) \, dt. \tag{10}$$

Since the order of integration of an ordinary and a Cauchy principal value integral can be interchanged [see Ref. 13 and also Eq. (26)], we find that

$$(L^*\phi)(t) = a(t)\phi(t) - (1/\pi i) \int_{\mathscr{C}} [b(\tau)\phi(\tau)/(\tau - t)] \, d\tau + \lambda \int_{\mathscr{C}} k(\tau, t)\phi(\tau) \, d\tau. \tag{11}$$

Let $N(L)$ denote the null-space of L, being that space of functions ϕ such that $L\phi = 0$. Suppose $n = \dim N(L)$ and $n^* = \dim N(L^*)$; then Noether's theorems tell us that n, n^* are finite and furthermore that

$$\kappa(L) = n - n^*. \tag{12}$$

It is perhaps worth noting at this point that for Fredholm equations $\kappa = 0$, and it is for this reason that singular equations with zero index are sometimes known as *quasi-Fredholm equations*. Such singular integral equations satisfy the Fredholm theorems. It can also be shown that if L_1, L_2 are two singular operators then

$$\kappa(L_1 L_2) = \kappa(L_1) + \kappa(L_2), \tag{13}$$

so that if $\kappa(L_1) = -\kappa(L_2)$, then obviously $\kappa(L_1 L_2) = 0$, and the equation $L_1 L_2 \phi = f$ is quasi-Fredholm. We shall make use of this observation later when we "regularize" the equation $L\phi = f$.

Let us return to the Riemann boundary value problem (8) and in particular consider the homogeneous case in which we put $g \equiv 0$. The *canonical function* X is defined for all z except the points of $\mathscr{C}$ by

$$X(z) = \begin{cases} \exp\left\{(1/2\pi i) \int_{\mathscr{C}} \{\log[\tau^{-\kappa} G(\tau)]/(\tau - z)\} \, d\tau\right\}, & z \in D^+, \\[2ex] z^{-\kappa} \exp\left\{(1/2\pi i) \int_{\mathscr{C}} \{\log[\tau^{-\kappa} G(\tau)]/(\tau - z)\} \, d\tau\right\}, & z \in D^-, \end{cases} \tag{14}$$

where we have assumed that the origin is in D^+. This function has the following properties: (1) It is sectionally analytic in the complex plane with the contour $\mathscr{C}$ deleted; (2) $X(z) = 0(z^{-\kappa})$, for large $|z|$; (3) $X^+(t) = G(t)X^-(t)$ for $t \in \mathscr{C}$; (4) it has no zeros in the finite complex plane with the contour $\mathscr{C}$ deleted.

Closely related to the canonical function X is what we shall call the *fundamental function* Z, which is defined at all points t on $\mathscr{C}$ by

$$Z(t) = (a(t) + b(t))X^+(t)/r(t)$$
$$= (a(t) - b(t))X^-(t)/r(t)$$
$$= \sqrt{X^+(t)X^-(t)}, \tag{15}$$

where $r = \sqrt{a^2 - b^2}$ and, as we have noted before, we have assumed $r(t) \neq 0$, $t \in \mathscr{C}$. From Eqs. (14) and (15), it follows that Z is given explicitly by

$$Z(t) = t^{-\kappa/2} \exp\left\{(1/2\pi i) \oint_{\mathscr{C}} \{\log[\tau^{-\kappa} G(\tau)]/(\tau - t)\} \, d\tau\right\}. \tag{16}$$

From these definitions and the Sokhotski–Plemelj formulas (7), we can obtain some results which will be of considerable use in our subsequent analysis. If Φ is sectionally analytic in the complex plane with $\mathscr{C}$ deleted, and if Φ vanishes at ∞, then from (7) it follows that

$$S(\Phi^+ - \Phi^-) = \Phi^+ + \Phi^- \tag{17}$$

at all points $t \in \mathscr{C}$. Suppose Q, R are analytic functions in the finite complex plane such that $\Phi = QX - R$ is zero at infinity. Then from (15) and (17), we find

$$S(bZQ/r) = -(aQZ/r) + R \qquad \text{for all } t \in \mathscr{C}. \tag{18}$$

Again, if Q, U are analytic in the finite plane and if $\Phi = QX^{-1} - U$ is zero at infinity, then

$$S(bQ/rZ) = (aQ/rZ) - U \qquad \text{for all } t \in \mathscr{C}. \tag{19}$$

We shall now state the explicit solution of the dominant equation $A\phi = f$ (for proof, see Ref. 10 or 13). Carleman (Ref. 9) first gave this solution for the particular case when $\mathscr{C}$ is the arc $(-1, 1)$ and $\kappa = 1$. This was later generalized by Vekua (Ref. 18) to the case of arbitrary κ and $\mathscr{C}$ a simple, smooth, closed contour. If $A\phi = f$, then

$$\phi(t) = [a(t)f(t)/r^2(t)] - [b(t)Z(t)/r(t)\pi i] \int_{\mathscr{C}} [f(\tau)/r(\tau)Z(\tau)(\tau - t)] \, d\tau$$

$$+ [b(t)Z(t)/r(t)]P_{\kappa-1}(t) \tag{20}$$

for $t \in \mathscr{C}$, where $P_{\kappa-1}$ denotes an arbitrary polynomial of degree $\leq \kappa - 1$. (If $\kappa - 1 < 0$, then we define $P_{\kappa-1}$ to be identically zero.) However, if $\kappa \leq -1$, the equation $A\phi = f$ will have the solution given by (20) only if f satisfies the *consistency conditions*,

$$\int_{\mathscr{C}} [\tau^{j-1}f(\tau)/r(\tau)Z(\tau)] \, d\tau = 0, \qquad j = 1(1)(-\kappa). \tag{21}$$

From (20) we see that when $\kappa \geq 1$, the null-space of A is spanned by the functions $t^{j-1}b(t)Z(t)/r(t)$ for $j = 1(1)\kappa$. It is not difficult to show that when $\kappa \leq -1$ the null-space of A^* is spanned by the functions $t^{j-1}/r(t)Z(t)$, $j = 1(1)(-\kappa)$, so that the consistency conditions (21) are equivalent to saying that f is orthogonal to the null-space of A^*.

Let us rewrite (20) as

$$\phi = A^I f + (bZ/r)P_{\kappa-1}, \tag{22}$$

where the operator A^I is defined by

$$A^I f = (af/r^2) - (bZ/r)S(f/rZ) = (Z/r)\{(af/rZ) - bS(f/rZ)\}. \tag{23}$$

From the latter form, we see that the index of A^I is $-\kappa$. Again, although we have used the symbol A^I, care must be taken in its manipulation. If f is Hölder continuous on $\mathscr{C}$, then

$$AA^I f = f \text{ when } \kappa \geq 0, \text{ but } A^I Af \neq f \text{ in general when } \kappa > 0;$$
$$A^I Af = f \text{ when } \kappa \leq 0, \text{ but } AA^I f \neq f \text{ in general when } \kappa < 0. \qquad (24)$$

Thus when $\kappa > 0$, A^I is a right inverse of A; when $\kappa < 0$, A^I is a left inverse. Only when $\kappa = 0$ do we have that A^I is the (unique) inverse of A. The proof of (24) follows on using (18), (19), and the *Poincaré–Bertrand formula*, given by

$$S(\phi_1 S\phi_2) + S(\phi_2 S\phi_1) = (S\phi_1)(S\phi_2) + \phi_1 \phi_2 \qquad (25)$$

for arbitrary ϕ_1, ϕ_2 which are Hölder continuous on $\mathscr{C}$ (see Ref. 10).

In the case when $\kappa \geq 1$ there is an interesting characterization of the particular solution $A^I f$ of the equation $A\phi = f$ which we shall need later (see Section 3.3). To show this we shall need *Parseval's equation* that

$$\int_{\mathscr{C}} \{\phi(t)(S\psi)(t) + \psi(t)(S\phi)(t)\} \, dt = 0, \qquad (26)$$

where ϕ, ψ are any Hölder continuous functions on $\mathscr{C}$. For each solution ϕ of $A\phi = f$ we shall consider the order of the zero of its Cauchy integral Φ at infinity. Since, for z large enough and $\tau \in \mathscr{C}$, we can write $1/(\tau - z) = -(1/z) \sum_{j=0}^{\infty} (\tau/z)^j$, the sum being absolutely and uniformly convergent, we can, from (6), write $\Phi(z) = -(1/z) \sum_{j=0}^{\infty} \mu_j z^{-j}$, where $\mu_j = (1/2\pi i) \int_{\mathscr{C}} \tau^j \phi(\tau) \, d\tau$. Now, for the function $A^I f$, on using (18) and (26) we find that $\mu_j = 0$ for $j = 0(1)(\kappa - 1)$, so that the corresponding Φ has a zero of order $\kappa + 1$ at infinity. On the other hand, if we use (18) with $Q(\tau) = (\tau - t)\tau^{j-1}$, we find that none of the solutions of the homogeneous equation $A\phi = 0$ can give a corresponding Φ with a zero of order greater than κ at infinity. Thus of all the solutions of $A\phi = f$, the Cauchy integral of that given by $A^I f$ has the highest-order zero at infinity.

Let us now consider the *Carleman–Vekua regularization* of the complete equation $L\phi = f$ (see either Ref. 10 or 13). We rewrite (5) as

$$A\phi = f - \lambda K\phi, \qquad (27)$$

and, in the first instance, we shall assume that the right-hand side is known. When $\kappa \geq 0$, the general solution of this equation is given by [see (20)]

$$\phi = A^I (f - \lambda K\phi) + (bZ/r)P_{\kappa-1}, \qquad (28)$$

which can be rewritten as

$$(I + \lambda A^I K)\phi = A^I f + (bZ/r)P_{\kappa-1}. \qquad (29)$$

Again, when $\kappa < 0$, we have

$$\phi = A^{\mathrm{I}}(f - \lambda K\phi), \tag{30}$$

provided that $f - \lambda K\phi$ is orthogonal to the null-space of A^*, a condition that can be verified *a posteriori*. With the convention that $P_{\kappa-1} \equiv 0$ when $\kappa \leq 0$, it is immediately apparent that (30) can also be rewritten as (29). The consistency conditions for $\kappa < 0$ are given by

$$\int_{\mathscr{C}} [\tau^{j-1}/r(\tau)Z(\tau)][f(\tau) - \lambda (K\phi)(\tau)] \, d\tau = 0, \qquad j = 1(1)(-\kappa). \tag{31}$$

The significance of (29) is that it is a Fredholm integral equation of the second kind, for if we write $(A^{\mathrm{I}}K\phi)(t)$ as $(1/\pi i) \int_{\mathscr{C}} N(t, \tau)\phi(\tau) \, d\tau$, then

$$N(t, \tau) = [a(t)k(t, \tau)/r^2(t)] - [b(t)Z(t)/r(t)\pi i] \left\{ \int_{\mathscr{C}} [k(\xi, \tau)/r(\xi)Z(\xi) \right.$$
$$\left. \times (\xi - t)] \, d\xi. \right.$$

Furthermore, the function $N(t, \tau)$ can in turn be written as $N_H(t, \tau)/|t - \tau|^\lambda$, $0 \leq \lambda < 1$, where N_H is Hölder continuous in both variables, so that at worst the function N is weakly singular.

Since (29) is a Fredholm integral equation, we can use the theory of such equations to draw further conclusions about the solutions of the singular equation $L\phi = f$. We know that there will be at most a countable number of values of λ for which the homogeneous equation $(I + \lambda A^{\mathrm{I}}K)\phi = 0$ has a nontrivial solution. Such values of λ are known as *eigenvalues*, and we shall assume throughout this chapter that λ is not an eigenvalue. In this case we can write the solution of (29) as

$$\phi = R[A^{\mathrm{I}}f + (bZ/r)P_{\kappa-1}], \tag{32}$$

where the operator R is such that

$$(I + \lambda A^{\mathrm{I}}K)R = R(I + \lambda A^{\mathrm{I}}K) = I. \tag{33}$$

We observe that when $\kappa > 0$, (32) gives a general solution of $L\phi = f$ containing κ arbitrary constants. The null-space of L in this case has the same dimension as the null-space of A, the null-space of L being spanned by the functions $R(bZt^{j-1}/r)$, $j = 1(1)\kappa$. When $\kappa < 0$, (32) gives us $\phi = RA^{\mathrm{I}}f$, and we must verify that the $-\kappa$ consistency conditions (31) are satisfied by this solution. Finally, we note that when $\kappa = 0$ the function $RA^{\mathrm{I}}f$ is the unique solution of $L\phi = f$ for any arbitrary function f that is Hölder continuous on $\mathscr{C}$.

The above reduction of $L\phi = f$ to an equivalent Fredholm equation of the second kind is known as an *equivalent regularization* of the singular

integral equation. Since, as we remarked earlier, the problem of finding
approximate solutions of Fredholm integral equations has been extensively
studied and is now well understood, the above theory provides us with a
possible method of finding approximate solutions of singular integral equations. We propose to take a closer look at the implementation of this method
in Section 4. However, it is apparent that the use of such methods is likely to
involve a considerable amount of computation so that we must also consider
direct methods for the approximate solution of $L\phi = f$ which do not require
regularization and in which, for example, we replace the integrals by
appropriate quadrature sums. Direct methods will be considered in the next
section and again in Sections 4 and 5.

Before leaving this outline of the theory of singular integral equations,
it should be noted the theory can also be developed in the space $L_p(\mathscr{C})$,
$1 < p < \infty$, of functions that are pth-order integrable on $\mathscr{C}$. To be more
specific, if a, b are continuous on $\mathscr{C}$, $f \in L_p(\mathscr{C})$, and K is a compact operator
in $L_p(\mathscr{C})$, then we can look for solutions $\phi \in L_p(\mathscr{C})$. We have, for example,
that if $\phi \in L_p(\mathscr{C})$ then $S\phi \in L_p(\mathscr{C})$ also. A summary of this theory is to be
found in Zabreyko (Ref. 1), where further references are given.

3. Approximate Solutions: $\mathscr{C}$, A Simple, Smooth, Closed Contour

3.1. A General Strategy. In this and the following two sections, we
shall consider various ways of finding approximate solutions of (3), assuming
that λ is not an eigenvalue. First, let us consider our strategy, in the broadest
possible terms, for finding an approximate solution of an equation $Ax = y$,
where x, y are elements of Banach spaces X, Y, respectively, and A is a
linear operator from X into Y. We must, in some way, *discretize* this
equation. Suppose that X and Y are replaced by sequences of simpler
spaces $\{X_n\}, \{Y_n\}$, respectively, where n is a positive integer. We assume that
$y_n \in Y_n$ is, in some sense, an approximation to $y \in Y$. Again, suppose that for
each n there exists an operator A_n such that A_n maps X_n into Y_n and A_n is,
in some way, an approximation to the operator A. Then, for each n, we
suppose that there exists a unique element $x_n \in X_n$ such that $A_n x_n = y_n$. Our
discretization must be such that $A_n x_n = y_n$ represents a system of linear
algebraic equations which we can solve on a digital computer. Thus in this
way we obtain a sequence of solutions $\{x_n\}$, and if we are to have a
satisfactory algorithm, then we must show that x_n converges in some
appropriate norm to the solution x of the original equation $Ax = y$, in the
limit as $n \to \infty$.

This then is our overall strategy, and we must now consider ways of implementing it. Throughout Section 3 we shall choose $\mathscr{C}$ to be the unit circle with center at the origin, which we shall denote by γ.

3.2. Solution by Piecewise Polynomial Approximation.
This method is perhaps conceptually the simplest of them all. Let $t_1, t_2, \ldots, t_n \equiv t_0$ denote n distinct points on γ described in the positive (anticlockwise) sense, and let $t_{j-1}t_j$ denote that arc of γ from the point t_{j-1} to the point t_j. Suppose that on each arc $t_{j-1}t_j$ we approximate to ϕ by a polynomial of degree m, say. Here we shall consider only the case when $m = 1$, so that ϕ is approximated by a set of piecewise linear polynomials that are continuous at each point t_j. This method has been discussed by Ivanov (Ref. 16) (who says that it was first considered by Lavrentiev in 1932) and more recently by Atkinson (Ref. 19), who has also considered the case in which $m = 2$. If, for $j = 0(1)n$, we define functions s_j by

$$s_j(t) = \begin{cases} (t_{j+1} - t)/(t_{j+1} - t_j), & t \in t_j t_{j+1}, \\ (t - t_{j-1})/(t_j - t_{j-1}), & t \in t_{j-1} t_j, \\ 0, & \text{otherwise,} \end{cases} \tag{34}$$

with $s_0(t) \equiv s_n(t)$, an approximation ϕ_n to ϕ on γ is given by

$$\phi_n(t) = \sum_{j=1}^{n} \phi(t_j) s_j(t). \tag{35}$$

If ϕ is Hölder continuous on γ, then it can be shown (see Atkinson, Ref. 19) that both ϕ_n and $S\phi_n$ converge uniformly to ϕ and $S\phi$, respectively. With these encouraging convergence properties for this form of approximation, let us consider a discretization of (5), in the case when $\kappa = 0$, by choosing

$$\phi_n(t) = \sum_{j=1}^{n} \alpha_j s_j(t), \tag{36}$$

where the constants $\alpha_1, \alpha_2, \ldots, \alpha_n$ have to be determined in some way. We cannot, in general, expect any ϕ_n to satisfy $L\phi = f$ exactly, so we define the *residual function* r_n by $r_n = L\phi_n - f$. If $w_1, w_2, \ldots, w_n \equiv w_0$ is another set of n distinct points described positively on γ, then one way of determining ϕ_n is to suppose that $r_n(w_i) = 0$ for $i = 1(1)n$. In this way we obtain a system of n linear equations in the n unknowns $\alpha_1, \alpha_2, \ldots, \alpha_n$ given by

$$(L\phi_n)(w_i) = f(w_i), \qquad i = 1(1)n. \tag{37}$$

When L is of the form $a + bS$, we can evaluate $L\phi_n$ analytically (see Ref. 19) so that there is no need to approximate to L in this case. However, when L also contains a regular part it might not be possible to evaluate $K\phi_n$ analytically, and some approximation $K_n\phi_n$, where K_n denotes a quadrature sum, may have to be used instead. Atkinson (Ref. 19) reports that when $w_i = t_i$, $i = 1(1)n$, the resulting system of equations in (37) tends to be ill-conditioned as n is increased. Some light has been shed on this problem by Ivanov (Ref. 16), who has analyzed in detail the particular case when the points t_i are equally spaced around γ (with $t_n = 1$) and the equation $L\phi = f$ is chosen as

$$a\phi + S\phi = f, \tag{38}$$

a being a constant. This equation has zero index, and Ivanov has shown that even for values of the constant a for which (38) possesses a solution, the system of equations $a\phi_n(t_i) + (S\phi_n)(t_i) = f(t_i)$ may not necessarily have a solution. With this particular example in mind it is perhaps not surprising that Atkinson observed ill-conditioning. However, on a more optimistic note, Atkinson reports that when the points w_i, $i = 1(1)n$, are chosen to be midway between the points t_i, $t = 1(1)n$, on γ, then for the examples he considered the resulting system of linear algebraic equations appeared to be well conditioned. McInnes (Ref. 20) has also made some interesting comments on a similar method proposed by Gabdulhaev (Ref. 21) for the equation with Hilbert kernel

$$a(t)\phi(t) + (1/2\pi) \int_0^{2\pi} b(t, \tau) \cot[(t - \tau)/2]\phi(\tau)\, d\tau = f(t),$$

$0 < t < 2\pi$, which can be transformed to an equation of the form of Eq. (3), taken over γ.

3.3. Use of a Rational Approximation.

Another direct method for finding an approximate solution of $L\phi = f$ has been described by Gabdulhaev (Ref. 22). For an arbitrary κ, Gabdulhaev looks for an approximation to that solution of $L\phi = f$ for which the Cauchy integral $\Phi(z) = (1/2\pi i)\int_\gamma [\phi(\tau)/(\tau - z)]\, d\tau$ has the zero of highest order at infinity. An approximation ϕ_n to ϕ is sought in the form

$$\phi_n(t) = \sum_{k=0}^{n} \alpha_k t^k + t^{-\kappa} \sum_{k=-n}^{-1} \alpha_k t^k, \tag{39}$$

and, as in Section 3.2, the $2n + 1$ coefficients α_k, $k = -n(1)n$, are found by collocation at the $2n + 1$ points t_j, where $t_j = \exp[2\pi i j/(2n + 1)]$, $j = -n(1)n$.

This gives rise to a system of linear algebraic equations of the form

$$(L_n\phi_n)(t_j) = f(t_j), \qquad j = -n(1)n, \tag{40}$$

where L_n is an approximation to L chosen so that $L_n(t^k) = L(t^k)$ for $\kappa = -(2n+1)(1)(2n-1)$. For this discretization, Gabdulhaev is able to prove a very satisfactory convergence theorem. Let $H_\alpha^{(r)}$, $r = 0, 1, 2, \ldots$, $0 < \alpha \le 1$, denote the space of functions whose rth-order derivative on γ is in H_α. We assume that a, b, f, and k (with respect to both variables) are in $H_\alpha^{(r)}$. For an arbitrary g, the *Hölder norm* of g in H_β is denoted and defined by

$$\|g\|_{H_\beta} = \max_{t \in \gamma} |g(t)| + \sup_{t_1 \ne t_2; t_1, t_2 \in \gamma} [|g(t_1) - g(t_2)|/|t_1 - t_2|^\beta]. \tag{41}$$

Gabdulhaev (Ref. 22) has shown that if λ is not an eigenvalue of L, then, for all $n \ge n_0$, the system of equations (40) possesses a unique solution, and furthermore

$$\|\phi - \phi_n\|_{H_\beta} \le (A \ln n + B)n^{-r-\alpha+\beta} \tag{42}$$

for $r \ge 0$ and $0 < \beta < \alpha \le 1$. This is a good result, although, as Ivanov has remarked, there is a problem in determining *a priori* the value of n_0.

4. Equations Taken over the Arc $(-1, 1)$

4.1. Some Preliminaries. Throughout this section we shall consider the case when $\mathscr{C}$ is an arc which we shall choose to be $(-1, 1)$. For the approximate solution of such equations, Ivanov (Ref. 16) advocates making appropriate transformations so that the equation over the arc becomes an equation over a closed contour. However, equations over arcs occur with sufficient frequency to investigate them in their own right, and we shall do this in this section. We shall now suppose that the singular integral equation is written as

$$a(t)\phi(t) + (b(t)/\pi) \int_{-1}^{1} [\phi(\tau)/(\tau - t)] \, d\tau + \lambda \int_{-1}^{1} k(t, \tau)\phi(\tau) \, d\tau = f(t), \tag{43}$$

$-1 < t < 1$, where a, b, f, and k are assumed to be real. Following Tricomi (Ref. 23), we define the operator T by

$$(T\phi)(t) = (1/\pi) \int_{-1}^{1} [\phi(\tau)/(\tau - t)] \, d\tau, \qquad -1 < t < 1, \tag{44}$$

so that we shall write (43) symbolically as

$$L\phi \equiv a\phi + bT\phi + \lambda K\phi \equiv A\phi + \lambda K\phi = f, \tag{45}$$

the operator A throughout Section 4 being defined by $A\phi = a\phi + bT\phi$. With $\mathscr{C}$ being the finite arc $(-1, 1)$ instead of a simple, smooth, closed contour some modifications to the theory given in Section 2 are required. Many, but not all, of the modifications are minor. At the risk of being a little repetitious we shall, in the remainder of the section, review the theory; it will minimize confusion in the later sections.

The Cauchy integral Φ of a function ϕ is still defined by (6) with $\mathscr{C}$ as $(-1, 1)$, and the Sokhotski–Plemelj formulas (7) are valid across $(-1, 1)$. However, the definition of index as given by (9) is now meaningless, and we must give an alternative one. Equation (43) may have different values of the index depending on conditions imposed on ϕ at the end points ± 1 (see Ref. 10). However, we shall find it convenient not to prescribe conditions on ϕ but always to choose the largest numerical value of the index as follows. First we adjust (43) so that $b(-1) \le 0$. Next, we introduce the real function θ, continuous on $[-1, 1]$, such that $-1 < \theta(-1) \le 0$ and $\theta(t) = (1/\pi) \arg[a(t) + ib(t)]$, $-1 \le t \le 1$. The function $\arg z$ is many-valued; the condition at -1 prescribes which branch is chosen at that point, and as we go from -1 to $+1$ we choose that branch of the argument so that θ remains continuous on $[-1, 1]$. Then the index of L is defined by

$$\kappa(L) = -[\theta(1)], \qquad (46)$$

where $[x]$ denotes the largest integer not exceeding x. (For the computation of κ, see Ref. 24.) Again κ depends only on the dominant part of L, and its significance insofar as dimensions of the null-spaces of L, L^* is concerned is as before. As an example, for the equation with $a = 0$, $b = -1$, we find that $\kappa = 1$.

The canonical function X is defined in the complex plane, with $[-1, 1]$ deleted, as

$$X(z) = (1-z)^{-\kappa} \exp\left\{-\int_{-1}^{1} [\theta(\tau)/(\tau - z)] \, d\tau\right\}, \qquad z \notin [-1, 1], \quad (47)$$

and satisfies properties (1)–(4) following Eq. (14). The fundamental function Z is defined on $(-1, 1)$ by

$$Z(t) = (1-t)^{-\kappa} \exp\left\{-\int_{-1}^{1} [\theta(\tau)/(\tau - t)] \, d\tau\right\}, \qquad t \in (-1, 1), \quad (48)$$

and from the Sokhotski–Plemelj formulas we find

$$X^{\pm}(t) = [a(t) \mp ib(t)][Z(t)/r(t)], \qquad Z(t) = \sqrt{X^+(t)X^-(t)}, \qquad (49)$$

$-1 < t < 1$, where the function r is now defined by $r(t) = \sqrt{a^2(t) + b^2(t)}$. We shall assume throughout that $r(t) > 0$ for $t \in [-1, 1]$.

If Q, R are analytic in the finite plane and such that $QX - R$ is zero at infinity, then on $(-1, 1)$

$$T(bZQ/r) = -(aQZ/r) + R. \tag{50}$$

Similarly if Q, U are analytic in the finite plane and $QX^{-1} - U$ is zero at infinity, then on $(-1, 1)$ we have

$$T(bQ/rZ) = (aQ/rZ) - U. \tag{51}$$

If a, b, and f are Hölder continuous on $[-1, 1]$, the solution of the dominant equation $A\phi = f$ is given by

$$\phi = (af/r^2) - (bZ/r)T(f/rZ) + (bZ/r)P_{\kappa-1} = A^1 f + (bZ/r)P_{\kappa-1}, \tag{52}$$

this solution existing only when $\kappa \le -1$ if f satisfies the consistency conditions

$$\int_{-1}^{1} [\tau^{j-1} f(\tau)/r(\tau)Z(\tau)] \, d\tau = 0; \qquad j = 1(1)(-\kappa). \tag{53}$$

There is, however, an important difference compared with the equation taken over a closed contour where if a, b, $f \in H_\alpha$ ($0 < \alpha < 1$), then $\phi \in H_\alpha$. From (52) it can be shown that the function ϕ may be in the space of functions that Muskhelishvili calls H^* (Ref. 10). We have $\phi \in H^*$ on $[-1, 1]$ if ϕ is Hölder continuous on every closed subinterval of $(-1, 1)$ and, near an end point c, we have $\phi(t) = \phi^*(t)(t - c)^{-\beta}$, $0 \le \beta < 1$, where ϕ^* is Hölder continuous on $[-1, 1]$. Thus ϕ, although absolutely integrable over $(-1, 1)$, may have a complicated algebraic/logarithmic singularity at either of the end points. MacCamy (Ref. 25) has shown that if a, b, f are analytic in the domain D defined by $D = \{z : 0 \le |z + 1| < 2\}$, with k analytic in $D \times D$, and if $b(z) \ne 0$, $b(z)/a(z) \ne \pm i$ for all $z \in D$, then as $t \to -1$ the solution of $L\phi = f$ is of the form

$$\phi(t) \sim (1 + t)^{\theta(-1)} \sum_{m=0}^{\infty} \sum_{n=0}^{\infty} c_{m,n}(1 + t)^m [(1 + t)\log(1 + t)]^n. \tag{54}$$

The symbol $\sim$ means "asymptotically equal to" as $t \to -1$. For our approximate methods, rather than attempt to approximate directly to the function ϕ, we shall approximate instead to a function ψ, where

$$\psi = r\phi/Z \quad \text{or} \quad \phi = Z\psi/r. \tag{55}$$

It follows, in fact, that the fundamental function $Z \in H^*$ (although $1/Z \in H$), and if a, b, f, and k are Hölder continuous on $[-1, 1]$, then ψ is also

Hölder continuous on $[-1, 1]$. This transformation of the dependent variable has been used by many authors in the particular case when $a = 0$, $b = -1$. Then $r = 1$, $Z = 1/\sqrt{1-t^2}$, and we have simply that $\phi = \psi/\sqrt{1-t^2}$.

Finally in this section we shall note the modifications needed to the Poincaré–Bertrand and Parseval formulas. We now have

$$T(\phi_1 T\phi_2) + T(\phi_2 T\phi_1) = (T\phi_1)(T\phi_2) - \phi_1\phi_2,$$

$$\int_{-1}^{1} [\phi_1(t)(T\phi_2)(t) + \phi_2(t)(T\phi_1)(t)] \, dt = 0, \qquad (56)$$

replacing (25) and (26), respectively.

4.2. Approximate Solution by Regularization.

We shall outline an algorithm, described in greater detail by Dow and Elliott (Ref. 24), in which the Carleman–Vekua regularization is used to find an approximate solution ψ_n to ψ where $L(Z\psi/r) = f$. First we rewrite (43) as,

$$A(Z\psi/r) = f - \lambda K(Z\psi/r) = g, \quad \text{say}, \qquad (57)$$

and we shall approximate to g by a polynomial g_n of degree $\leq n$. Let $\{q_j\}$ denote a sequence of polynomials, where q_j is of exact degree j, for $j = 0, 1, 2, \ldots$. Suppose we approximate to f by a polynomial f_n of degree $\leq n$, such that

$$f_n(t) = \sum_{j=0}^{n} F_{j,n} q_j(t). \qquad (58)$$

Similarly, we approximate to the operator K by the operator K_n, where

$$(K_n h)(t) = \int_{-1}^{1} k_n(t, \tau) h(\tau) \, d\tau,$$

$$k_n(t, \tau) = \sum_{j=0}^{n} N_{j,n}(\tau) q_j(t), \qquad (59)$$

the function k_n being some approximation to k on $[-1, 1] \times [-1, 1]$. The function ψ_n will be defined as the solution of the equation

$$A(Z\psi_n/r) = f_n - \lambda K_n(Z\psi_n/r)$$

$$= \sum_{j=0}^{n} \left\{ F_{j,n} - \lambda \int_{-1}^{1} [N_{j,n}(\tau)Z(\tau)\psi_n(\tau)/r(\tau)] \, d\tau \right\} q_j. \qquad (60)$$

To solve (60) it is convenient to introduce the functions u_j, $j = 0, 1, 2, \ldots$, defined by

$$u_j = (aq_j/rZ) - bT(q_j/rZ) = (r/Z)A^1 q_j. \qquad (61)$$

It is worth noting that in the particular case when b is a polynomial the functions u_j will also be polynomials (see Section 4.3). On solving (60) using (52) we find

$$\psi_n(t) = \sum_{j=0}^{n} \left\{ F_{j,n} - \lambda \int_{-1}^{1} [N_{j,n}(\tau)Z(\tau)\psi_n(\tau)/r(\tau)]\, d\tau \right\} u_j(t) + b(t)P_{\kappa-1}(t). \quad (62)$$

The consistency conditions required when $\kappa \le -1$ will be discussed later. Rearranging (62), we can write it as

$$\psi_n(t) + \lambda \int_{-1}^{1} \left\{ \sum_{j=0}^{n} [u_j(t)Z(t)N_{j,n}(\tau)/r(\tau)] \right\} \psi_n(\tau)\, d\tau$$

$$= \sum_{j=0}^{n} F_{j,n}u_j(t) + b(t)P_{\kappa-1}(t), \quad (63)$$

which we observe is a Fredholm integral equation of the second kind with a degenerate kernel, so that it can be reduced to an equivalent system of linear algebraic equations in a standard way. Let ξ_n denote that $(n+1) \times 1$ column vector $(\xi_{0,n}, \xi_{1,n}, \dots, \xi_{n,n})^T$, where

$$\xi_{j,n} = \int_{-1}^{1} [N_{j,n}(\tau)Z(\tau)\psi_n(\tau)/r(\tau)]\, d\tau. \quad (64)$$

Let $\mathbf{f}_n$ denote the $(n+1) \times 1$ column vector $(F_{0,n}, F_{1,n}, \dots, F_{n,n})^T$ and define $\mathbf{g}_n = \mathbf{f}_n - \lambda \xi_n$. Suppose A_n is the square matrix of order $n+1$ with elements $a_{i,j}^{(n)}$ defined by

$$a_{i,j}^{(n)} = \delta_{i,j} + \lambda \int_{-1}^{1} [N_{i,n}(\tau)Z(\tau)u_j(\tau)/r(\tau)]\, d\tau \quad (65)$$

for $i, j = 0(1)n$. When $\kappa \ge 1$, we assume that the arbitrary polynomial $P_{\kappa-1}$ can be written as a linear combination of polynomials l_j, where l_j is of exact degree j, so that $P_{\kappa-1} = \sum_{j=0}^{\kappa-1} \rho_j l_j$. Finally, let $H_{n,\kappa-1}$ denote the $(n+1) \times \kappa$ matrix with elements $h_{i,j}^{(n)}$, where

$$h_{i,j}^{(n)} = \int_{-1}^{1} [b(\tau)N_{i,n}(\tau)Z(\tau)l_j(\tau)/r(\tau)]\, d\tau \quad (66)$$

for $i = 0(1)n$, $j = 0(1)(\kappa-1)$, and let $\rho_{\kappa-1}$ denote the $\kappa \times 1$ column vector $(\rho_0, \rho_1, \dots, \rho_{\kappa-1})^T$. Then on multiplying each side of (63) by $N_{i,n}(t)Z(t)/r(t)$ and integrating with respect to t over $(-1, 1)$ for $i = 0(1)n$, we obtain

$$A_n \mathbf{g}_n = \mathbf{f}_n - \lambda H_{n,\kappa-1}\rho_{\kappa-1}, \quad (67)$$

so that if A_n possesses an inverse $A_n^I = (a_{i,j}^{(n)I})$ then formally

$$\mathbf{g}_n = A_n^I \mathbf{f}_n - \lambda A_n^I H_{n,\kappa-1}\rho_{\kappa-1}.$$

Substituting this solution back into (62) gives

$$\psi_n(t) = \sum_{j=0}^{n} u_j(t) \left(\sum_{k=0}^{n} a_{j,k}^{(n)\mathrm{I}} F_{k,n} \right) + \sum_{s=0}^{\kappa-1} \rho_s \left[b(t) l_s(t) - \lambda \sum_{j=0}^{n} p_j(t) \left(\sum_{k=0}^{n} a_{j,k}^{(n)\mathrm{I}} h_{k,s}^{(n)} \right) \right]$$

(68)

for any $t \in [1, 1]$. When $\kappa \le -1$, (60) will possess a solution only if the consistency conditions are satisfied. For $k = 0(1)(-\kappa - 1)$ we require that the numbers $\delta_k^{(n)}$ defined by

$$\delta_k^{(n)} = \int_{-1}^{1} \left\{ f_n(t) - \lambda \int_{-1}^{1} [k_n(t, \tau) Z(\tau) \psi_n(\tau) / r(\tau)] \, d\tau \right\} [t^k / r(t) Z(t)] \, dt$$

be zero. In terms of the quantities we have computed we have

$$\delta_k^{(n)} = \sum_{j=0}^{n} g_{j,n} \int_{-1}^{1} [t^k q_j(t) / r(t) Z(t)] \, dt,$$

(69)

where $\mathbf{g}_n = (g_{0,n}, g_{1,n}, \ldots, g_{n,n})^T$. It is suggested that one can check that the coefficients $\delta_k^{(n)}$ are zero (or approximately zero) *a posteriori*.

The above paragraphs have only outlined a possible algorithm using the Carleman–Vekua regularization procedure. Many points of detail have been omitted such as, for example, the computation of f_n, k_n; the choice of the polynomials q_j; the computation of the coefficients $F_{j,n}$ and the functions $N_{j,n}$; the evaluation of integrals arising in (65), (66), and (68); etc. For some discussion on these points the reader is referred to Dow and Elliott (Ref. 24). That paper also contains an analysis of the convergence of the algorithm. If we define the operator $K': C \to H$ by

$$(K'g)(t) = \int_{-1}^{1} [Z(\tau) / r(\tau)] k(t, \tau) g(\tau) \, d\tau,$$

where g is any continuous function on $[-1, 1]$ and k is Hölder continuous on $[-1, 1] \times [-1, 1]$, then a norm for K' is defined by $\|K'\| = \sup_{g \in C} \|K'g\|_{H_\alpha} / \|g\|_\infty$, where the Hölder norm $\| \cdot \|_{H_\alpha}$ is defined in (41), provided we replace γ by $[-1, 1]$. Similarly, we define the operator K'_n in terms of k_n and a norm $\|K'_n\|$. If a, b, f, f_n, k, and k_n are Hölder continuous and if for some α, $0 < \alpha \le 1$, we have $\lim_{n \to \infty} \|f - f_n\|_{H_\alpha} = 0$ and $\lim_{n \to \infty} \|K' - K'_n\| = 0$, then $\lim_{n \to \infty} \|\psi - \psi_n\|_\infty = 0$. Although no estimate of the rate of convergence is given, a computable upper bound for $\|\psi - \psi_n\|_\infty$ is given which can be used once ψ_n has been computed (Ref. 24, Theorem 7.1).

In the above discussion we have assumed that λ is not an eigenvalue of L, in which case it can be shown, under the same conditions as those for which convergence is established, that A_n^{I} exists for large enough n. From (68) we obtain an approximate solution of the inhomogeneous equation

$L\phi = f$, together with κ approximate solutions of the homogeneous equation $L\phi = 0$. It is proposed that with these solutions explicitly exhibited, one can then impose any extra conditions on ϕ (such as its vanishing at an end point or its normalization) as may be required in a particular problem.

4.3. The Galerkin Method.

Before we discuss the Galerkin method we shall introduce two sets of orthogonal polynomials that are closely associated with (43). As we have remarked previously, the function $Z \in H^*$, and since r is continuous and strictly positive on $[-1, 1]$, it follows that Z/r is positive and integrable on $[-1, 1]$. Hence, there exists [see Szegö (Ref. 26)] a sequence of orthogonal polynomials $\{p_k\}$ such that

$$p_k(x) = \alpha_k x^k + \cdots, \qquad \text{with } \alpha_k > 0,$$

$$\int_{-1}^{1} [Z(\tau)/r(\tau)]p_k(\tau)p_l(\tau)\, d\tau = h_k \delta_{k,l}. \tag{70}$$

Similarly, since from (48) it can be shown that $1/Z$ is continuous and nonnegative on $[-1, 1]$, it follows that $1/(Zr)$ is positive and integrable on $(-1, 1)$. Again, there exists a sequence of orthogonal polynomials $\{q_k\}$ such that

$$q_k(x) = \beta_k x^k + \cdots, \qquad \text{with } \beta_k > 0,$$

$$\int_{-1}^{1} [q_k(\tau)q_l(\tau)/Z(\tau)r(\tau)]\, d\tau = j_k \delta_{k,l}. \tag{71}$$

There is an interesting relationship between these two sets of polynomials. If we assume that b is a polynomial of degree μ, then for $n \geq \max(\kappa, \mu)$ we find on using (50), (51), and (56) that

$$(aZp_n/r) + bT(Zp_n/r) = (-1)^\kappa (h_n\beta_m/j_m\alpha_n)q_m \tag{72}$$

and

$$(aq_m/rZ) - bT(q_m/rZ) = (-1)^\kappa (\alpha_n j_m/\beta_m h_n)p_n, \tag{73}$$

where $m = n - \kappa$. These results are generalizations of the well-known results in the special case when $a = 0$, $b = -1$, $\kappa = 1$, so that $Z = 1/\sqrt{1-t^2}$, $p_k = T_k$, and $q_k = U_k$, where T_k, U_k are the Chebyshev polynomials of the first and second kind, respectively.

Let us now consider applying the Galerkin method to find an approximate solution of (43) in the particular case when b is a polynomial of degree μ, and let us suppose that $\kappa \geq 0$. We approximate to ψ by ψ_n, where

$$\psi_n(t) = \sum_{j=0}^{n} \gamma_{j,n} p_j(t), \tag{74}$$

the coefficients $\gamma_{j,n}$ having to be determined. As in Section 3.2, we let $r_n = L(Z\psi_n/r) - f$ denote the residual function, so that

$$r_n(t) = \sum_{j=0}^{n} \gamma_{j,n} [A(Zp_j/r)](t) + \lambda \sum_{j=0}^{n} \gamma_{j,n} [K(Zp_j/r)](t) - f(t). \qquad (75)$$

To determine $\gamma_{j,n}$, $j = 0(1)n$, we shall impose the $n - \kappa + 1$ conditions

$$\int_{-1}^{1} [r_n(t)q_i(t)/r(t)Z(t)] \, dt = 0, \qquad i = 0(1)(n - \kappa). \qquad (76)$$

In the next section we shall discuss the reason for imposing only $n - \kappa + 1$ conditions, but accepting for the time being that this is reasonable, we shall obtain a rectangular system of $n - \kappa + 1$ linear algebraic equations in $n + 1$ unknowns which can be written in matrix form as

$$(A_n + \lambda K_n)\gamma_n = \mathbf{f}_m, \qquad m = n - \kappa. \qquad (77)$$

Here γ_n is the column vector $(\gamma_{0,n}, \gamma_{1,n}, \ldots, \gamma_{n,n})^T$, and $\mathbf{f}_m$ is the $(m+1) \times 1$ column vector $(f_0, f_1, \ldots, f_m)^T$, where each element f_i is defined by $f_i = \int_{-1}^{1} [f(t)q_i(t)/r(t)Z(t)] \, dt$. The matrices A_n, K_n are of order $(m+1) \times (n+1)$, and if $A_n = (a_{i,j}^{(n)})$, then

$$a_{i,j}^{(n)} = \int_{-1}^{1} [q_i(t)/r(t)Z(t)][A(Zp_j/r)](t) \, dt, \qquad i = 0(1)m, j = 0(1)n. \qquad (78)$$

Again if $K_n = (k_{i,j}^{(n)})$, then

$$k_{i,j}^{(n)} = \int_{-1}^{1} [q_i(t)/r(t)Z(t)]\left\{ \int_{-1}^{1} [k(t, \tau)Z(\tau)p_j(\tau)/r(\tau)] \, d\tau \right\} dt, \qquad (79)$$

where $i = 0(1)m$, $j = 0(1)n$. Let us consider the structure of the matrix A_n a little more closely. From (78) on using the second of (56) and (51), we find

$$a_{i,j}^{(n)} = \int_{-1}^{1} [Z(t)/r(t)]p_j(t)U(t) \, dt, \qquad (80)$$

where U is the polynomial of degree $i + \kappa$ such that $q_i X^{-1} - U$ is zero at infinity. From (70) if $i + \kappa \le j - 1$, then $a_{i,j}^{(n)} = 0$. On the other hand, since we have assumed that b is a polynomial of degree μ, from (70) we have $bT(Zp_j/r) = T(bZp_j/r)$, provided $j \ge \mu$. Substituting this into (78) and using (50), we find

$$a_{i,j}^{(n)} = \int_{-1}^{1} [q_i(t)/Z(t)r(t)]R(t) \, dt \qquad \text{for } j \ge \mu, \qquad (81)$$

where R is the polynomial of degree $j - \kappa$ such that $p_j X - R$ is zero at infinity. Again by the orthogonality of the polynomials q_i, we have $a_{i,j}^{(n)} = 0$, provided $\mu \leq j \leq i + \kappa - 1$. Thus we have that $a_{i,j}^{(n)} = 0$ whenever $\mu \leq j \leq i + \kappa - 1$ or $j \geq i + \kappa + 1$, so that the matrix A_n has a comparatively simple structure with many of its elements being zero.

Although the implementation of such an algorithm for the general equation $L\phi = f$ does not appear to have been discussed in the literature, it has been considered in particular cases. Erdogan and Gupta (Ref. 27) have solved (43) in the case when $a = 0$, $b = -1$, so that $\kappa = 1$. In this case, based, as we have noted earlier, on the Chebyshev polynomials, we obtain a system of n equations for $n + 1$ unknowns, and the coefficients $a_{i,j}^{(n)}$ are nonzero only when $j = i + 1$. To this system an extra normalizing condition on ϕ has been added to give a system of $n + 1$ equations in $n + 1$ unknowns for which a unique solution is found. A discussion of the evaluation of the integrals which arise in the elements of the matrices A_n, K_n and the vector $\mathbf{f}_m$ is found in Ref. 27. This appears to be a satisfactory algorithm, based on a direct method of solution, and recently an error analysis has been given by Linz (Ref. 28). He has proved that if f possesses a continuous derivative of order $p_1 + 1$, $p_1 \geq 1$ on $[-1, 1]$, and if k possesses continuous derivatives in each variable of order $p_2 + 1$, $p_2 \geq 1$ on $[-1, 1]$, then $\|\psi - \psi_n\|_\infty \leq c n^{-p}$, where c is constant and $p = \min(p_1, p_2)$. This can be considered to be a very satisfactory result.

Karpenko (Ref. 29) has considered (43) in the case when a and b are constants with a not necessarily zero. In this case $\kappa = 1$, and we find $Z(t) = (1-t)^\alpha (1+t)^\beta$, where α, β are constants depending on a, b and such that $\alpha + \beta = -1$. The polynomials p_k are the Jacobi polynomials $P_k^{(\alpha,\beta)}$, and the polynomials q_k are also Jacobi polynomials $P_k^{(-\alpha,-\beta)}$. For this case, Karpenko has proved convergence under the assumption that f, k are Hölder continuous, but he has not given orders of convergence.

5. Conclusions

In this final section we propose to give a brief summary of what has gone before and then to outline areas in which further investigation needs to be done. It is worth reiterating that of necessity we have had to be very selective in what we have presented, but it is hoped that sufficient information has been given to point out the particular problems of singular integral equations. As well as illustrating the differences between the solutions of integral equations when taken over closed contours or open arcs, we have also looked at a few basic methods for their approximate solution. We have considered various representations for the unknown function ϕ, from

piecewise polynomial approximation in Section 3.2 to rational approximation in Section 3.3 and to its representation in Section 4 as the product of an
"unpleasant" function Z, whose value at a point t may be found by
quadrature, and a polynomial. We have also considered various ways in
which a system of linear algebraic equations can be set up by the use of
collocation in Sections 3.2 and 3.3, regularization to give a Fredholm
equation with degenerate kernel in Section 4.2, and a Galerkin-type method
in Section 4.3. Where possible, we have indicated convergence properties of
the proposed algorithm. Ivanov (Ref. 16) has discussed other methods, and
his book must surely be the standard one at the present time on the
approximate solution of such equations.

Among the many problems that remain, there is still that of the use of
the so-called direct methods. In a direct method, we eschew the process of
regularization but replace directly the integrals occurring in $L\phi = f$ by
suitable quadrature sums. Quadrature rules have been discussed extensively
by Davis and Rabinowitz (Ref. 30), and for Cauchy principal-value integrals
particular attention is drawn to the papers by Hunter (Ref. 31) and Elliott
and Paget (Ref. 32). Can direct methods be used in general? McInnes (Ref.
20) concludes. "The preliminary analysis reducing the singular equation to a
Fredholm equation before the application of approximate methods seems to
be an unfortunate but necessary condition." This appears to be unduly
pessimistic. Ivanov (Ref. 16) observes some advantages in trying to develop
direct methods, "firstly, . . . the process of regularization can be computationally complicated; secondly, a direct solution of the problem will allow
one to consider the application of such methods to a wider class of
equations . . . . In particular, we can consider singular integral equations for
which the problem of equivalent regularization has not yet been solved (for
example, some systems of singular integral equations) or equations for
which regularization is not possible (for example, some forms of singular
integral equations of convolution type, or the special case . . . with $a^2 - b^2$
equal to zero on some set of points)." In Sections 1 to 4, we carefully avoided
the case where $r(t) = 0$, although such equations are not without importance
(see Ref. 13). When considering possible direct methods it is the author's
opinion that insufficient emphasis has been placed on the similarity of the
theory of the equation $L\phi = f$ with that of the rectangular system of linear
algebraic equations which we write as $L_{m,n}\phi_n = f_m$. Here $L_{m,n}$ denotes an
$m \times n$ matrix, ϕ_n is an $n \times 1$ column vector, and f_m is an $m \times 1$ column
vector. Now the index of $L_{m,n}$ can be defined very simply. The dimension of
the null-space of $L_{m,n}$ is $n - \rho$, where ρ is the rank of $L_{m,n}$, and the dimension
of the null-space of $L_{m,n}^T$ is $m - \rho$. Following (12), we can define the index of
$L_{m,n}$ as $\kappa(L_{m,n}) = (n - \rho) - (m - \rho) = n - m$. Thus if the index of L is known
to be κ, then we set up a system of $n - \kappa$ equations in n unknowns in some

way, as a discretization of $L\phi = f$. We require $L_{n-\kappa,n}$ to converge in some sense to L as $n \to \infty$. Similarly, we choose $\mathbf{f}_m$ as some approximation to f, and we want ϕ_n to tend toward ϕ in some norm as $n \to \infty$. Again, we note that if the rank of $L_{m,n} = \min(m, n)$ then $L_{m,n}$ possesses a right inverse for $\kappa \geq 0$ and a left inverse for $\kappa \leq 0$. This can be compared with (24). Just as it proved profitable to look upon Fredholm integral equations as the limit as $n \to \infty$ of systems of n equations in n unknowns, so it might be equally profitable to look upon a singular equation with index κ as the limit as $n \to \infty$ of $n - \kappa$ equations in n unknowns. Although it seems likely that one might be able to prove convergence of direct methods for regularizable equations in this way, the case of equations which are not regularizable appears to be far from resolved.

In concluding this section, we would like to draw the reader's attention to the survey paper published in 1973 by Erdogan *et al.* (Ref. 33). Since that time, in addition to papers already mentioned, there have been papers by Fromme and Golberg (Ref. 34), Krenk (Ref. 35), Šeško (Ref. 36), and Theocaris and Ioakimidis (Ref. 37). Finally, a recent paper by Ikebe *et al.* (Ref. 38) considers approximate solution of the Riemann boundary-value problem, which, as we have seen in Section 2, is closely related to the equation $L\phi = f$. Considerations of solving approximately the Riemann problem might well shed further light on solving $L\phi = f$. But there are many aspects of singular integral equations which to date have attracted very little attention. Among these we might list finding approximate solutions of (1). the eigenvalue problem, (2) systems of one-dimensional singular integral equations, (3) multidimensional singular integral equations (see Ref. 15 for the theory), and (4) nonlinear singular integral equations. Ivanov (Ref. 16) has touched briefly on each of these topics, but much remains to be done.

References

1. ZABREYKO, P. P., KOSHELEV, A. I., KRASNOSEL'SKII, M. A., MIKHLIN, S. G., RAKOVSHCHIK, L. S., and STETSENKO, Y., *Integral Equations—A Reference Text*, Noordhoff International Publishing Company, Leyden, Holland, 1975.
2. ATKINSON, K. E., *A Survey of Numerical Methods for the Solution of Fredholm Integral Equations of the Second Kind*, Society for Industrial and Applied Mathematics, Philadelphia, Pennsylvania, 1976.
3. HILBERT, D., *Über eine Anwendung der Integralgleichungen auf ein Problem der Funktiontheorie*, Verhandl, des III International Mathematische Kongress, Heidelberg, Germany, 1904.
4. HILBERT, D., *Grundzüge einer Allgemeinen Theorie der Linearen Integralgleichungen*, B. G. Teubner, Leipzig/Berlin, Germany, 1912.

5. POINCARÉ, H., *Lecons de Mécanique Céleste*, Vol. 3, Gauthier-Villars, Paris, France, 1910.
6. FREDHOLM, I., *Sur une Classe d'Équations Functionelles*, Acta Mathematica, Vol. 27, pp. 365–390, 1903.
7. FREDHOLM, I., *Les Équations Intégrales Lineaires*, Comptes Rendues du Congresse de Stockholm, Sweden, 1909, pp. 92–100; B. G. Teubner, Leipzig, 1910.
8. NOETHER, F., *Über eine Klasse Singulärer Integralgleichungen*, Mathematische Annalen, Vol. 82, pp. 42–63, 1921.
9. CARLEMAN, T., *Sur la Resolution de Certaines Équations Intégrales*, Arkiv fur Mathematik, Vol. 16, No. 26, 1922.
10. MUSKHELISHVILI, N. I., *Singular Integral Equations*, Noordhoff, Groningen, Holland, 1953.
11. PRIVALOV, I., *Boundary Properties of Analytic Functions*, Second Edition, Gosudarstv. Izdat. Tehn.-Teor. Lit., Moscow/Leningrad, U.S.S.R., 1950.
12. MUSKHELISHVILI, N. I., *Some Basic Problems of the Mathematical Theory of Elasticity*, Third Edition, Noordhoff, Groningen, Holland, 1953.
13. GAKHOV, F. D., *Boundary Value Problems*, Pergamon Press, Oxford, England, 1966.
14. VEKUA, N. P., *Systems of Singular Integral Equations*, Noordhoff International Publishing Company, Groningen, Holland, 1967.
15. MIKHLIN, S. G., *Multidimensional Singular Integrals and Integral Equations*, Pergamon Press, Oxford, England, 1965.
16. IVANOV, V. V., *The Theory of Approximate Methods and Their Application to the Numerical Solution of Singular Integral Equations*, Translated by A. Ideh, translation edited by R. S. Anderssen and D. Elliott, Noordhoff International Publishing Company, Leyden, Holland, 1976 (Russian edition published in 1968).
17. HYERS, D. H., *A Review of "The Theory of Approximate Methods and Their Applications to the Numerical Solution of Singular Integral Equations" by V. V. Ivanov*, Bulletin of the American Mathematical Society, Vol. 83, pp. 964–967, 1977.
18. VEKUA, I. N., *On Linear Singular Integral Equations Containing Integrals in the Sense of Cauchy Principal Value*, Doklady Akademia Nauk, U.S.S.R., Vol. 26, pp. 335–338, 1940.
19. ATKINSON, K. E., *The Numerical Evaluation of the Cauchy Transform on Simple Closed Curves*, Society for Industrial and Applied Mathematics Journal on Numerical Analysis, Vol. 9, pp. 284–299, 1972.
20. MCINNES, A. W., *On the Uniform Approximation of a Class of Singular Integral Equations in a Hölder Space*, University of Illinois, PhD Thesis, 1972.
21. GABDULHAEV, B. G., *A Direct Method for Solving Integral Equations*, American Mathematical Society Translation, Series 2, Vol. 91, pp. 213–224, 1970.
22. GABDULHAEV, B. G., *Approximate Solution of Singular Integral Equations by the Method of Mechanical Quadratures*, Soviet Mathematics Doklady, Vol. 9, pp. 239–332, 1968.

23. TRICOMI, F. G., *Integral Equations*, Interscience Publishers, New York, New York, 1957.
24. DOW, M. L., and ELLIOTT, D., *The Numerical Solution of Singular Integral Equations over* [−1, 1], Society for Industrial and Applied Mathematics Journal on Numerical Analysis, Vol. 16, pp. 115–134, 1979.
25. MACCAMY, R. C., *On Singular Integral Equations with Logarithmic or Cauchy Kernels*, Journal of Mathematics and Mechanics, Vol. 7, pp. 355–376, 1958.
26. SZEGÖ, G., *Orthogonal Polynomials*, American Mathematical Society Colloquium Publications, Vol. 23, 1967.
27. ERDOGAN, F., and GUPTA. G. D., *On the Numerical Solution of Singular Integral Equations*, Quarterly of Applied Mathematics, Vol. 30, pp. 525–534, 1972.
28. LINZ, P., *An Analysis of a Method for Solving Singular Integral Equations*, BIT, Vol. 17, pp. 329–337, 1977.
29. KARPENKO, L. N., *Approximate Solution of a Singular Integral Equation by Means of Jacobi Polynomials*, Journal of Applied Mathematics and Mechanics, Vol. 30, pp. 668–675, 1966.
30. DAVIS, P. J., and RABINOWITZ, P., *Methods of Numerical Integration*, Academic Press, New York, New York, 1975.
31. HUNTER, D. B., *Some Gauss-Type Formulas for the Evaluation of Cauchy Principal Value Integrals*, Numerische Mathematik, Vol. 19, pp. 419–424, 1972.
32. ELLIOTT, D., and PAGET, D. F., *Gauss-Type Quadrature Rules for Cauchy Principal Value Integrals*, Mathematics of Computation, Vol. 33, pp. 301–309, 1979.
33. ERDOGAN, F., GUPTA, G. D., and COOK, T. S., *Numerical Solution of Singular Integral Equations*, Mechanics of Fracture, Vol. 1, pp. 368–425, 1973.
34. FROMME, J., and GOLBERG, M., *Numerical Solution of a Class of Integral Equations Arising in Two-Dimensional Aerodynamics*, this volume, Chapter 4.
35. KRENK, S., *On the Integration of Singular Integral Equations*, Danish Center for Applied Mathematics and Mechanics, Report No. 65, 1974.
36. ŠEŠKO, M. A., *On the Numerical Solution of a Singular Integral Equation on an Open Contour*, Akademia Nauk BSSR, Vestsi, Ser. Fiz.-Mat. Vol. 1, pp. 29–36, 1975.
37. THEOCARIS, P. S., and IOAKIMIDIS, N. I., *Numerical Solution of Cauchy-Type Singular Integral Equations*, Transactions of the Athens Academy, Vol. 40, pp. 1–39, 1977.
38. IKEBE, Y., LI, T. Y., and STENGER, F., *The Numerical Solution of the Hilbert Problem*, The Theory of Approximation with Applications, Edited by A. G. Law and B. N. Sahney, Academic Press, New York, New York, 1976.

4

Numerical Solution of a Class of Integral Equations Arising in Two-Dimensional Aerodynamics[1,2]

J. A. FROMME[3] AND M. A. GOLBERG[4]

Abstract. We consider the numerical solution of a class of integral equations arising in the determination of the compressible flow about a thin airfoil in a ventilated wind tunnel. The integral equations are of the first kind with kernels having a Cauchy singularity. Using appropriately chosen Hilbert spaces, it is shown that the kernel gives rise to a mapping which is the sum of a unitary operator and a compact operator. This enables us to study the problem in terms of an equivalent integral equation of the second kind. Using Galerkin's method, we are able to derive a convergent numerical algorithm for its solution. It is shown that this algorithm is numerically equivalent to Bland's collocation method, which is then used as our method of computation. Extensive numerical calculations are presented establishing the validity of the theory.

1. Introduction

Recent interest in aerodynamics testing at high subsonic and transonic speeds has increased the need for improved understanding of the phenomena involved and for improved computational methods to guide experiments, as well as to serve as a means for extrapolating wind-tunnel test data to free-flight conditions. While fundamental differences exist

[1] This paper was prepared with support of the National Aeronautics and Space Administration, Grant No. NSG-2140.
[2] The authors would like to acknowledge the help of Messrs. Tuli Haromy, Charles Doughty, Karl Kuopus, and Steven Sedlacek in the preparation of this paper.

[3] Associate Professor of Mathematics, Department of Mathematics, University of Nevada at Las Vegas, Las Vegas, Nevada.
[4] Professor of Mathematics, University of Nevada at Las Vegas, Las Vegas, Nevada.

between two-dimensional and three-dimensional flows, solutions of two-dimensional problems provide significant insight into three-dimensional flow phenomena.

This chapter treats the problem of predicting unsteady airloads on oscillating thin planar airfoils in ventilated wind tunnels. Essentially, we have extended the work of Bland (Refs. 1–2) in two directions. First, we have developed a flexible computer program called TWODI (Ref. 3), which calculates a wide variety of quantities of aerodynamic interest. Second, we have extended the mathematical analysis of the singular integral equation used by Bland by presenting a rigorous existence theory and using this to give a theoretical analysis of the convergence of the numerical method described in Refs. 1–2.

From a mathematical point of view, we are primarily concerned with the numerical solution of a class of integral equations of the first kind, where the kernel has a Cauchy-type singularity. Such equations seem to fall into a *gray* area of numerical analysis. Equations of the second kind are well investigated, both theoretically and numerically, as this and other volumes attest (Refs. 4–5). On the other hand, equations of the first kind with integrable singularities have become an increasingly well-studied subject. Here, the source of interest lies in the fact that these kernels generate compact operators, and thus have unbounded inverses; thus, the resulting numerical problem is ill-posed (Ref. 6). In our case, the nonintegrable singularity generates a bounded but noncompact operator, so that the problem turns out to be well-posed. As such, it falls into neither of the well-investigated categories of integral equations. Fortunately, there are additional special properties that these equations possess which allow us to show that they are equivalent to equations of the second kind, thus greatly facilitating their numerical solution. We would hope that the analysis presented here would promote the understanding of similar, but more complex, equations occurring in three-dimensional aerodynamics and other areas as well.

The chapter is divided into eight sections. In Section 2, we give a brief description of the boundary-value problem leading to the integral equations that we study. Section 3 is devoted to a brief exposition of Bland's approach to the solution of these equations. Section 4 presents a theoretical analysis of Bland's equation including a Fredholm-type existence theory. In Section 5, we develop a projection method of solution analogous (in fact, equivalent) to Galerkin's method for equations of the second kind. It is shown, under appropriate conditions, that the method is numerically equivalent to Bland's collocation technique. This then justifies the use of his method, since we are able to appeal to well-known results for equations of the second kind to establish convergence and numerical stability. In Section 6, we present a discussion of some of the computational aspects of the problem, particularly that of evaluating the kernel. In Section 7, we give a collection of numerical examples and we close with a discussion of future lines of research.

2. Boundary-Value Problem

As stated in the introduction, the physical problem that we wish to solve is that of finding the pressure distribution around a thin oscillating planar airfoil in a ventilated wind tunnel. In Ref. 1, Bland showed that this required the solution of the following boundary-value problem:

$$\Psi_{xx}(x, y, t) + \Psi_{yy}(x, y, t) - (1/c^2)[v(\partial/\partial x) + (\partial/\partial t)]^2 \Psi(x, y, t) = 0, \quad (1)$$

$$\Psi(x, 0^+, t) = \begin{cases} -\tfrac{1}{2}\Delta p(x, t), & |x| \le b, \\ 0, & |x| > b, \end{cases} \quad (2)$$

$$\Psi(x, \pm H, t) \pm C_w \Psi_y(x, \pm H, t) = 0, \quad -\infty < x < \infty, \quad (3)$$

where Ψ is fluid pressure, x and y are Cartesian coordinates as shown in Fig. 1, t is time, v is free-stream velocity, c is the speed of sound, Δp is the unknown pressure jump across the airfoil, $2H$ is the tunnel height, $2b$ is the airfoil chord, and C_w is the wall ventilation coefficient. A detailed discussion of these equations can be found in Refs. 1, 3. In addition, the downwash w is given by

$$w(x, t) = [v(\partial/\partial x) + (\partial/\partial t)]h(x, t), \quad (4)$$

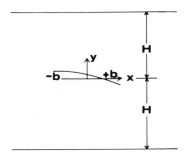

Fig. 1.   Coordinate system for an airfoil in a wind tunnel.

where h is the known airfoil profile function. The downwash $w(x, t)$ is related to p by

$$w(x, t) = -(1/\rho v) \int_{-\infty}^{x} \Psi_y(\lambda, 0^+, t - (x - \lambda)/v)\, d\lambda, \tag{5}$$

where ρ is the free-stream fluid density. Normalizing all lengths by dividing by b, all velocities by v, all pressures by $\frac{1}{2}\rho v^2$, and separating variables letting

$$\Psi(x, y, t) = \Psi(x, y)\exp(i\omega t), \qquad h(x, t) = h(x)\exp(i\omega t), \tag{6}$$

gives the time-independent problem

$$\Psi_{xx}(x, y) + \Psi_{yy}(x, y) - M^2[(\partial/\partial x) + ik]^2\Psi(x, y) = 0, \tag{7}$$

$$\Psi(x, 0^+) = \begin{cases} -\frac{1}{2}\Delta p(x), & |x| \le 1, \\ 0, & |x| > 1, \end{cases} \tag{8}$$

$$\Psi(x, \pm\eta_H) \pm C_w \Psi_y(x, \pm\eta_H) = 0, \qquad -\infty < x < \infty, \tag{9}$$

where

$$M = v/c < 1$$

is the Mach number,

$$\beta = \sqrt{(1 - M^2)}$$

is the co-Mach number,

$$k = \omega b/v$$

is the dimensionless reduced frequency,

$$c_w = C_w/b$$

is the dimensionless ventilation coefficient, and

$$\eta_H = H/b$$

is the tunnel height-to-chord ratio. Using Fourier transforms, Bland obtained $\Psi(x, y)$ in terms of $\Delta p(x)$ (see Ref. 1). This resulted in the following integral equation for the nondimensionalized pressure jump $\Delta p(x)$:

$$[(d/dx) + ik]h(x) = \int_{-1}^{1} \Delta p(\xi)K(x - \xi, k, M, \eta_H, c_w)\, d\xi, \tag{10}$$

in terms of the now nondimensionalized downwash w given by the left-hand side of (10). The kernel of (10) has the form

$$K(x, k, M, \eta_H, c_W)$$

$$= \beta/4\pi x - (ik/4\pi\beta) \log|x|$$

$$+ [(1 + \text{sign } x)/8][1 + c_W k \tanh k\eta_H]/[c_W + (1/k) \tanh k\eta_H]$$

$$- (1/4\eta_H)[\text{sign } x \, F'(|x|/\beta\eta_H)$$

$$- (ik\eta_H/\beta)F(|x|/\beta\eta_H)] \exp(ikM^2x/\beta^2)$$

$$+ (1/8\eta_H)[\text{csch}(\pi x/2\beta\eta_H) - 2\beta\eta_H/\pi x$$

$$+ \{\exp(ikM^2x/\beta^2) - 1\} \text{csch}(\pi x/2\beta\eta_H)]$$

$$- (ik/4\pi\beta)[\log\{(1/x) \tanh(\pi x/4\beta\eta_H)\}$$

$$+ \{\exp(ikM^2x/\beta^2) - 1\} \log \tanh(\pi|x|/4\beta\eta_H)], \qquad (11)$$

where

$$F(x) = \sum_{n=1}^{\infty} \{(\alpha_n/\hat{\lambda}_n) \exp(-\hat{\lambda}_n x) - [1/\pi(n - \tfrac{1}{2})] \exp[-(n - \tfrac{1}{2})\pi x]\}, \quad (12)$$

$$\alpha_n = \frac{1}{[1 + \gamma/(1 + (\gamma\lambda_n)^2)][1 + (k\eta_H/\gamma)^2]}, \qquad (13)$$

$$\hat{\lambda}_n = \lambda_n\sqrt{(1 - \zeta_n^2)}, \qquad \zeta_n = Mk\eta_H/\beta\lambda_n, \qquad (14)$$

$$\tan \lambda_n + \gamma\lambda_n = 0, \qquad (15)$$

$$\gamma = c_W/\eta_H, \qquad \eta_H = H/b. \qquad (16)$$

Thus, we are led to the study of equations of the form

$$w(x) = \int_{-1}^{1} \Delta p(\xi)K(x - \xi) \, d\xi, \qquad (17)$$

where [from here on, we write the kernel as in (18)]

$$K(x - \xi) = (\beta/4\pi)[1/(x - \xi)] - (ik/4\pi\beta) \log|x - \xi| + K_b(x - \xi), \quad (18)$$

and $K_b(x - \xi)$ is bounded. If

$$K_b = 0 \quad \text{and} \quad k = 0,$$

then (18) reduces to the classical airfoil equation

$$w(x) = (1/4\pi) \int_{-1}^{1} [1/(x - \xi)]\Delta p(\xi) \, d\xi, \qquad (19)$$

discussed by Söhngen, Carleman, Tricomi, and others (Refs. 7–12). In this case, solving (17) corresponds to the inversion of the finite Hilbert transform (Ref. 10). From the work of Söhngen and Tricomi, it is known that, for physically reasonable solutions to exist, it is necessary that one impose the auxiliary Kutta condition

$$\lim_{\xi \to 1} \Delta p(\xi) = 0. \tag{20}$$

In this case, it has been shown (Refs. 8, 10) that a unique solution exists for all

$$w(x) \in L_p, \qquad p > 1.$$

Consequently, we look for a solution of (17) satisfying (20). A useful observation of Bland was to reformulate (17), so that (20) automatically holds. Using the Söhngen inversion formula for (19), one obtains the solution

$$\Delta p(\xi) = (4/\pi)\sqrt{[(1-\xi)/(1+\xi)]} \int_{-1}^{1} \sqrt{[(1+x)/(1-x)]}[w(x)/(\xi-x)]\, dx$$

$$= \sqrt{[(1-\xi)/(1+\xi)]}p(\xi), \tag{21}$$

which satisfies the Kutta condition if

$$w(x) \in L_p, \qquad p > 1,$$

(see Ref. 10). Using this observation, Bland was led to look for solutions of (17) in the form

$$\Delta p(\xi) = \sqrt{[(1-\xi)/(1+\xi)]}p(\xi), \tag{22}$$

where $p(\xi)$ is termed the pressure factor (Ref. 1). From this, it is easily seen that $p(\xi)$ solves the integral equation

$$w(x) = \int_{-1}^{1} \sqrt{[(1-\xi)/(1+\xi)]}K(x-\xi)p(\xi)\, d\xi. \tag{23}$$

It is on this equation that we focus for the remainder of the chapter. Hereafter, (23) will be referred to as Bland's integral equation.

3. Collocation and Airfoil Polynomials

Preliminary to the discussion of the numerical solution of (23), we present some results on a class of special functions studied by Bland in Ref. 1. Define

$$\chi_n(x) = \cos[(n - \tfrac{1}{2})\cos^{-1}x]/\cos(\tfrac{1}{2}\cos^{-1}x), \qquad -1 \le x \le 1, \qquad n = 1, 2, \ldots, \tag{24}$$

$$\Psi_n(\xi) = \sin[(n - \tfrac{1}{2})\cos^{-1}\xi]/\sin(\tfrac{1}{2}\cos^{-1}\xi), \qquad -1 \le \xi \le 1, \qquad n = 1, 2, \ldots.$$
(25)

These functions may be identified as renormalized Jacobi polynomials (Ref. 1) and can be obtained computationally by solving the recursion relations

$$\chi_1(x) = 1, \qquad \chi_2(x) = -1 + 2x, \qquad \chi_{n+2}(x) = 2x\chi_{n+1}(x) - \chi_n(x), \quad (26)$$

$$\Psi_1(\xi) = 1, \qquad \Psi_2(\xi) = 1 + 2\xi, \qquad \Psi_{n+2}(\xi) = 2\xi\Psi_{n+1}(\xi) - \Psi_n(\xi). \quad (27)$$

It is easy to show that (26) and (27) are stable. We refer to the functions $\{\chi_n(x)\}$ as downwash polynomials and to $\{\Psi_n(\xi)\}$ as pressure polynomials (Ref. 3). Collectively, they are called airfoil polynomials (Ref. 1).

The basic properties of these functions that we need are summarized below:

$$(1/\pi) \int_{-1}^{1} \sqrt{[(1+x)/(1-x)]}\chi_n(x)\chi_m(x)\,dx = \delta_{nm},$$

$$(1/\pi) \int_{-1}^{1} \sqrt{[(1-\xi)/(1+\xi)]}\Psi_n(\xi)\Psi_m(\xi)\,d\xi = \delta_{nm},$$
(28)

where δ_{nm} is the Kronecker delta,

$$(H_\chi \chi_n)(\xi) = (1/\pi) \int_{-1}^{1} \sqrt{[(1+x)/(1-x)]}[\chi_n(x)/(x-\xi)]\,dx = \Psi_n(\xi),$$
(29)

$$(H_\Psi \Psi_n)(x) = (1/\pi) \int_{-1}^{1} \sqrt{[(1-\xi)/(1+\xi)]}[\Psi_n(\xi)/(x-\xi)]\,d\xi = -\chi_n(x).$$
(30)

Let spaces of complex-valued, measurable functions be defined as follows:

$$L_\chi^2 = \left\{ f: [-1, 1] \to \mathbb{C} \,\middle|\, \int_{-1}^{1} \sqrt{[(1+x)/(1-x)]}|f|^2\,dx < \infty \right\},$$
(31)

$$L_\Psi^2 = \left\{ f: [-1, 1] \to \mathbb{C} \,\middle|\, \int_{-1}^{1} \sqrt{[(1-\xi)/(1+\xi)]}|f|^2\,d\xi < \infty \right\}.$$
(32)

Introducing inner products on L_χ^2 and L_Ψ^2 by

$$\langle f, g \rangle_\chi = (1/\pi) \int_{-1}^{1} \sqrt{[(1+x)/(1-x)]}f(x)\bar{g}(x)\,dx,$$
(33)

$$\langle f, g \rangle_\Psi = (1/\pi) \int_{-1}^{1} \sqrt{[(1-\xi)/(1+\xi)]}f(\xi)\bar{g}(\xi)\,d\xi,$$
(34)

it follows that $\{\chi_n(x)\}$ and $\{\Psi_n(\xi)\}$ are orthonormal in L_Ψ^2 and L_χ^2, respectively. In fact, they are complete.

L_χ^2 and L_Ψ^2 are Hilbert spaces, and we denote their norms by $\|\ \|_\chi$ and $\|\ \|_\Psi$, respectively. That is,

$$\|f\|_\chi = \sqrt{\langle f, f \rangle_\chi}, \qquad \|g\|_\Psi = \sqrt{\langle g, g \rangle_\Psi}.$$

As will be seen later, the relations (29) and (30) are crucial to our study in that they imply that the operators H_χ and H_Ψ can be extended to define unitary operators.

The zeros of the airfoil polynomials can be expressed as:

$$\chi_n(x_k^n) = 0, \qquad x_k^n = -\cos[2k\pi/(2n-1)], \qquad k = 1, 2, \ldots, n-1, \qquad n \geq 2, \tag{35}$$

$$\Psi_n(\xi_k^n) = 0, \qquad \xi_k^n = \cos[2k\pi/(2n-1)], \qquad k = 1, 2, \ldots, n-1, \qquad n \geq 2. \tag{36}$$

These zeros are interdigitated according to

$$-1 < \xi_{n-1}^n < x_1^n < \cdots < \xi_1^n < x_{n-1}^n < 1.$$

We also need the logarithmic transform

$$(1/\pi) \int_{-1}^{1} \sqrt{[(1-\xi)/(1+\xi)]} \Psi(\xi) \log|x-\xi| \, d\xi$$

$$= \begin{cases} x + \frac{1}{2}\log 2, & n = 1, \\ [(\chi_{n+1}(x)+\chi_n(x))/2n] - [(\chi_n(x)+\chi_{n-1}(x))/2(n-1)], & n \geq 2, \end{cases} \tag{37}$$

and the N-point Jacobi-Gauss quadrature formulas

$$\int_{-1}^{1} \sqrt{[(1+x)/(1-x)]} f(x) \, dx \cong [2\pi/(2N+1)] \sum_{k=1}^{N} (1+x_k^N)f(x_k^N), \tag{38}$$

$$\int_{-1}^{1} \sqrt{[(1-\xi)/(1+\xi)]} g(\xi) \, d\xi \cong [2\pi/(2N+1)] \sum_{k=1}^{N} (1-\xi_k^{N+1})g(\xi_k^{N+1}), \tag{39}$$

where $\{x_k^N\}$ and $\{\xi_k^N\}$ are defined in (35) and (36).

Using the above results, we now review briefly Bland's collocation method for the solution of (23). Since L_Ψ^2 is spanned by the pressure polynomials, $p \in L_\Psi^2$ can be expanded as

$$p = \sum_{n=1}^{\infty} \langle p, \Psi_n \rangle_\Psi \Psi_n. \tag{40}$$

(We omit writing the arguments of functions when convenient.) Because of this, one seeks an approximate solution p_N of the form

$$p_N = \sum_{n=1}^{N} a_n \Psi_n, \qquad (41)$$

where a_n, $n = 1, 2, \ldots, N$, are to be determined. Letting

$$r_N(x) = w(x) - \int_{-1}^{1} \sqrt{[(1-\xi)/(1+\xi)]} K(x-\xi) p_N(\xi)\, d\xi, \qquad (42)$$

and setting $r_N(x)$ equal to zero at the zeros of $\chi_{N+1}(x)$ gives the N linear equations

$$\sum_{n=1}^{N} a_n \left[\int_{-1}^{1} \sqrt{[(1-\xi)/(1+\xi)]} K(x_k^{N+1} - \xi) \Psi_n(\xi)\, d\xi \right] = w(x_k^{N+1}),$$

$$k = 1, 2, \ldots, N, \qquad (43)$$

where $\{x_k^{N+1}\}$ are given by (35). This is the method that we use ultimately to solve (23). The numerical examples given by Bland (Ref. 1) and by us (Ref. 3) indicate that p_N converges to p.

To establish this, it is convenient to regard (23) as an operator equation acting between the Hilbert spaces L_Ψ^2 and L_x^2 defined in this section. Making use of the properties of the airfoil polynomials given here enables us to show that (23) has interesting analytic properties which permit a detailed analysis of the above numerical method.

4. Analytical Properties of Bland's Integral Equation

To set the stage for our theoretical treatment of (23), we make some simple preliminary observations. From (11), the kernel can be written as

$$K(x-\xi) = (\beta/4\pi)[1/(x-\xi)] - (ik/4\pi\beta) \log|x-\xi| + K_b(x-\xi), \qquad (44)$$

where $K_b(x-\xi)$ is the bounded part. Multiplying both sides of (44) by $4/\beta$ allows us to rewrite it as

$$\tilde{w}(x) = \int_{-1}^{1} \sqrt{[(1-\xi)/(1+\xi)]} \tilde{K}(x-\xi) p(\xi)\, d\xi, \qquad (45)$$

where

$$\tilde{w}(x) = (4/\beta)w(x) \qquad \text{and} \qquad \tilde{K}(x-\xi) = (4/\beta)K(x-\xi).$$

This transformation obviously has no effect on the solutions of (23). The key observation, and one that was made by Bland in his thesis, is that (45) is actually equivalent to an equation of the second kind. To see this, decompose $\tilde{K}(x - \xi)$ as

$$\tilde{K}(x - \xi) = 1/\pi(x - \xi) + K_S(x - \xi). \tag{46}$$

The first term in (46) can be used to define the integral operator

$$H_\Psi p(x) = (1/\pi) \int_{-1}^{1} \sqrt{[(1 - \xi)/(1 + \xi)]} [p(\xi)/(x - \xi)] \, d\xi,$$

which is just the Hilbert transform (30). From (29) and (30), it is seen that H_Ψ maps the orthonormal basis $\{\Psi_n\}$, $n = 1, 2, \ldots$, onto the orthonormal basis $\{-\chi_n\}$, $n = 1, 2, \ldots$; thus, it can be extended to a bounded invertible operator from L_Ψ^2 to L_χ^2. If one then operates on both sides of (45) by H_Ψ^{-1}, it takes the form of an equation of the second kind in L_Ψ^2. This leads one to suspect that the well-developed numerical methods for such equations would be useful in solving (23). As we shall see, this is the case.

To proceed further, it is necessary to develop some of the analytical properties of (45). For this, we digress slightly and state some definitions and theorems from functional analysis.

Notation. Let H_1 and H_2 be complex Hilbert spaces. The inner products on H_i, $i = 1, 2$, will be denoted by $\langle , \rangle_i$, and the corresponding norm by $\|\cdot\|_i$. The set of bounded linear operators from H_1 to H_2 will be denoted by $[H_1, H_2]$.

Definition 4.1. Let H_1 and H_2 be Hilbert spaces. Let

$$H \in [H_1, H_2].$$

If

$$\|Hx\|_2 = \|x\|_1$$

for all $x \in H_1$, we say that H is an isometry. In addition, if H has a bounded inverse, we say that H is unitary.

Theorem 4.1. Let H_1 be a Hilbert space with a complete orthonormal basis $\{\Psi_n\}$, $n = 1, 2, \ldots$. A necessary and sufficient condition that $H \in [H_1, H_2]$ be unitary is that the set $\{H\Psi_n\}$, $n = 1, 2, \ldots$, be a complete orthonormal basis in H_2.

Proof. See Ref. 13. $\qquad\qquad\qquad\qquad\qquad\qquad\qquad\qquad\qquad\qquad\square$

Definition 4.2. Let $H \in [H_1, H_2]$. The adjoint of H, denoted by H^*, is the bounded linear operator defined by

$$\langle H^* x, y \rangle_1 = \langle x, Hy \rangle_2, \tag{47}$$

where x and y are arbitrary vectors in H_2 and H_1, respectively.

Theorem 4.2. Let $H \in [H_1, H_2]$. Then, H has a unique adjoint $H^* \in [H_2, H_1]$.

Proof. See Ref. 13. □

Theorem 4.3. Let H be a unitary operator from H_1 to H_2. Then, $H^* = H^{-1}$.

Proof. First of all, note that, since H is an isometry, it preserves inner products; i.e.,

$$\langle Hx, Hy \rangle_2 = \langle x, y \rangle_1 \qquad \text{for all } (x, y) \in H_1.$$

In fact,

$$\|H(x + y)\|_2^2 = \|(x + y)\|_1^2.$$

But

$$\|H(x + y)\|_2^2 = \langle Hx, Hx \rangle_2 + 2 \operatorname{Re}\langle Hx, Hy \rangle_2 + \langle Hy, Hy \rangle_2$$

$$= \langle x, x \rangle_1 + 2 \operatorname{Re}\langle Hx, Hy \rangle_2 + \langle Hy, y \rangle_1. \tag{48}$$

Similarly,

$$\|(x + y)\|_1^2 = \langle x, x \rangle_1 + 2 \operatorname{Re}\langle x, y \rangle_1 + \langle y, y \rangle_1. \tag{49}$$

Equating (48) and (49) and similar equations obtained by replacing y by iy gives the result. Now,

$$\langle Hx, Hy \rangle_2 = \langle H^*Hx, y \rangle_1 = \langle x, y \rangle_1, \tag{50}$$

by the definition of adjoint and the above observation. Since (50) holds for all $y \in H_1$, we get that

$$H^*Hx = x, \qquad \forall x \in H_1. \tag{51}$$

Let $y \in H_2$. Then, since H^{-1} exists, $y = Hx$ for some $x \in H_1$. Therefore,

$$y = Hx = H(H^*Hx) = (HH^*)Hx = HH^*y.$$

Thus,

$$y = HH^*y \qquad \text{for all} \qquad y \in H_2;$$

and this with (51) show that H^* is both a left and right inverse of H, so that

$$H^{-1} = H^*. \qquad □$$

Definition 4.3. Let $T \in [H_1, H_2]$. Let $\{\Psi_n\}$, $n = 1, 2, \ldots$, be a complete orthonormal basis for H_1. We say that T is a Hilbert–Schmidt operator if

$$\sum_{n=1}^{\infty} \|T\Psi_n\|_2^2 < \infty.$$

A straightforward adaptation of Theorem (12.11) of Ref. 14 shows that, if T is Hilbert–Schmidt, then T is compact. Note that Hilbert–Schmidt operators are sometimes called operators of finite double norm (Ref. 14).

The results of the above theorems are now applied to (23) or, equivalently, (45). As was stated in Section 3, we wish to regard (23) as an operator equation acting between L_Ψ^2 and L_x^2. Using the equivalent form (45), the kernel decomposition gives rise to three operators H, K_1, K_2, defined by

$$Hp(x) = (1/\pi) \int_{-1}^{1} \sqrt{[(1-\xi)/(1+\xi)]}[p(\xi)/(x-\xi)]\,d\xi, \qquad (52)$$

$$K_1 p(x) = -(ik/\pi\beta^2) \int_{-1}^{1} \sqrt{[(1-\xi)/(1+\xi)]} \log|x-\xi| p(\xi)\,d\xi, \qquad (53)$$

$$K_2 p(x) = \int_{-1}^{1} \sqrt{[(1-\xi)/(1+\xi)]} \tilde{K}_c(x-\xi) p(\xi)\,d\xi, \qquad (54)$$

where the integral in (52) is taken in the sense of a Cauchy principal value. The basic properties of these operators are summarized in Theorem 4.4 below.

Theorem 4.4. H, K_1, K_2 define bounded linear operators from L_Ψ^2 to L_x^2. In addition, H is unitary and K_1, K_2 are compact.

Proof. As stated above, H is just the finite Hilbert transform (30). Since

$$H\Psi_k = -\chi_k,$$

where $\{\Psi_k\}$ and $\{\chi_k\}$ are defined by (24) and (25), respectively, H can be uniquely extended as a bounded operator from L_Ψ^2 to L_x^2 by

$$Hp(x) = \sum_{k=1}^{\infty} \langle p, \Psi_k \rangle H\Psi_k(x) = -\sum_{k=1}^{\infty} \langle p, \Psi_k \rangle \chi_k(x). \qquad (55)$$

Since $\{\chi_k\}$ is a complete orthonormal basis in L_x^2, it follows from Theorem 4.1 that H is unitary.

From (37), we see that

$$K_1\Psi = -(ik/\beta^2)\begin{cases} x + \frac{1}{2}\log 2, & n=1, \\ [(\chi_{n+1}+\chi_n)/2n] - [(\chi_n+\chi_{n-1})/2(n-1)], & n \geq 2, \end{cases} \qquad (56)$$

so that we extend K_1 as an operator from L_Ψ^2 to L_x^2 by

$$K_1 p = \sum_{n=1}^{\infty} \langle p, \Psi_n \rangle_x K_1 \Psi_n, \tag{57}$$

provided that the sum in (57) converges in the L_x^2 norm. To verify this, observe that

$$\|K_1 \Psi_n\|_x^2 = \langle K_1 \Psi_n, K_1 \Psi_n \rangle_x = (k^2/\beta^4)[(n^2 - n + 1)/2n^2(n-1)^2], \qquad n \geq 2, \tag{58}$$

$$\|K_1 \Psi_n\|_x^2 = \text{const}, \qquad n = 1.$$

Thus, (57) converges; since $\|K_1 \Psi_n\|_x^2$ is of order $1/n^2$, the series $\sum_{n=1}^{\infty} \|K_1 \Psi_n\|_x^2$ converges; thus, by the comment following Definition 4.3, K_1 is compact.

Now,

$$|K_2 p(x)| \leq \int_{-1}^{1} \sqrt{[(1-\xi)/(1+\xi)]} K_b(x - \xi) \|p(\xi)\| \, d\xi$$

$$\leq \left[\int_{-1}^{1} \sqrt{[(1-\xi)/(1+\xi)]} |p(\xi)|^2 \, d\xi \right]^{1/2}$$

$$\times \left[\int_{-1}^{1} \sqrt{[(1-\xi)/(1+\xi)]} |K_b(x-\xi)|^2 \, d\xi \right]^{1/2},$$

by the Cauchy–Schwarz inequality. Thus,

$$\|K_2 p\|_x^2 \leq \pi \|p\|_\Psi^2 \int_{-1}^{1} \sqrt{[(1+x)/(1-x)]}$$

$$\times \left[\int_{-1}^{1} \sqrt{[(1-\xi)/(1+\xi)]} |K_b(x-\xi)|^2 \, d\xi \right] dx, \tag{59}$$

where the integral in (59) exists, since $K_b(x - \xi)$ is bounded and the function $\sqrt{[(1+x)/(1-x)]}$ is integrable on $[-1, 1]$. From this, it is seen that

$$\|K_2 p\|_x \leq C \|p\|_\Psi,$$

and so K_2 is bounded. To establish compactness, we again resort to showing that

$$\sum_{k=1}^{\infty} \|K_2 \Psi_k\|_x^2$$

converges. For this, observe that

$$\sum_{k=1}^{\infty} |K_2 \Psi_k|^2 = \sum_{k=1}^{\infty} \left| \int_{-1}^{1} \sqrt{[(1-\xi)/(1+\xi)]} K_b(x - \xi) \Psi_k(\xi) \, d\xi \right|^2. \tag{60}$$

But

$$\int_{-1}^{1} \sqrt{[(1-\xi)/(1+\xi)]} K_b(x-\xi) \Psi_k(\xi) \, d\xi = \pi \langle \eta_x, \Psi_k \rangle_{\Psi}, \tag{61}$$

where

$$\eta_x(\xi) = K_b(x-\xi).$$

Thus, the right-hand sum in (60) can be written as

$$\pi^2 \sum_{k=1}^{\infty} |\langle \eta_x, \Psi_k \rangle_{\Psi}|^2. \tag{62}$$

By Parseval's theorem, this is equal to

$$\pi \int_{-1}^{1} \sqrt{[(1-\xi)/(1+\xi)]} |\eta_x(\xi)|^2 \, d\xi.$$

Multiplying both sides of (62) by $\sqrt{[(1+x)/(1-x)]}$ and integrating with respect to x gives

$$\sum_{k=1}^{\infty} \int_{-1}^{1} \sqrt{[(1+x)/(1-x)]} |K_2 \Psi_k|^2 \, dx = \pi \sum_{k=1}^{\infty} \|K_2 \Psi_k\|_x^2$$

$$= \pi \int_{-1}^{1} \sqrt{[(1+x)/(1-x)]} \left[\int_{-1}^{1} \sqrt{[(1+\xi)/(1-\xi)]} |K_b(x-\xi)|^2 \, d\xi \right] dx. \tag{63}$$

The integral in (63) is finite, since $K_b(x-\xi)$ is bounded and $\sqrt{[(1+x)/(1-x)]}$ is integrable. Thus, K_2 is compact, since it has finite double norm.   □

From the result of Theorem 4.3, Eq. (23) can be written in operator form as

$$(H+K)p = \tilde{w}, \tag{64}$$

where

$$H+K \in [L_{\Psi}^2, L_x^2]$$

and

$$K = K_1 + K_2$$

is compact. A solution to (64) will now mean an element

$$p \in L_{\Psi}^2$$

solving (64). In general, such a solution will satisfy (23) or (45) almost everywhere. If the solution is sufficiently smooth, it will satisfy (45) in the usual pointwise sense. Since physically one is interested in weighted integrals of p, such generalized solutions are perfectly reasonable. In fact, in much aerodynamic work, the pressures themselves are not of primary importance; rather, such things as lift and aeroelastic forces are the usual quantities of interest (Refs. 1–3). As will be shown in Section 7, one obtains convergence of these quantities, even though one cannot in general guarantee pointwise convergence results for either $\Delta p(\xi)$ or $p(\xi)$.

From Theorem 4.4, we know that H has an inverse. Applying H^{-1} to both sides of (64) gives the equivalent equation

$$p + H^{-1}Kp = H^{-1}\tilde{w}. \tag{65}$$

From Theorem 4.4, $H^{-1}K$ is a compact operator and (65) has the form

$$(I + L)p = H^{-1}\tilde{w}, \tag{66}$$

where I is the identity operator on L_Ψ^2 and

$$L = H^{-1}K.$$

Equation (66) is now in the standard form of equations of the second kind. From this, it is possible to obtain solvability theorems by appealing to the Fredholm theory. We state only one such theorem.

Theorem 4.5. Bland's integral equation (23) has a unique solution in L_Ψ^2 iff the homogeneous equation

$$(H + K)p = 0$$

has only the zero solution.

Proof. From the above discussion, it is seen that (23) has a unique solution iff (66) has a unique solution. By the Fredholm alternative for operators of the form $I + L$, L compact, (66) has a unique solution in L_Ψ^2 iff

$$(I + L)p = 0 \tag{67}$$

has only the zero solution (Ref. 5). Applying H to both sides of the above relation shows that this is equivalent to the condition that

$$(H + K)p = 0$$

have only the zero solution. $\qquad\square$

Throughout the rest of our discussion, it will be assumed that the solvability condition in Theorem 4.5 is satisfied, and we now proceed to a discussion of the numerical solution of (23).

5. Projection Method for a Class of Operator Equations

Motivated by the results in Section 4, we now consider the numerical solution of the class of equations

$$Tp = w, \tag{68}$$

where

$$T \in [H_1, H_2] \quad \text{and} \quad H_i, i = 1, 2,$$

are Hilbert spaces,

$$T = H + K,$$

where H is unitary and K is compact. From Theorem 4.3, Eq. (68) can be written in the equivalent form

$$(I + H^*K)p = H^*w. \tag{69}$$

Equation (69) is of the second kind; since there are many well-understood methods for solving such equations (Ref. 5), we consider the possibility of using them to find numerical algorithms for it. For our purposes, Galerkin's method appears to make best use of the theoretical structure of (69), and we now give a brief discussion of this popular technique.

Assume that $\{\Psi_n\}$ is a complete orthonormal basis for H_1, and look for approximate solutions of (69) of the form

$$p_N = \sum_{n=1}^{N} a_n \Psi_n, \tag{70}$$

where the a_n's are constants to be determined. Since p_N in general will not solve (69) exactly, we consider the residual R_N given by

$$R_N = p_N + H^*Kp_N - H^*w. \tag{71}$$

If p_N were the true solution of (69), then

$$R_N = 0.$$

However, this will not be the case in general, and we try to pick p_N so as to make R_N small. Galerkin's method attempts to do this by making R_N orthogonal to Ψ_n, $n = 1, 2, \ldots, N$. That is, we require

$$\langle R_N, \Psi_n \rangle_1 = 0, \qquad n = 1, 2, \ldots, N. \tag{72}$$

Substitution of (71) into (72) gives the following set of linear equations to determine the a_n's:

$$a_n + \sum_{m=1}^{N} \langle H^*K\Psi_m, \Psi_n \rangle_1 a_m = \langle H^*w, \Psi_n \rangle_1, \qquad n = 1, 2, \ldots, N, \tag{73}$$

where the orthogonality of the Ψ_n's has been used in the derivation of (73). If the equations in (73) have a unique solution, then a_n, $n = 1, 2, \ldots, N$, can be determined and the approximation p_N is well defined. The following theorems, taken from Atkinson (Ref. 5), justify this procedure. First, we recast Galerkin's method into a slightly more abstract form.

Let

$$U_N = \text{span}\{\Psi_n\}, \qquad n = 1, 2, \ldots, N,$$

and let Q_N be the operator of orthogonal projection onto U_N. The Galerkin equations (70) and (73) are equivalent to the operator equation

$$Q_N R_N = 0, \qquad p_N \in U_N, \tag{74}$$

or, written out in full,

$$Q_N(I + H^*K)p_N = Q_N H^* w, \qquad p_N \in U_N. \tag{75}$$

Assume that (69) has a unique solution. Then, from the Fredholm alternative (Ref. 5), the operator $I + H^*K$ has a bounded inverse, since H^*K is compact.

Theorem 5.1. Let

$$K_N = Q_N H^* K.$$

Then,

$$\lim_{N \to \infty} \|K - K_N\| = 0.$$

Proof. See Atkinson (Ref. 5). □

Theorem 5.2. Let K_N be as in Theorem 5.1, and assume that N is large enough, so that

$$\|K - K_N\| < 1/\|I + H^*K\|.$$

Then, the operator $(I + K_N)^{-1}$ exists, is bounded, and

$$\|(I + K_N)^{-1}\| \leq \|(I + H^*K)^{-1}\|/[1 - \|I + H^*K\| \cdot \|K - K_N\|]. \tag{76}$$

From this, it follows that the Galerkin equations (73) have a unique solution and

$$\|p - p_N\|_1 \leq \|(I + K_N)^{-1}\| \cdot \|p - Q_N p\|_1. \tag{77}$$

Inequality (77) implies that p_N converges in norm to p, the solution of (69).

Proof. See Atkinson (Ref. 5, pp. 51–52) for details of (76) and (77). The convergence follows from (77) and the fact that $\{\Psi_n\}$ is complete in H_1, so that

$$\|p - Q_N p\|_1 = \left[\sum_{N+1}^{\infty} |\langle p, \Psi_n \rangle_1|^2 \right]^{1/2},$$

which converges to zero.   □

Theorem 5.2 is the basic convergence result that we will use in discussing Bland's method. Note that it has the immediate consequence of showing that the application of Galerkin's method to Bland's equation (45) is a convergent numerical method. That this is true follows from the fact that the equivalent operator version (65) satisfies all of the hypotheses needed to prove Theorem 5.2.

As pointed out in Atkinson, there exists a *dual* to Theorem 5.2. That is, if one can establish the existence of a unique solution to the Galerkin equations for some sufficiently large N, then it follows that one can prove the existence of a solution to (69). From a practical point of view, this is important, since we have the numerical information available from the TWODI program; thus, the solvability of the numerical problem can be used to infer the existence of solutions to the original problem.

Although we have established that Galerkin's method is theoretically a reasonable procedure to use numerically, it does initially appear to have several drawbacks, the most important of which is the necessity of performing the complicated integrations required to evaluate the inner products in (73). Since practically these integrals must be done numerically, it is important to examine the effect of this on the Galerkin equations. Surprisingly, due to the structure of T, particularly the unitarity of H, considerable simplification results; as will be shown, under appropriate conditions, Galerkin's method becomes equivalent to the collocation method.

Relation Between Galerkin's Method and Collocation Method. We now proceed to examine the above-mentioned relation between the collocation method and Galerkin's method. Recall from (73) that the Galerkin approximation for p is obtained by solving the linear equations

$$a_n + \sum_{m=1}^{N} \langle H^* K \Psi_m, \Psi_n \rangle_1 a_m = \langle H^* w, \Psi_n \rangle_1, \qquad n = 1, 2, \ldots, N. \quad (78)$$

Using the definition of adjoint, these equations become

$$a_n + \sum_{m=1}^{N} \langle K \Psi_m, H \Psi_n \rangle_2 a_m = \langle w, H \Psi_n \rangle_2, \qquad n = 1, 2, \ldots, N. \quad (79)$$

Since H is unitary,

$$\{H\Psi_n\} = \{\chi_n\}$$

is a complete orthonormal basis for H_2. Using this and the fact that

$$\langle \chi_n, \chi_m \rangle_2 = \delta_{mn},$$

(79) becomes

$$\sum_{m=1}^{N} \{\langle H\Psi_m, \chi_n \rangle_2 + \langle K\Psi_m, \chi_n \rangle_2\} a_m = \langle w, \chi_n \rangle_2$$

$$= \sum_{m=1}^{N} \langle (H+K)\Psi_m, \chi_n \rangle_2 a_m$$

$$= \langle w, \chi_n \rangle_2, \qquad n = 1, 2, \ldots, N. \qquad (80)$$

Or, since

$$H + K = T,$$

we have

$$\sum_{m=1}^{N} \langle T\Psi_m, \chi_n \rangle_2 a_m = \langle w, \chi_n \rangle_2, \qquad n = 1, 2, \ldots, N. \qquad (81)$$

Note that (81) is formulated directly in terms of the original equation of the first kind

$$Tp = w$$

and can be regarded as a projection method in its own right. To see this, again look for an approximate solution to (68) in the form

$$p_N = \sum_{n=1}^{N} a_n \Psi_n.$$

Using the same argument as for Galerkin's method leads us to consider the residual

$$r_N = Tp_N - w. \qquad (82)$$

In order to make r_N small, we pick a_n, $n = 1, 2, \ldots, N$, to satisfy the orthogonality conditions[5]

$$\langle r_N, \chi_n \rangle = 0, \qquad n = 1, 2, \ldots, N. \qquad (83)$$

Writing (83) out in full gives (81). If one knows $\{\chi_n\}$, $n = 1, 2, \ldots, N$, explicitly, then numerically it is more efficient to use (81) than (78).

---

[5] Sometimes, this is referred to as the method of weighted residuals.

Since our main interest is in Bland's equation (23), we now specialize (68) to the case where H_i, $i = 1, 2$, are taken to be Hilbert spaces of functions on the interval $[a, b]$ and the operator T is an integral operator of the form

$$Tp(x) = \int_b^a k(x, \xi) p(\xi) \, d\xi, \tag{84}$$

where the integral in (84) is generally taken in the sense of a Cauchy principal value. Let

$$L_m(x) = T\Psi_m(x). \tag{85}$$

Then,

$$\langle T\Psi_m, \chi_n \rangle_2 = \langle L_m, \chi_n \rangle_2. \tag{86}$$

Since the inner products in (86) are integrals, we assume that it can be approximated by a quadrature rule $\mathcal{Q}_N$ having N nodes, denoted by x_k, $k = 1, 2, \ldots, N$, and corresponding weights W_k, $k = 1, 2, \ldots, N$. Thus, (76) can be approximated by the finite sum

$$\sum_{k=1}^{N} W_k L_m(x_k) \chi_n(x_k).$$

A similar approximation applied to $\langle w, \chi_n \rangle_2$ gives

$$\langle w, \chi_n \rangle_2 \simeq \sum_{k=1}^{N} W_k w(x_k) \chi_n(x_k).$$

Thus, the numerical solution to the Galerkin equations (81) is obtained by solving

$$\sum_{m=1}^{N} \hat{a}_m \left(\sum_{k=1}^{N} W_k L_m(x_k) \chi_n(x_k) \right) = \sum_{k=1}^{N} W_k w(x_k) \chi_n(x_k), \qquad n = 1, 2, \ldots, N. \tag{87}$$

Define vectors and matrices by

$$a = [\hat{a}_n], \qquad w = [w(x_k)], \tag{88}$$

$$\chi = [\chi_n(x_k)], \qquad L = [L_n(x_k)], \qquad W = \operatorname{diag}[W_k], \qquad n, k = 1, 2, \ldots, N. \tag{89}$$

Using (88) and (89), (87) takes the form

$$(\chi W)La = (\chi W)w. \tag{90}$$

If it is assumed that the matrix χ is nonsingular, then χW is nonsingular and (90) is equivalent to

$$La = w. \tag{91}$$

Referring back to Section 3, it is easily seen that, for Bland's equation, (91) are just the collocation equations resulting from (73). Thus, it is seen that, if the quadrature errors are ignored in the solution of the Galerkin equations, they will then give exactly the same numerical solution as the collocation method, provided that the matrix χ is invertible, as will be established below. Thus, from a numerical standpoint, it makes little difference whether or not one uses the collocation method or Galerkin's method for the solution of (23): the result will be the same numerical approximation to p. However, from a theoretical point of view Galerkin's method appears to be more desirable; as we have seen, it gives convergence along with computable error bounds.[6]

To complete our proof of the equivalence of the collocation method and Galerkin's method, we now establish the nonsingularity of χ.

Definition 5.1. Let $\{f_n(x)\}$, $n = 1, 2, \ldots, N$, be functions defined on the interval $[a, b]$. We say that $\{f_n(x)\}$, $n = 1, 2, \ldots, N$, are unisolvent if the matrix $[f_n(x_k)]$, $n, k = 1, 2, \ldots, N$, is nonsingular for every set of distinct points $\{x_k\}$, $k = 1, 2, \ldots, N$, in $[a, b]$.

Theorem 5.3. Let $\{p_n(x)\}$, $n = 1, 2, \ldots, N$, be a basis for the polynomials of degree $\leq N - 1$ on $[a, b]$. Then, $\{p_n(x)\}$ are unisolvent.

Proof. Let

$$\pi = [p_n(x_k)], \qquad n, k = 1, 2, \ldots, N.$$

Since

$$p(x) = \sum_{k=0}^{N-1} p_{nk} x^k,$$

it follows that

$$\pi = PV,$$

where

$$P = [p_{nk}]$$

---

[6] Note added in proof. The authors have obtained recently an independent convergence proof of the collocation method.

and V is the Vandermonde matrix

$$V(x_1, x_2, \ldots, x_N) = \begin{bmatrix} 1 & 1 & \ldots & 1 \\ x_1 & x_2 & \ldots & x_N \\ x_1^2 & x_2^2 & \ldots & x_N^2 \\ \cdots\cdots\cdots\cdots\cdots\cdots\cdots\cdots \\ x_1^{N-1} & x_2^{N-1} & \ldots & x_N^{N-1} \end{bmatrix}$$

Thus,

$$\det \Pi = \det P \det V.$$

But $\det P$ is nonzero, since $\{p_n(x)\}$ is a basis and

$$\det V = (-1)^N \prod_{\substack{i<j \\ i,j=1,\ldots,N}} (x_i - x_j) \neq 0,$$

since the points x_k are distinct. Thus,

$$\det \pi \neq 0,$$

and π is nonsingular.   □

Corollary 5.1. Let $\{p_n(x)\}$, $n = 1, 2, \ldots, N$, be polynomials of degree $\leq N - 1$ orthogonal with respect to some inner product on the set of functions on $[a, b]$. Then, they are unisolvent.

Proof. The orthogonality implies that $p_n(x)$ are linearly independent, and thus a basis for the polynomials of degree $\leq N - 1$ on $[a, b]$. The corollary now follows from the theorem.   □

Corollary 5.2. Let $\{\chi_n\}$, $n = 1, 2, \ldots, N$, be the first N downwash polynomials as defined in Section 3. Then, $\{\chi_n\}$ are unisolvent.

Proof. Since $\{\chi_n\}$ are of degree $\leq N - 1$ and orthogonal on $[-1, 1]$ with respect to the inner product $\langle \, , \, \rangle_x$, by Corollary 5.2 they are unisolvent.
                                                                                                           □

Corollary 5.3. Let $\{\chi_n\}$ be as above, and let $\{x_k\}$, $k = 1, 2, \ldots, N$, be the zeros of χ_{N+1}. Then, the matrix

$$\chi = [\chi_n(x_k)]$$

is nonsingular.

Proof. From (35) we see that χ_{N+1} has N distinct zeros on $[-1, 1]$. Since $\{\chi_n\}$ are unisolvent, the result follows.   □

From Theorem 5.3 and the above discussion, we arrive at our main equivalence result for the collocation method and Galerkin's method for Bland's equation.

Theorem 5.4. Let p_N be the approximate solution to (23) given by using Galerkin's method based on the pressure polynomials as a complete orthonormal basis for L_Ψ^2. Then, if the inner products in (81) are evaluated using the Jacobi–Gauss quadrature rule (88), the resulting numerical approximation to (23) is the same as one obtains using the collocation method with the same basis and collocating at the zeros of $\chi_{N+1}(x)$.

Proof. From (90), it suffices to prove invertibility of χ, which was done in Theorem 5.3. □

The importance of Theorem 5.4 is that it enables us to regard the collocation method as numerically equivalent to Galerkin's method. As we have seen, Galerkin's method is convergent; since neglecting the quadrature errors in the evaluation of the inner product in (81) gives Bland's collocation equations, we can conclude that this method is convergent if we neglect these errors (see below). We summarize this, our main result, as Theorem 5.5.

Theorem 5.5. Let p_N^C be the numerical approximation to the solution of (23) as described in Section 3. Let p_N^G be the Galerkin approximation to p as described in Section 5. Let $\hat{p}_N^G$ be the approximation to p_N^G obtained by evaluating the inner products by the Jacobi–Gauss rule (81). Then,

$$\hat{p}_N^G = p_N^C;$$

thus, neglecting these quadrature errors, $\hat{p}_N^G$ converges in the norm of L_Ψ^2 to p, and so does p_N^C. The error in this approximation is given by (77), and so

$$\|p_N^C - p\|_\Psi \leq C_N \|p - Q_N p\|_\Psi = C_N \left[\sum_{n=N+1}^\infty |\langle p, \Psi_n \rangle_\Psi|^2 \right]^{1/2} \tag{92}$$

Note that (92) shows that the rate of convergence of p_N^C to p depends on the smoothness of p, and thus requires a knowledge of the behavior of the generalized Fourier coefficients of functions expanded in airfoil polynomials. In addition, it requires an estimate of the smoothness of the solution p. We expect to pursue these points in future work.

As we stated at the beginning of this section, Bland's starting point for the solution of (23) was a least-square procedure which he claimed was equivalent to the collocation method. His proof of this fact (Ref. 1) resided in the assumption that the collocation matrix $[L_n(x_k)]$ was nonsingular. As a further consequence of the Galerkin theory, we establish the validity of this proposition.

Let G denote the matrix $[\langle T\Psi_m, \chi_n\rangle_2]$ given in (86). From the properties of Jacobi–Gauss quadrature,

$$\langle T\Psi_m, \chi_n\rangle_2 = g_{nm} + e_{nm},$$

where $[g_{nm}]$ is the matrix χWL as defined in (90) and

$$[e_{nm}] = E$$

is the matrix of quadrature errors. Thus,

$$G = \chi WL + E. \tag{93}$$

By Theorem 5.3

$$L = (\chi W)^{-1}(G - E). \tag{94}$$

From (94), it is seen that L has an inverse iff $G - E$ does. From Theorem 5.2, we know that, for all sufficiently large N, G has an inverse. Thus,

$$G - E = G(I - G^{-1}E),$$

and it follows from Banach's lemma (Ref. 5) that $I - G^{-1}E$ has an inverse provided that

$$\|G^{-1}E\| < 1,$$

where $\|\cdot\|$ is any matrix norm on the set of $N \times N$ complex matrices. Since

$$\|G^{-1}E\| \leq \|G^{-1}\| \cdot \|E\|,$$

it suffices to show that

$$\|E\| < 1/\|G^{-1}\|.$$

Under the assumption that the quadrature errors can be made arbitrarily small if N is large enough, we can pick N so that G^{-1} exists and

$$\|E\| < 1/\|G^{-1}\|.$$

Thus, we conclude that, for N large enough, the collocation matrix is nonsingular; consequently, Bland's observation that the collocation method is equivalent to a least-square method is valid. Theorem 5.6 states this result.

Theorem 5.6. Provided that N is sufficiently large, Bland's least-square method, his collocation method, and our Galerkin method all give the same numerical approximation to the solution of (23).

Conditioning of the Collocation Matrix. As a final application of the Galerkin approach to (23), we offer a brief discussion of the conditioning of the collocation matrix L. In Refs. 1–2, Bland remarks on the fact that L is well conditioned without offering any proof. The accuracy of our own numerical results also supports this observation. It also appears to be part of the folklore of singular integral equations of the first kind that the strong Cauchy singularity leads to numerical methods which are well conditioned (Refs. 12 and 15). Although this is intuitively reasonable, we are unaware of any mathematical proof of this fact. For equations of the second kind, there exist fairly complete results on conditioning. A summary of these may be found in Atkinson (Ref. 5). Since we know that (23) is equivalent to an equation of the second kind, these results should be of use here. That this is the case is demonstrated below.

Definition 5.2. Let A be an $N \times N$ complex matrix. Let $\|\cdot\|$ be a matrix norm on C^N. The condition number of A relative to $\|\cdot\|$ is given by

$$C(A) = \|A\| \cdot \|A^{-1}\|. \tag{95}$$

A matrix is said to be well conditioned if $C(A)$ is of order 1 and poorly conditioned if $C(A)$ is large (Ref. 5). It follows immediately from (95), that, if A and B are matrices, then

$$C(AB) \leq C(A)C(B), \tag{96}$$

$$C(A^{-1}) = C(A). \tag{97}$$

Since

$$L = (\chi W)^{-1}(G - E) = (\chi W)^{-1} G^{-1}(I - G^{-1}E),$$

(96) and (97) give

$$C(L) \leq C(\chi)C(W)C(G)C(I - G^{-1}E). \tag{98}$$

Since W is a diagonal matrix, one can show that

$$C(W) \leq c[\max |W_k| / \min |W_k|],$$

where c is a constant independent of W. Since W_k, $k = 1, 2, \ldots, N$, are the quadrature weights,

$$C(W) \sim \tilde{c}N.$$

Now, for N large,

$$I - G^{-1}E \sim I,$$

so that

$$C(I - G^{-1}E) \sim C(I) \sim O(1).$$

Thus,

$$C(L) \sim cNC(\chi)C(G).$$

From Atkinson's results (Ref. 5), one can show that

$$\|G\| \leq k$$

independently of N for a suitably chosen norm, so that

$$C(L) \sim cNC(\chi),$$

which gives a reasonable estimate of the conditioning of L. It appears that, if N is not too large, $C(L)$ is well conditioned. A more complete analysis will have to await future work.

From the above results, it is seen that the collocation method put forward in Ref. 1 is theoretically attractive. In the following section, we present some of the details of its implementation.

6. Some Computational Considerations

From the results of Section 5, it is seen that, for computational purposes, one can implement Galerkin's method as Bland's collocation method as given by Eq. (43). Numerically, this is carried out by a program called TWODI, which is described in considerable detail in Ref. 3. In this section, we give a brief outline of the main computational components of TWODI.

The majority of the work in solving (43) goes into the computation of the collocation matrix

$$L = \left[\int_{-1}^{1} \sqrt{[(1-\xi)/(1+\xi)]} K(x_k^{N+1} - \xi) \Psi_n(\xi) \, d\xi \right], \qquad k, n = 1, 2, \ldots, N, \tag{99}$$

and particularly into the computation of the kernel. Using the decomposition of the kernel given in (44), it is seen that L can be written as the sum of three terms:

$$L = L_1 + L_2 + L_3,$$

where L_1 involves integrals of the form

$$\int_{-1}^{1} \sqrt{[(1-\xi)/(1+\xi)]} [\Psi_n(\xi)/(x_k^{N+1} - \xi)] \, d\xi, \qquad k, n = 1, 2, \ldots, N, \tag{100}$$

L_2 involves calculating

$$\int_{-1}^{1} \sqrt{[(1-\xi)/(1+\xi)]} \log|x_k^{N+1} - \xi| \Psi_n(\xi)\, d\xi, \qquad k, n = 1, 2, \ldots, N,$$

(101)

and

$$L_3 = \int_{-1}^{1} \sqrt{[(1-\xi)/(1+\xi)]} K_b(x_k^{N+1} - \xi) \Psi_n(\xi)\, d\xi, \qquad k, n = 1, 2, \ldots, N.$$

(102)

From (30) and (37), it is seen that L_1 and L_2 can be given in closed form, so that we concentrate on the evaluation of L_3. Using the Jacobi–Gauss quadrature rule, the integrals in (102) are approximated by

$$[2\pi/(2N+1)] \sum_{j=1}^{N} (1 - \xi_j^{N+1}) K_b(x_k^{N+1} - \xi_j^{N+1}) \Psi_n(\xi_j^{N+1}), \qquad (103)$$

where the number of quadrature points ξ_j^{N+1} is taken equal to the number of collocation points. This procedure is referred to as Hsu's interdigitation method in the aerodynamics literature (Ref. 16), and is quite reasonable, since it tends to maximize the minimum argument at which $K_b(x)$ needs to be evaluated (Ref. 3). This is a highly desirable property, since $K_b(x)$ is determined by the series $F(x)$ and $F'(x)$ defined in (12) and (13), which converge slowly near $x = 0$. Since for many problems the number of basis elements can be kept quite small, frequently under ten (Ref. 3), evaluating $K_b(x)$ and in particular $F(x)$ and $F'(x)$ consumes the bulk of the computing time.

From (13)–(15), it is seen that $F(x)$ and $F'(x)$ are determined, once the roots λ_n of the eigenvalue equation

$$\tan \lambda_n + \gamma \lambda_n = 0, \qquad \gamma > 0, \qquad (104)$$

are known. It was shown in Ref. 3 that to achieve D digit accuracy for $K_b(x)$ requires approximately $2\beta\eta_H(DN)^2$ terms of $F(x)$ and $F'(x)$. For example, a typical case with

$$M = 0, \quad \eta_H = 15, \quad D = 10, \quad N = 10$$

(N is the number of pressure polynomials) gives this number as 3000. For this reason, an efficient solution method for (104) is needed. We have found that the following iteration scheme works well in practice:

$$\lambda_n^0 = (n - \tfrac{1}{2})\pi, \qquad \lambda_n^{k+1} = n\pi - \tan^{-1}(\gamma \lambda_n^k). \qquad (105)$$

It is straightforward to show using the contraction mapping theorem (Ref. 3), that

$$\lim_{k\to\infty} \lambda_n^k = \lambda_n.$$

The convergence is geometric with convergence rate ρ_n for λ_n given by

$$\rho_n \le \min\{1/\pi(2n-1), 1/\gamma\pi^2(2n-1)^2\}. \tag{106}$$

Although Newton's method can be shown to converge slightly faster for large λ_n, it requires the evaluation of two functions at each iteration, so that (105) will generally be superior.

Solution of the collocation equations is carried out by Gaussian elimination. As the results in the next section show, we are able to obtain accurate solutions of (23) for a wide variety of cases of engineering interest.

As we pointed out in Section 4, it is often the case that various integrals of $p(\xi)$ have engineering significance. Typical of these are lift, pitching moment, and aerodynamic work (Ref. 2). These are defined by

$$C_L = \tfrac{1}{2}\int_{-1}^{1} \Delta p(\xi)\, d\xi, \tag{107}$$

$$C_M = \tfrac{1}{2}\int_{-1}^{1} (\xi+\tfrac{1}{2})\Delta p(\xi)\, d\xi, \tag{108}$$

$$A_{rs} = \int_{-1}^{1} h_r(\xi)\Delta p_s(\xi)\, d\xi, \tag{109}$$

respectively, where $h_r(\xi)$ are appropriate structural basis functions and Δp_s is the pressure corresponding to the downwash w_s (Ref. 2). In general, then, we consider calculating, for $w(\xi)$ real, integrals of the form

$$I = \int_{-1}^{1} w(\xi)\Delta p(\xi)\, d\xi. \tag{110}$$

Since

$$\Delta p(\xi) = \sqrt{[(1-\xi)/(1+\xi)]}p(\xi),$$

we take our approximation to $\Delta p(\xi)$ as

$$\Delta p_N(\xi) = \sqrt{[(1-\xi)/(1+\xi)]}p_N(\xi), \tag{111}$$

and to I as

$$I_N = \int_{-1}^{1} \sqrt{[(1-\xi)/(1+\xi)]}w(\xi)p_N(\xi)\, d\xi. \tag{112}$$

From the definition of the inner product on L_Ψ^2, we see that I_N has the form

$$I_N = \pi \langle p_N, w \rangle_\Psi. \tag{113}$$

Since p_N converges to p in the norm of L_Ψ^2, it follows immediately from the Cauchy–Schwarz inequality that, if

$$w \in L_\Psi^2,$$

I_N converges to I. Since continuous functions are in L_Ψ^2, it follows that the approximations to lift, pitching moment, etc., converge to their true values. The error in this approximation is of the same order as in the approximation to p. This result is similar to (but stronger than) the one given by Milne in Ref. 12.

7. Numerical Results

In this section, we present a variety of numerical results that have been obtained using TWODI. An extensive account of these is contained in our NASA report (Ref. 3), and we refer the reader to that document for a more detailed discussion.

Since there exist known closed-form solutions to (23), we have used these as a check on the code's performance. Two of these cases, the Söhngen solution (Ref. 7) and the Küssner–Schwarz solution (Refs. 17–18) are discussed. We also give an extension of some of the early results presented by Bland (Refs. 1–2) and a comparison to some results of Milne (Ref. 12).

Example 7.1. Söhngen Solutions. The Söhngen solution corresponds to solving (23) in the steady incompressible case with no tunnel walls and is thus given by (21). Using the results of Section 4, the inversion formula (21) gives an exact solution to (19) for all downwash functions $w(x) \in L_x$. In this case, then, we are numerically inverting the finite Hilbert transform. For computational purposes, we have picked deflection functions

$$h_n(x) = \chi_n(x), \qquad n = 1, 2, 3, 4, 5.$$

From (9), the downwashes $w_n(x)$ are given by

$$w_n(x) = (d/dx)h_n(x),$$

since for steady flow

$$k = 0.$$

138   J. A. Fromme and M. A. Golberg

Since (19) corresponds to the infinite atmosphere case (no tunnel walls), we chose

$$\eta_H = 10^4 \quad \text{and} \quad c_w = 10^{100}$$

as appropriate parameters for Bland's kernel to approximate that in (19). The exact pressures are given in Table 1, and the results given by TWODI using 6 basis functions agree in all cases to six decimal accuracy. This is the maximal accuracy that we have required, since it generally exceeds by orders ot magnitude the accuracy achieved in engineering practice (Ref. 3).

In Ref. 3, results for lift, pitching moment, and aerodynamic forces are given. Again, the results were accurate to within the six-decimal-digit accuracy requested of TWODI; thus, they confirm in this case the theoretical predictions made in Sections 5 and 6.

Example 7.2. Küssner–Schwarz Solution. This solution of (23) provides an exact closed-form representation of $\Delta p(\xi)$ for unsteady incompressible flow in an infinite atmosphere (Ref. 3). As in the previous example, we have chosen downwashes

$$w_n = [(d/dx) + ik]\chi_n, \quad n = 1, 2, 3, 4, 5. \tag{114}$$

Table 1.   Pressures for the Söhngen comparison.

ξ	$\Delta p_1(\xi)$	$\Delta p_2(\xi)$	$\Delta p_3(\xi)$	$\Delta p_4(\xi)$	$\Delta p_5(\xi)$
−0.9	0.0000	−34.8712	20.9227	−87.8754	−2.51073
−0.8	0.0000	−24.0000	4.80000	−30.7200	−67.5840
−0.7	0.0000	−19.0438	−3.80876	−5.33227	−84.2498
−0.6	0.0000	−16.0000	−9.60000	7.68000	−81.4080
−0.5	0.0000	−13.8564	−13.8564	13.8564	−69.2820
−0.4	0.0000	−12.2202	−17.1083	15.6419	−53.1823
−0.3	0.0000	−10.9022	−19.6239	14.3909	−36.3696
−0.2	0.0000	−9.79796	−21.5555	10.9737	−21.0068
−0.1	0.0000	−8.84433	−22.9953	6.01415	−8.56131
0.0	0.0000	−8.00000	−24.0000	0.00000	0.00000
0.1	0.0000	−7.23627	−24.6033	−6.65737	4.11020
0.2	0.0000	−6.53197	−24.8215	−13.5865	3.55339
0.3	0.0000	−5.87040	−24.6557	−20.4290	−1.54978
0.4	0.0000	−5.23723	−24.0913	−26.8146	−10.7258
0.5	0.0000	−4.61880	−23.0940	−32.3316	−23.0940
0.6	0.0000	−4.00000	−21.6000	−36.4800	−37.2480
0.7	0.0000	−3.36067	−19.4919	−38.5805	−51.0016
0.8	0.0000	−2.66667	−16.5333	−37.5467	−60.7573
0.9	0.0000	−1.83533	−12.1132	−31.0537	−59.3324
1.0	0.0000	0.00000	0.00000	0.00000	0.00000

In this case, Bland's equation reduces to

$$\int_{-1}^{1} \sqrt{[(1-\xi)/(1+\xi)]} K(x-\xi) p(\xi)\, d\xi = w(x), \qquad (115)$$

where

$$K(x) = 1/4\pi x - (ik/4\pi) \exp(-ikx)[\text{Ci}(k|x|) + i\text{Si}(kx) + i\pi/2], \qquad (116)$$

and $\text{Ci}(x)$ and $\text{Si}(x)$ are the cosine and sine integral functions (Ref. 19). Equation (115) has the solution

$$p(\xi) = [-H_x w - ikI_\Lambda w] + [1 - C(k)]I_T w, \qquad (117)$$

where $H_x w$ is given by (29) and

$$I_\Lambda w = (1/\pi)\sqrt{[(1+\xi)/(1-\xi)]} \int_{-1}^{1} \Lambda(x, \xi) w(x)\, dx, \qquad (118)$$

where

$$\Lambda(x, \xi) = (1/2) \log\left[\frac{1 - x\xi + \sqrt{(1-x^2)}\sqrt{(1-\xi^2)}}{1 - x\xi - \sqrt{(1-x^2)}\sqrt{(1-\xi^2)}}\right], \qquad (119)$$

$$2C(k) - 1 = [H_1^{(2)}(k) - iH_0^{(2)}(k)]/[H_1^{(2)}(k) + iH_0^{(2)}(k)], \qquad (120)$$

$H_i^{(2)}(k)$ are Hankel functions of the second kind, and

$$I_T(w) = (1/\pi)\left[\int_{-1}^{1} \sqrt{[(1+x)/(1-x)]} w(x)\, dx\right].$$

Using methods similar to those in Section 4, it follows that $p(\xi)$ is well defined if $w \in L_x^2$. Using the results in Ref. 3, the pressures corresponding to the downwashes w_n, $n = 1, 2, 3, 4, 5$, are given by

$$\Delta p_1(\xi) = \sqrt{[(1-\xi)/(1+\xi)]}\{[-4ikC(k) + 2k^2]\Psi_1(\xi) + 2k^2\Psi_2(\xi)\}, \qquad (121)$$

$$\Delta p_2(\xi) = \sqrt{[(1-\xi)/(1+\xi)]}\{[-8C(k) - 4ik - 2k^2]\Psi_1(\xi)$$
$$- (k^2 + 8ik)\Psi_2(\xi) + k^2\Psi_3(\xi)\}, \qquad (122)$$

$$\Delta p_3(\xi) = \sqrt{[(1-\xi)/(1+\xi)]}\{[-8C(k) + 4ik]\Psi_1(\xi) - (16 + k^2)\Psi_2(\xi)$$
$$- (8ik + \tfrac{1}{3}k^2)\Psi_3(\xi) + \tfrac{2}{3}k^2\Psi_4(\xi)\}, \qquad (123)$$

$$\Delta p_4(\xi) = \sqrt{[(1-\xi)/(1+\xi)]}\{[-16C(k) - 4ik]\Psi_1(\xi) - 8\Psi_2(\xi) - (24 + \tfrac{2}{3}k^2)\Psi_3(\xi)$$
$$- (8ik + \tfrac{1}{6}k^2)\Psi_4(\xi) + \tfrac{1}{2}k^2\Psi_5(\xi)\}, \qquad (124)$$

$$\Delta p_5(\xi) = \sqrt{[(1-\xi)/(1+\xi)]}\{[-16C(k) + 4ik]\Psi_1(\xi) - 24\Psi_2(\xi) - 8\Psi_3(\xi)$$
$$- (32 + \tfrac{1}{2}k^2)\Psi_4(\xi) - (8ik + \tfrac{1}{10}k^2)\Psi_5(\xi) + \tfrac{2}{5}k^2\Psi_6(\xi)\}. \qquad (125)$$

From these, the lift and moment coefficients follow:

$$C_{L_1} = \pi[-2ikC(k)+k^2], \qquad C_{M_1} = \pi(\tfrac{1}{2}k^2), \tag{126}$$

$$C_{L_2} = \pi[-4C(k)-2ik-k^2], \qquad C_{M_2} = \pi(\tfrac{1}{4}k^2-2ik), \tag{127}$$

$$C_{L_3} = \pi[-4C(k)+2ik], \qquad C_{M_3} = \pi(-4-\tfrac{1}{4}k^2), \tag{128}$$

$$C_{L_4} = \pi[-8C(k)-2ik], \qquad C_{M_4} = \pi(-2), \tag{129}$$

$$C_{L_5} = \pi[-8C(k)+2ik], \qquad C_{M_5} = \pi(-6). \tag{130}$$

See Tables 2–4 for values of the pressure and the lift and moment coefficients based on the above expressions.

Using TWODI, (115) was solved for

$$h_n(x) = \chi_n(x), \qquad n = 1, 2, 3, 4, 5.$$

The numerical results for the pressures and the lift and moment coefficients using eight basis functions are given in Tables 5–7. Since the present form of TWODI calculates the kernel in its most general form, it will not accept $\eta_H = \infty$, which gives the limiting kernel (116). This evaluation is most efficient for small values of η_H and k. Therefore, as a good test of the code's

Table 2. In-phase pressures for the Küssner–Schwarz comparison, $k = 1$.

ξ	$\mathrm{Re}(\Delta p_1)$	$\mathrm{Re}(\Delta p_2)$	$\mathrm{Re}(\Delta p_3)$	$\mathrm{Re}(\Delta p_4)$	$\mathrm{Re}(\Delta p_5)$
-0.9	-0.00476	-22.1235	39.8542	-58.0293	31.2030
-0.8	1.19673	-17.2664	19.0216	-10.3985	-45.0000
-0.7	1.90178	-15.1291	7.87586	11.3355	-67.3412
-0.6	2.39782	-13.7510	0.15838	22.4864	-68.0052
-0.5	2.76939	-12.6708	-5.74258	27.4860	-58.0773
-0.4	3.05338	-11.7245	-10.4536	28.3354	-43.2973
-0.3	3.26916	-10.8415	-14.2720	26.1859	-27.2281
-0.2	3.42795	-9.98838	-17.3565	21.8174	-12.2751
-0.1	3.53653	-9.14889	-19.7975	15.8246	-0.10282
0.0	3.59891	-8.31548	-21.6488	8.70238	8.16904
0.1	3.61715	-7.48545	-22.9421	0.89173	11.8675
0.2	3.59169	-6.65892	-23.6943	-7.19346	10.7190
0.3	3.52144	-5.83773	-23.9107	-15.1462	4.83414
0.4	3.40348	-5.02478	-23.5855	-22.5477	-5.27874
0.5	3.23253	-4.22359	-22.6988	-28.9431	-18.6663
0.6	2.99945	-3.43774	-21.2084	-33.8024	-33.8236
0.7	2.68808	-2.66984	-19.0296	-36.4390	-48.4592
0.8	2.26630	-1.91849	-15.9772	-35.8007	-58.8993
0.9	1.65154	-1.16439	-11.5352	-29.6921	-57.9949
1.0	0.00000	0.00000	0.00000	0.00000	0.00000

Table 3.   Out-of-phase pressures for the Küssner–Schwarz comparison, $k = 1$.

ξ	$\mathrm{Im}(\Delta p_1)$	$\mathrm{Im}(\Delta p_2)$	$\mathrm{Im}(\Delta p_3)$	$\mathrm{Im}(\Delta p_4)$	$\mathrm{Im}(\Delta p_5)$
−0.9	−9.40537	13.9580	5.58891	−10.7213	40.2743
−0.8	−6.47322	4.80655	15.3666	−23.1229	41.3507
−0.7	−5.13645	0.00520	19.8108	−25.0513	32.0497
−0.6	−4.31548	−3.19563	21.7644	−22.5833	20.3991
−0.5	−3.73731	−5.53878	22.1740	−18.0058	9.70705
−0.4	−3.29600	−7.32879	21.5109	−12.5557	1.50242
−0.3	−2.94050	−8.71875	20.0630	−7.01504	−3.63101
−0.2	−2.64268	−9.79529	18.0309	−1.91505	−5.69315
−0.1	−2.38547	−10.6108	15.5684	2.37511	−5.06828
0.0	−2.15774	−11.1978	12.8022	5.60437	−2.39563
0.1	−1.95175	−11.5761	9.84331	7.61651	1.52646
0.2	−1.76179	−11.7558	6.79503	8.33836	5.81963
0.3	−1.58335	−11.7392	3.75865	7.77561	9.59778
0.4	−1.41257	−11.5205	0.83939	6.01520	12.0401
0.5	−1.24577	−11.0839	−1.84626	3.23568	12.4733
0.6	−1.07887	−10.3989	−4.15891	−0.26982	10.4758
0.7	−0.90643	−9.40897	−5.91387	−4.04442	6.03222
0.8	−0.71925	−7.99927	−6.82594	−7.34788	−0.18414
0.9	−0.49502	−5.87254	−6.31302	−8.75717	−6.07319
1.0	0.00000	0.00000	0.00000	0.00000	0.00000

performance we have chosen

$$\eta_H = 300 \qquad \text{and} \qquad k = 1$$

for comparison purposes to the true solution. As can be seen from Tables 5–7, our results are generally correct to errors of the order 10^{-3}. This agreement is well within the accuracy usually achieved in engineering practice (Ref. 3).[7]

Table 4.   Lift and moment coefficients for the Küssner–Schwarz comparison, $k = 1$.

Mode	C_L	C_M
1	2.51156 − 3.38937i	1.57080
2	−9.92033 − 5.02312i	−0.785398 − 6.28319i
3	−6.77874 + 7.54325i	−13.3518
4	−13.5575 − 3.76305i	−6.28319
5	−13.5575 + 8.80332i	−18.8496

[7] We have recently shown that changing the method of calculating L_2 and L_3 yields six or more figure accuracy.

Table 5.   Computed in-phase pressures for Küssner–Schwarz comparison,
$\eta_H = 300$, $k = 1$, $M = 0$, number of basis elements = 8.

ξ	$\text{Re}(\Delta p_1)$	$\text{Re}(\Delta p_2)$	$\text{Re}(\Delta p_3)$	$\text{Re}(\Delta p_4)$	$\text{Re}(\Delta p_5)$
−0.9	0.00029	−22.1166	39.8107	−57.9898	31.1405
−0.8	1.20011	−17.2555	18.9831	−10.3804	−44.9802
−0.7	1.90365	−15.1143	7.84448	11.3234	−67.2518
−0.6	2.39819	−13.7333	0.13703	22.4443	−67.8793
−0.5	2.76832	−12.6514	−5.75172	27.4200	−57.9520
−0.4	3.05101	−11.7045	−10.4495	28.2548	−43.2034
−0.3	3.26565	−10.8222	−14.2549	26.1017	−27.1849
−0.2	3.42353	−9.77075	−17.3276	21.7403	−12.2885
−0.1	3.53144	−9.13379	−19.7591	15.7637	−0.16613
0.0	3.59343	−8.30359	−21.6039	8.66429	8.07255
0.1	3.61156	−7.47705	−22.8940	0.87952	11.7605
0.2	3.58625	−6.65414	−23.6464	−7.18029	10.6249
0.3	3.51642	−5.83641	−23.8664	−15.1117	4.77267
0.4	3.39913	−5.02650	−23.5477	−22.4988	−5.29617
0.5	3.22903	−4.22769	−22.6696	−28.8890	−18.6395
0.6	2.99694	−3.44334	−21.1890	−33.7529	−33.7647
0.7	2.68663	−2.67593	−19.0199	−36.4031	−48.3910
0.8	2.26586	−1.92400	−15.9758	−35.7843	−58.8496
0.9	1.65186	−1.16821	−11.5389	−29.6948	−57.9853

Example 7.3. Extension of Bland's Results. In Ref. 1, Bland presented numerical results for a flat plate oscillating about the 42.5% chord at $M = 0.85$ in a closed tunnel with $\eta_H = 7.5$. His results for the values of the magnitudes of the lift coefficients were

$$|C_L| = 12.2351, \quad k = 0,$$
$$|C_L| = 7.99420, \quad k = 0.1,$$
$$|C_L| = 5.43549, \quad k = 0.2,$$

and have been verified with TWODI.

Bland's results were based on a program written only for the closed-tunnel condition $c_w = \infty$. We have extended these results to include the effects of ventilating the wind tunnel. Typical values of the lift coefficients for various values of c_w and η_H are presented in Table 8. These results are new.

The above results indicate continuous behavior of the lift coefficients as a function of c_w and η_H. Although this is reasonable on physical

Table 6.   Computed out-of-phase pressures for Küssner–Schwarz comparison, $\eta_H = 300$, $k = 1$, $M = 0$, number of basis elements = 8.

ξ	$\mathrm{Im}(\Delta p_1)$	$\mathrm{Im}(\Delta p_2)$	$\mathrm{Im}(\Delta p_3)$	$\mathrm{Im}(\Delta p_4)$	$\mathrm{Im}(\Delta p_5)$
−0.9	−9.40155	13.9362	5.58925	−10.7160	40.2407
−0.8	−6.46884	4.78967	15.3537	−23.0932	41.2824
−0.7	−5.13165	−0.00675	19.7834	−25.0051	31.9825
−0.6	−4.31049	−3.20222	21.7247	−22.5341	20.3629
−0.5	−3.73237	−5.53986	22.1258	−17.9662	9.71526
−0.4	−3.29132	−7.32451	21.4587	−12.5350	1.55265
−0.3	−2.93627	−8.70957	20.0112	−7.01818	−3.55274
−0.2	−2.63906	−9.78191	17.9838	−1.94244	−5.60663
−0.1	−2.38258	−10.5941	15.5292	−2.32700	−4.99352
0.0	−2.15564	−11.1789	12.7733	5.54207	−2.34813
0.1	−1.95048	−11.5560	9.82604	7.54837	−1.53882
0.2	−1.76132	−11.7358	6.78949	8.27314	5.79789
0.3	−1.58363	−11.7203	3.76377	7.72111	9.55125
0.4	−1.41348	−11.5038	0.85301	5.97701	11.9839
0.5	−1.24716	−11.0703	−1.82716	3.21609	12.4243
0.6	−1.08055	−10.3890	−4.13799	−0.27243	10.4476
0.7	−0.90820	−9.40307	−5.89500	−4.03566	6.02970
0.8	−0.72086	−7.99723	−6.81261	−7.33642	−0.16842
0.9	−0.49621	−5.87344	−6.30739	−8.75209	−6.05846

grounds, no analytic proof has been attempted in this study. We expect to pursue this matter further in future work.

Comparison with Milne's Results.   In Ref. 12, Milne examined the use of a projection method for the solution of steady flow in an infinite atmosphere. He did not make use of Bland's transformation (22); thus, he worked in the L_p-setting used by Tricomi (Ref. 10). For numerical purposes, the pressures were expanded in a basis consisting of piecewise linear splines. For the airfoil equation with downwash

$$w = (1/4\pi)\chi_1,$$

Table 7.   Computed lift and moment for the Küssner–Schwarz comparison.

Mode	C_L	C_M
1	2.50990 − 3.38758i	1.56898 − 0.00052i
2	−9.91313 − 5.01876i	−0.786022 − 6.27628i
3	−6.76970 + 7.52965i	−13.3370 + 0.00224i
4	−13.5648 + 3.77175i	−6.27720 − 0.01926i
5	−13.5511 + 8.80183i	−18.8494 + 0.00921i

Table 8.   Lift coefficients versus η_H and c_w for $M = 0.85$, $k = 0$.

c_w	$\eta_H = 1$	$\eta_H = 7.5$	$\eta_H = 10$	$\eta_H = 100$	$\eta_H = 1000$
0	1.99486(5)*	8.22740(5, 10)	8.98389(5)	11.5788(5)	11.8920(3)
10^{-4}	1.99506(5)	8.22744(5)	8.98391(5)	11.5788(4)	11.8920(3)
10^{-2}	2.01449(5)	8.23118(5)	8.98630(5)	11.5788(4)	11.8920(3)
1	3.83187(5)	8.57219(5)	9.20734(5)	11.5822(4)	11.8920(3)
10^4	20.1757(5)	11.8449(5)	11.7403(5)	11.7513(3)	11.8952(3)
10^6	21.8952(5)	12.2308(5)	12.0980(5)	11.9257(3)	11.9243(3)
∞		12.2351(5)	12.1021(5)	11.9292(3)	11.9275(3)‡
∞		12.2442(1)†			
		12.2351(2–20)†			

* Numbers in parentheses indicate number of pressure basis functions used in the calculations.
† These results are those of Bland's, which were programmed only for $c_w = \infty$.
‡ Compare with $\eta_H = \infty$, $C_{L_\alpha} = 2\pi/\beta = 11.9275$.

he obtained 2-digit accuracy of the coefficient with 20 basis elements and 3-digit accuracy for the pitching moment with 10 basis elements. Our results, using pressure polynomials, were correct to six-digit accuracy using one basis function.

The above results constitute only a small fraction of the numerical studies that we have carried out using TWODI. In Ref. 3, we have given a fairly comprehensive parametric analysis of the effects of the depth-to-chord ratio η_H and the ventilation coefficient on the section coefficients. In addition, we have made the first extensive analysis of the effect of tunnel ventilation on acoustic resonance between the airfoil and the wind tunnel. For further illustration of these results, the interested reader is referred to Ref. 3.

8. Conclusions

We have presented a successful technique for the numerical solution of a large class of singular integral equations arising in two-dimensional aerodynamics. As stated in the introduction, we have given a detailed theoretical examination of Bland's collocation method and developed a flexible code capable of calculating a wide variety of quantities of engineering interest. Although the results given in Section 7 are generally accurate for practical purposes, there are several areas where we feel improvement can be made. In the present form of TWODI, the calculation of the kernel frequently requires taking large numbers of terms in the series for $F(x)$ and $F'(x)$, often thousands, to obtain acceptable accuracy. Methods for accelerating their convergence are being investigated, and we hope that they will prove successful. In fact, we believe that this illustrates a point sometimes overlooked in studying solution methods for integral equations, that is, the kernel computation is frequently the most expensive and time-consuming aspect of the whole process (Ref. 5).

Since we view this work as a preliminary study, there are many things that we have not examined in detail which will be taken up shortly. Among these are: the effect of the numerical integration used in evaluating the collocation matrix L, particularly for large values of k; *a priori* estimates of the number of basis elements needed (the results of Sections 3 and 4 are the groundwork for this); and methods for speeding up convergence. The fact that (23) is equivalent to an equation of the second kind and our adoption of Galerkin's method as the primary theoretical vehicle leads one to believe that the recent results of Sloan (Refs. 5, 20) for improving the convergence of Galerkin's method could be of value here.

Throughout this work, we have taken as given the formulation of the boundary-value problem (1)–(3), in particular the boundary condition (3). In Bland's thesis (Ref. 1), this condition was presented as a purely phenomenological formulation of the flow condition across the porous tunnel wall. Comparison of the results of this study with actual wind-tunnel test data should give some indication of the degree to which it represents the actual flow conditions across the boundary. One of the authors (Fromme) has recently developed a *first principle derivation* of a boundary condition which appears to have physical substance. We expect to explore the solution of this boundary-value problem in future work.

From a mathematical point of view, one of the most interesting aspects of this problem is the role of the unitarity of the finite Hilbert transform (30). Note that Bland's transformation (22) gives a transform which differs from the classical Hilbert transform (Ref. 10) by the weight factor $\sqrt{[(1-\xi)/(1+\xi)]}$. This is an exact parallel to the unitarity of the standard Hilbert transform. We feel that a more thorough understanding of this property is essential for the further application of Galerkin-type methods to other integral equations in aerodynamics, particularly for three-dimensional flows.

References

1. BLAND, S. R., *The Two-Dimensional Oscillating Airfoil in a Wind Tunnel in Subsonic Compressible Flow*, University of North Carolina, PhD Thesis, 1968.
2. BLAND, S. R., *The Two-Dimensional Oscillating Airfoil in a Wind Tunnel in Subsonic Flow*, SIAM Journal on Applied Mathematics, Vol. 18, pp. 830–848, 1970.

3. FROMME, J., and GOLBERG, M., *Unsteady Two-Dimensional Airloads Acting on Thin Oscillating Airfoils in Subsonic Ventilated Wind Tunnels*, NASA Contractor Draft Report, NSG-2140, 1977.
4. GOLBERG, M. A., Chapter 1, this volume; ELLIOT, D., Chapter 3, this volume.
5. ATKINSON, K., *A Survey of Numerical Methods for the Solution of Fredholm Integral Equations of the Second Kind*, Society for Industrial and Applied Mathematics, Philadelphia, Pennsylvania, 1976.
6. NASHED, M. Z., *Aspects of Generalized Inverses in Analysis and Regularization*, Generalized Inverses and Applications, Edited by M. Z. Nashed, Academic Press, New York, New York, 1976.
7. SÖHNGEN, H., *Die Lösungen der Integralgleichung* $g(x) = (1/2\pi)\int_{-a}^{a}(f(\xi)/x-\xi)\,d\xi$ *und Deren Anwendung in der Traflügeltheorie*, Matematische Zeitschrift, Vol. 45, pp. 245–264, 1939.
8. SÖHNGEN, H., *Zur Theorie der Endlichen Hilbert-Transformation*, Matematische Zeitschrift, Vol. 60, pp. 31–51, 1954.
9. MUSHKELISHVILI, N. I., *Singular Integral Equations*, Wolters-Noordhoff Publishing Company, Groningen, Holland, 1953.
10. TRICOMI, F. G., *On the Finite Hilbert Transformation*, Quarterly of Applied Mathematics, Vol. 2, pp. 199–211, 1951.
11. TRICOMI, F. G., *Integral Equations*, John Wiley and Sons, (Interscience Publishers), New York, New York, 1957.
12. MILNE, R. D., *Application of Integral Equations to Fluid Flows in Unbounded Regions*, Finite Elements in Fluids, Vol. 2, Edited by R. H. Gallagher, J. T. Oden, C. Taylor, and O. C. Zienkiewicz, John Wiley and Sons, New York, New York, 1975.
13. DUNFORD, N., and SCHWARZ, J., *Linear Operators—Part I*, John Wiley and Sons (Interscience Publishers), New York, New York, 1967.
14. AGMON, S., *Lectures on Elliptic Boundary Value Problems*, D. Van Nostrand Company, New York, New York, 1965.
15. NOBLE, B., *Some Applications of the Numerical Solution of Integral Equations to Boundary Value Problems*, Proceedings of the Conference on the Applications of Numerical Analysis, Edited by J. Morris, Springer–Verlag, New York, New York, 1971.
16. HSU, P. T., *Some Recent Developments in the Flutter Analysis of Low Aspect Ratio Wings*, Proceedings of the National Specialists Meetings on Dynamics and Aeroelasticity, Fort Worth, Texas, 1958.
17. KÜSSNER, H. G., *Das Zweidimensionale Problem der Beliebig Bewegeten Tragfläche unter Bereucksichtigung von Partialbewegugung der Flussigkeit*, Luftfahrtforschung, Vol. 17, pp. 355–361, 1940.
18. SCHWARZ, L., *Beruchung der Druckverteilung einer Harmonisch sich Verformerden Tragfläche in Ebener Strömung*, Luftfahrtforschung, Vol. 17, pp. 379–386, 1940.
19. ABRAMOWITZ, M., and STEGUN, I. A., *Handbook of Mathematical Functions*, United States Government Printing Office, Washington, DC, 1964.
20. SLOAN, I., *Error Analysis for a Class of Degenerate Kernel Methods*, Numerische Mathematik, Vol. 25, pp. 231–238, 1976.

5

Numerical Solution of a Class of Integral Equations Arising in Two-Dimensional Aerodynamics—The Problem of Flaps[1]

J. A. FROMME[2] AND M. A. GOLBERG[3]

Abstract. Bland's collocation method is extended to calculate aerodynamic forces on airfoils with flaps. This requires the solution of Cauchy singular integral equations with discontinuous right-hand sides. Straightforward application of collocation is shown to yield slow convergence, of order $1/N$. Examination of a variety of methods that have been proposed to accelerate convergence leads us to select a simple form of extrapolation as an effective means of achieving engineering accuracy. Numerical results are presented supporting these assertions.

1. Introduction

In Chapter 4 we have considered the numerical solution of a class of Cauchy singular integral equations where it was demonstrated that acceptable accuracy could be obtained for smooth right-hand sides. Physically this corresponded to airfoils whose shape and elastic deflection modes were represented by highly differentiable functions. Although such problems are important (Refs. 1–2), recent investigations into the feasibility of developing automatic flutter suppression systems (Ref. 3) have led to renewed interest in predicting the aerodynamic behavior of airfoils with one or more flaps. Mathematically this leads to solving integral equations with discontinuous right-hand sides, and our purpose here is to describe an extension of the method used in Chapter 4 to deal with this class of problems.

Since many engineering problems require only a knowledge of moments of the solution rather than the solution itself, we shall concentrate on procedures for calculating them. As is shown in Section 4 it is possible to

[1] This work was supported by NASA Grant NSG-2140.
[2] Associate Professor of Mathematics, University of Nevada at Las Vegas, Las Vegas, Nevada.
[3] Professor of Mathematics, University of Nevada at Las Vegas, Las Vegas, Nevada.

147

obtain 3- to 4-figure accuracy (sufficient for most purposes) for such important quantities as lift and pitching moment using less than 20 basis elements.

The chapter is divided into five sections. In Section 2 we present the flap problem and show that Galerkin's method can be expected to yield a very slow rate of convergence. Numerical results for some simple free air problems bear this out. Section 3 is devoted to a discussion of various convergence acceleration techniques modeled after those that have been introduced for equations of the second kind. Theoretically we can demonstrate improved rates of convergence, but they are generally not large enough for our purposes. In Section 4 we develop the empirical basis for an extrapolation technique and present numerical results to support its effectiveness. Section 5 closes with a discussion of directions for future research.

2. The Flap Problem

We begin by stating that our notation will be the same as in Chapter 4. Consider the integral equation

$$Hp + Kp = w, \tag{1}$$

where

$$w(x) = (d/dx + ik)h(x) \tag{2}$$

and $h(x)$ is the profile of the airfoil given in Fig. 1 of Chapter 4. If $h(x)$ is highly differentiable, then $w(x)$ will also be highly differentiable, and Eq. (1) can be solved efficiently using Bland's collocation method. However, if $h(x)$ represents an airfoil with a simple trailing edge flap hinged at $x = x_1$, then

$$h(x) = \begin{cases} 0, & -1 \le x < x_1, \\ x - x_1, & x_1 \le x \le 1, \end{cases} \tag{3}$$

so that

$$w(x) = \begin{cases} 0, & -1 \le x < x_1, \\ 1 + ik(x - x_1), & x_1 \le x \le 1. \end{cases} \tag{4}$$

From Eq. (4) it is seen that $w(x)$ has a jump discontinuity at $x = x_1$. More generally if there are flaps hinged at points $-1 < x_1 < x_2 < x_3 < \cdots < x_n < 1$, then $w(x)$ will be discontinuous at each of them. We are thus led to the problem of solving Eq. (1) where $w(x)$ is the sum of a step function and a continuous function. Such functions are obviously elements of L_Ψ^2, so that the theory developed in Chapter 4 holds. Thus we can apply Galerkin's method, and the L^2_Ψ convergence of our approximations will be guaranteed. For practical purposes it is important to know the rate of convergence. It can be estimated using the fact that (Refs. 4–5)

$$\|p - p_N\|_\Psi \le \|(H + P_N K)^{-1}\| \, \|Hp - P_N Hp\|_x, \tag{5}$$

where P_N is the operator of orthogonal projection onto $\text{Span}(\{\chi_k\}^n_{k=1})$. In the simplest case with $K = 0$, Eq. (5) yields (recall $\|H\| = \|H^{-1}\| = 1$)

$$\|p - p_N\|_\Psi \le \|Hp - P_N Hp\|_x = \|w - P_N w\|_x. \tag{6}$$

If $w(x)$ has the form given by Eq. (4), then a straightforward calculation shows that

$$\|p - p_N\|_\Psi = 0(1/\sqrt{N}). \tag{7}$$

In the general case we have

$$\|p - p_N\|_\Psi \le C_N (\|Kp - P_N Kp\|_x + \|w - P_N w\|_x). \tag{8}$$

Since Kp is continuous (Ref. 5), the dominant error term in Eq. (8) is $C_N \|w - P_N w\|_x$, so that again we expect

$$\|p - p_N\|_\Psi = 0(1/\sqrt{N}). \tag{9}$$

The results of solving

$$Hp = w,$$

where $w(x)$ has the form in Eq. (4) with $x_1 = 0$, are shown in Tables 1 and 2. Examination of the third column in Table 1 shows that $\|p_N\|_\Psi$ appears to converge at a rate of $0(1/N)$, somewhat better than that predicted by Eq. (7). (Recall that in Chapter 4 it was shown that Galerkin's method was numerically equivalent to collocation and that is how it is computationally implemented.) Even so, the achievement of an accuracy of less than 1% error seems to require about 150 basis elements, a number which is generally impractical and which substantially exceeds our present capacity of 20. Although it is possible to increase this number, computing times can become excessive due to the difficulty of rapidly calculating the kernel. Consequently, we seek other strategies for improving the accuracy of Bland's collocation method in the presence of flaps.

3. Convergence Acceleration

In view of our comments above on the slow convergence of Bland's collocation method for discontinuous downwashes we shall now consider several alternate procedures for the solution of Eq. (1). In this respect, we

Table 1. Principal error term in norm for a midchord flap
$(M = 0, \ k = 0, \ \eta = \infty)$.

N	$\|p_N\|_\Psi$	% Error	$|\%\ \text{Error}| \times N$
1	1.77245	−10.545	10.5
2	1.50774	+5.964	11.9
3	1.67441	−4.430	13.3
4	1.54963	+3.352	13.4
5	1.64833	−2.843	14.2
6	1.56602	+2.330	14.0
7	1.63626	−2.051	14.4
8	1.57475	+1.785	14.3
9	1.62929	−1.617	14.5
10	1.58017	+1.447	14.5
11	1.62477	−1.335	14.7
12	1.58386	+1.217	14.6
13	1.62159	−1.136	14.8
14	1.58654	+1.049	14.7
15	1.61923	−0.989	14.8
16	1.58857	+0.923	14.8
17	1.61471	−0.876	14.9
18	1.59817	+0.823	14.8
19	1.61597	−0.786	14.9
20	1.59145	+0.743	14.9

$$\|p\|_\Psi \ \text{exact} = \sqrt{\pi/2 + 1} = 1.60337$$

have been guided by two principles: (1) minimization of the number of kernel evaluations and (2) utilization of the algorithm in TWODI to the maximum extent. This of course does not rule out the future introduction of new computational techniques. For example, the recent results of Nissim and Lottati on Possio's equation (Ref. 6) indicate that acceptable accuracy may be obtained using a piecewise polynomial basis.

By using the fact that Eq. (1) is equivalent to an equation of the second kind (Refs. 1–2) and referring to Section 3.3 of Chapter 1, the following procedures can be considered:

1. Singularity subtraction (Refs. 7–8)
2. Regularization (Refs. 8–10)
3. Iteration (Ref. 11)
4. Reverse flow theorems (Refs. 10, 12, 13)
5. Calculation of the resolvent (Refs. 14–16)
6. Extrapolation (Refs. 10, 17)

For completeness each of the above is examined, and our reasons for rejecting procedures 1–5 and accepting procedure 6 are given.

Table 2.   Principal error term in lift for a midchord flap
$(M=0, k=0, \eta=\infty)$.

N	C_L	% Error	$|\% \text{ Error}| \times N$
1	6.28319	+22.203	22.2
2	4.54656	−11.573	23.1
3	5.60723	+9.057	27.2
4	4.80272	−6.591	26.4
5	5.43401	+5.687	28.4
6	4.90481	−4.605	27.6
7	5.35469	+4.145	29.2
8	4.95964	−3.539	28.3
9	5.30922	+3.260	29.3
10	4.99386	−2.873	28.7
11	5.27974	+2.687	29.6
12	5.01725	−2.418	29.0
13	5.25908	+2.285	29.1
14	5.03424	−2.088	29.8
15	5.24379	+1.988	29.4
16	5.04715	−1.837	29.4
17	5.23202	+1.759	29.9
18	5.05729	−1.640	29.5
19	5.22269	+1.577	30.0
20	5.06546	−1.481	29.6

C_L exact $= 2 + \pi$

3.1. Singularity Subtraction.   It is well known that if $w(x)$ has a jump discontinuity only at $x = x_1$, then

$$p(x) = A \log|x - x_1| + p_c(x),$$

where $p_c(x)$ is continuous (Refs. 7, 13, 18). Using this, we might seek solutions to (1) of the form

$$p_N(x) = \sum_{i=1}^{n} A_i \log|x - x_i| + \sum_{i=1}^{N} a_i \Psi_i(x),$$

where $\{x_i\}_{i=1}^{n}$ are the locations of the hinge points. For the particular case of the Possio kernel (free air compressible flow) this has been tried in Ref. 7. However, Rowe et al. in Ref. 8, trying to extend this procedure to three-dimensional problems, indicated that the calculation of the coefficients $(\{A_i\}_{i=1}^{n}, \{a_i\}_{i=1}^{N})$ appeared to be extremely sensitive to the location of the collocation points, and thus the method appeared to be unreliable. This difficulty seems to indicate that the collocation matrix becomes ill-conditioned as N increases. This is not surprising in light of the fact that

$(\{\log|x - x_i|\}_{i=1}^n, \{\Psi_i\}_{i=1}^N)$ are increasingly linearly dependent for large N (Ref. 17). Because of its apparent instability, the above method has been ruled out.

3.2. Regularization; Landahl's Method. Since simple subtraction of the singular part of $p(x)$ appeared unreliable, Rowe *et al.* in Ref. 8 suggested a somewhat different approach. Unnoted by them, this technique was essentially a generalization of the well-known Kantorovich method for equations of the second kind (Ref. 19) and consists of replacing Eq. (1) with one that has a smoother right-hand side. Although many implementations are possible (Refs. 10, 20), we shall describe one that appears computationally feasible. A more complete discussion may be found in our forthcoming NASA report (Ref. 10).

Since the singular part of the solution is generally governed only by H, we shall consider the following procedure which makes use of this fact. Let p_0 be defined by

$$Hp_0 = w, \tag{10}$$

so that

$$p_0 = H^{-1}w, \tag{11}$$

and note that for most w's which occur in practice (Ref. 18) p_0 can be obtained analytically via the Söhngen inversion formula (Refs. 2, 21). Define the residual downwash (Ref. 7) by

$$w_R = w - (H + K)p_0 \tag{12}$$

and observe that

$$w_R = Kp_0 = -KH^{-1}w. \tag{13}$$

The residual pressure p_R is then obtained by solving

$$(H + K)p_R = w_R, \tag{14}$$

and it is easily shown that

$$p = p_0 + p_R. \tag{15}$$

Since Kp_0 is continuous (Ref. 5), Eq. (14) is a problem with a smoother downwash than in Eq. (1).

Let $p_{R,N}$ be the approximation to p_R computed as in Chapter 4. We take as our approximation to p the function $\tilde{p}_N$ given by

$$\tilde{p}_N = p_0 + p_{R,N}. \tag{16}$$

The following theorem governs the behavior of the sequence $\{\tilde{p}_N\}_{N=1}^{\infty}$.

Theorem 3.1. For all N sufficiently large, $\tilde{p}_N$ is well defined and converges in the norm of L_{Ψ}^2 to p. In addition the error estimate

$$\|p - \tilde{p}_N\|_{\Psi} \leq \|(H + P_N K)^{-1}\| \|Kp - P_N K p\|_{\chi} \qquad (17)$$

is valid.

Proof. Observe first that $p_{R,N}$ satisfies

$$Hp_{R,N} + P_N K p_{R,N} = P_N w_R. \qquad (18)$$

To obtain (18), start with the fact that the Galerkin approximation was defined in Chapter 4 by

$$p_{R,N} + Q_N H^{-1} K p_{R,N} = Q_N H^{-1} w_R, \qquad (19)$$

where Q_N is the operator of orthogonal projection onto $\mathrm{Span}(\{\Psi_k\}_{k=1}^N)$. Operating on both sides of (19) with H gives

$$Hp_{R,N} + HQ_N H^{-1} K p_{R,N} = HQ_N H^{-1} w_R.$$

Since $H\Psi_k = -\chi_k$, $k = 1, 2, \ldots, N$, it follows that $HQ_N H^{-1}$ is just the operator of orthogonal projection onto $\mathrm{Span}(\{\chi_k\}_{k=1}^N)$. Thus Eq. (18) holds. Now

$$\begin{aligned}
(H + P_N K)\tilde{p}_N &= (H + P_N K)p_0 + (H + P_N K)p_{R,N} \\
&= (H + P_N K)p_0 + P_N w_R \\
&= (H + P_N K)(H^{-1}w) + P_N w_R \\
&= w + P_N K H^{-1} w - P_N K H^{-1} w \\
&= w. \qquad (20)
\end{aligned}$$

Thus $\tilde{p}_N$ satisfies

$$(H + P_N K)\tilde{p}_N = w. \qquad (21)$$

For N sufficiently large it follows from Chapter 4 that $(H + P_N K)^{-1}$ exists, giving

$$\tilde{p}_N = (H + P_N K)^{-1} w. \qquad (22)$$

Subtracting Eq. (22) from Eq. (1) yields

$$\begin{aligned}
p - \tilde{p}_N &= (H + K)^{-1} w - (H + P_N K)^{-1} w \\
&= (H + P_N K)^{-1}(P_N K - K)p. \qquad (23)
\end{aligned}$$

Taking norms on both sides of Eq. (23) gives Eq. (7). $\qquad\qquad\square$

By comparing the estimate given above for $\tilde{p}_N$ to that for p_N it is seen that $\tilde{p}_N$ converges to p at a rate proportional to $\|Kp - P_N Kp\|_x$, while that for p_N is proportional to $\|Hp - P_N Hp\|_x$. Since Kp is continuous while Hp is not, we expect $\tilde{p}_N$ to converge more rapidly than p_N. For steady problems, where the logarithmic term is absent, the method should produce substantial improvement over calculating p_N. However, in unsteady problems, which are our primary concern, the presence of the logarithmic term in K will generally make Kp continuous but not differentiable. In view of this, the rate of convergence is not expected to be large enough to offer a meaningful reduction in computational effort.

3.3. Iteration. Recent work by Sloan *et al.* (Ref. 11) has shown that iterating on a Galerkin approximation for equations of the second kind gives improved rates of convergence. Motivated by this, we shall consider the possibility of extending their results to Bland's integral equation.

Here we let p_N^G denote the Galerkin approximation and define the first Picard iterate p_N^S of p_N^G by

$$Hp_N^S + Kp_N^G = w, \tag{24}$$

so that

$$p_N^S = H^{-1}(w - Kp_N^G). \tag{25}$$

The behavior of the sequence $\{p_N^S\}_{N=1}^{\infty}$ is governed by the following theorem.

Theorem 3.2. For all N sufficiently large, p_N^S is well defined and converges uniformly to p. In addition, we have the L_2 error estimate

$$\|p - p_N^S\|_\Psi \le \|(H + KQ_N)^{-1}\| \, \|K - KQ_N\| \, \|Hp - P_N Hp\|_x. \tag{26}$$

Proof. The uniform convergence of p_N^S is established first. From Eq. (1) we find that

$$p = -H^{-1}Kp - H^{-1}w, \tag{27}$$

and subtracting Eq. (25) from Eq. (27) gives

$$p - p_N^S = H^{-1}(Kp_N^G - Kp).$$

By using the Söhngen inversion formula, it follows that $H^{-1}K$ is an integral operator with a logarithmically singular kernel (Ref. 22). Let $\tilde{K}(x, \xi)$ denote the kernel of $H^{-1}K$; then

$$p(x) - p_N^S(x) = \int_{-1}^{1} \sqrt{(1-\xi)/(1+\xi)} \tilde{K}(x, \xi)(p_N^G(\xi) - p(\xi))\, d\xi.$$

By the Cauchy–Schwarz inequality

$$|p(x) - p_N^S(x)| \leq \left[\int_{-1}^{1} \sqrt{(1-\xi)/(1+\xi)} |\tilde{K}(x, \xi)|^2\, d\xi\right]^{1/2} \|p_N^G - p\|_\Psi. \quad (28)$$

Since

$$\sup_{x \in [-1,1]} \left[\int_{-1}^{1} \sqrt{(1-\xi)/(1+\xi)} |\tilde{K}(x, \xi)|^2\, d\xi\right]^{1/2} < \infty,$$

uniform convergence follows from Eq. (28) and the L_Ψ^2 convergence of p_N^G.

To obtain Eq. (26) we first show that p_N^S satisfies

$$Hp_N^S + KQ_N p_N^S = w. \quad (29)$$

To see this, apply P_N to both sides of Eq. (24), giving

$$P_N Hp_N^S + P_N Kp_N^G = P_N w. \quad (30)$$

But

$$P_N Kp_N^G = P_N w - Hp_N^G,$$

so that

$$P_N Hp_N^S = Hp_N^G. \quad (31)$$

Substitution of Eq. (31) into Eq. (24) yields

$$Hp_N^S + K(H^{-1}P_N H)p_N^S = w. \quad (32)$$

Now $Q_N = H^{-1}P_N H$, so that Eq. (29) is valid.

Since K is Hilbert–Schmidt, it follows that $\|K - KQ_N\| \to 0$, so that for N sufficiently large, $(H + KQ_N)^{-1}$ exists as a bounded operator. Hence

$$p_N^S = (H + KQ_N)^{-1}w. \quad (33)$$

Subtracting Eq. (33) from Eq. (1) gives

$$p - p_N^S = (H + K)^{-1}w - (H + KQ_N)^{-1}w$$
$$= (H + KQ_N)(KQ_N - K)p.$$

Now

$$p = Q_N p + (I - Q_N p) \quad (34)$$

and $Q_N^2 = Q_N$, so that

$$(KQ_N - K)p = (KQ_N - K)(p - Q_N p). \tag{35}$$

Thus

$$p - p_N^S = (H + KQ_N)^{-1}(KQ_N - K)(p - Q_N p). \tag{36}$$

Taking norms on both sides of Eq. (36) yields Eq. (26). □

Comparing Eqs. (5) and (26) shows that the rate of convergence of p_N^S to p is enhanced by the factor $\|KQ_N - K\|$ over that of p_N^G. For unsteady problems it follows from Chapter 4 that

$$\|K - KQ_N\| = O(1/\sqrt{N}),$$

so that for discontinuous downwashes

$$\|p - p_N^S\|_\Psi = O(1/N),$$

which is the observed numerical rate of convergence for collocation. At best we might expect a rate of $O(1/N^{3/2})$, one which is still too slow to obtain less than 1% error with fewer than 20 basis elements.

A further difficulty with iteration is the nontrivial calculations that have to be performed in order to obtain p_N^S. This arises from the difficulty in calculating the term $H^{-1}Kp_N^G$. Such problems do not occur for equations of the second kind (Refs. 11, 19), so at present we cannot recommend the above procedure for solving Eq. (1).

3.4. Reverse Flow Theorems. As stated in Section 1, a major concern is the computation of aerodynamic forces which can be obtained as weighted integrals of p. In this regard there is a classical technique, due originally to von Karman (Ref. 23) and generalized later by Flax (Ref. 12), for calculating such quantities without having to solve Eq. (1) directly. The procedure, usually called a *reverse flow theorem*,[4] is summarized in the following theorem.

Theorem 3.3. Let $l(p) = \langle l, p \rangle_\Psi$. Then

$$l(p) = \langle p^*, w \rangle_x, \tag{37}$$

---

[4] The terminology stems from the fact that $(H^* + K^*)p^* = l$ describes the pressure on the airfoil with downwash l in a free stream moving from right to left (the reverse flow) with the Kutta condition imposed at the leading rather than at the trailing edge.

where p^* is the solution of the adjoint equation

$$(H^* + K^*)p^* = l \qquad (38)$$

and H^* and K^* are the adjoints of H and K.

Proof.   Using the definition of adjoint, we get

$$\langle l, p \rangle_x = \langle (H^* + K^*)p^*, p \rangle_\Psi = \langle p^*, (H + K)p \rangle_x = \langle p^*, w \rangle_x. \qquad \square$$

Since it has been shown in Chapter 4 that the aerodynamic forces can be calculated in terms of the inner products $\langle l, p \rangle_\Psi$, the effect of Theorem 3.3 is to enable us to calculate these by solving Eq. (39) instead of Eq. (1). For moments such as the lift and pitching moment, l is a polynomial, so that numerically one anticipates no difficulties. However, for quantities such as the hinge moment (Ref. 13), l itself is a step function, so that the adjoint equation (38) is of the same form as Eq. (1) and no advantage is expected. Because of this we have presently not implemented this procedure. However, future work will be directed at determining its usefulness.

3.5. Williams' Method.   In a series of recent papers Williams has exploited the fact that (1) is a convolution equation to obtain a semianalytic formula for the resolvent kernel of $H + K$ (Refs. 14–16). His approach requires the numerical calculation of at least two solutions to (1) with *smooth* right-hand sides. These are then combined analytically to arrive at the above-mentioned formulas. The expression given in the original paper (Ref. 14) appeared to be computationally unwieldy, and this led us initially to reject the approach. However, in a current and as yet unpublished paper (Ref. 16) some of these difficulties appear to have been alleviated by the development of computationally more efficient procedures. Since we have not had time to make a complete assessment of this technique and to evaluate its competitiveness against the method we describe below, its use will have to await further investigation.

4. Extrapolation

Perhaps the most common method for accelerating the convergence of numerical algorithms is the use of an extrapolation procedure (Refs. 10, 17). Although such methods are commonplace in codes for solving boundary value problems (Refs. 24–25) and have been considered for use with quadrature methods for integral equations (Ref. 17), there seem to be few available results for projection methods, particularly for those using polynomial bases. Since asymptotic expansions have been worked out for procedures using "local" bases, one generally has the theoretical foundation available for the implementation of extrapolation when such techniques are used. Lacking this information, we have taken the approach of experimentally determining the form of these expansions.

As we remarked in Section 2, the third column of Table 1 indicates that

$$| \|p\|_\Psi - \|p_N\|_\Psi |/\|p\|_\Psi = (0.149/N) + O(1/N^2),$$

with similar behavior for the lift. Tables 3 and 4 show the effect of solving Eq. (1) with a nonzero kernel K. Again, the $O(1/N)$ convergence rate is apparent. The results of a large number of other calculations (Ref. 10) have also exhibited similar behavior.

In addition to the slow rate of convergence, two other important features have emerged from these studies: (1) The convergence is oscillatory, with the period of oscillation depending on the location of the hinge points, and (2) the period of oscillation appears to be independent of the kernel K for a fixed downwash. Tables 1–4 clearly exhibit this behavior.

Table 3. C_{L_δ} for an oscillating flap hinged at the 50% chord ($M = 0$, $k = 0.1$, $\eta = \infty$).

N	Real	Imaginary	Magnitude	Phase	% Error (magnitude)
1	5.24189	−0.50914	5.26656	−5.55	+20.9665
2	3.83875	−0.39049	3.85856	−5.81	−11.3735
3	4.71938	−0.46530	4.74226	−5.63	+8.9240
4	4.05062	−0.40762	4.07108	−5.75	−6.4922
5	4.57548	−0.45292	4.59785	−5.65	+5.6071
6	4.13538	−0.41479	4.15613	−5.73	−4.5387
7	4.50954	−0.44719	4.53166	−5.66	+4.0868
8	4.18095	−0.41868	4.20186	−5.72	−3.4883
9	4.47172	−0.44390	4.49370	−5.67	+3.2149
10	4.20940	−0.42113	4.23041	−5.72	−2.8326
11	4.44720	−0.44177	4.46909	−5.67	+2.6496
12	4.22884	−0.42281	4.24993	−5.71	−2.3842
13	4.43001	−0.44027	4.45184	−5.68	+2.2534
14	4.24298	−0.42403	4.26411	−5.71	−2.0585
15	4.41730	−0.43916	4.43907	−5.68	+1.9601
16	4.25371	−0.42495	4.27489	−5.71	−1.8109
17	4.40751	−0.43831	4.42925	−5.68	+1.7345
18	4.26214	−0.42568	4.28335	−5.70	−1.6166
19	4.39974	−0.43764	4.42145	−5.68	+1.5554
20	4.26894	−0.42627	4.29017	−5.70	−1.4600

Table 4. C_{L_δ} for an oscillating flap hinged at the 75% chord ($M = 0$, $k = 0.1$, $\eta = \infty$).

N	Real	Imaginary	Magnitude	Phase	% Error (magnitude)
1	5.20343	−0.76932	5.25999	−8.41	+62.2622
2	3.80393	−0.57961	3.84783	−8.66	+18.6994
3	2.86126	−0.44037	2.89495	−8.75	−10.6954
4	4.01530	−0.60740	4.06098	−8.60	+25.2747
5	3.45638	−0.52720	3.49635	−8.67	+7.8568
6	3.01247	−0.46186	3.04767	−8.72	−5.9843
7	3.70555	−0.56291	3.74806	−8.64	+15.6216
8	3.36402	−0.51342	3.40298	−8.68	+4.9765
9	3.07119	−0.47035	3.10700	−8.71	−4.1540
10	3.56680	−0.54280	3.60786	−8.65	+11.2967
11	3.32122	−0.50707	3.35971	−8.68	+3.6417
12	3.10241	−0.47489	3.13854	−8.70	−3.1811
13	3.48818	−0.53135	3.52842	−8.66	+8.8461
14	3.29653	−0.50341	3.33475	−8.68	+2.8717
15	3.12177	−0.47771	3.15811	−8.70	−2.5774
16	3.43758	−0.52397	3.47729	−8.67	+7.2688
17	3.28046	−0.50103	3.31850	−8.68	+2.3704
18	3.13496	−0.47963	3.17144	−8.70	−2.1662
19	3.40229	−0.51882	3.44162	−8.67	+6.1684
20	3.26916	−0.49936	3.30708	−8.68	+2.0181

Consequently, we have adopted the following heuristic procedure for extrapolating the aerodynamic moments. For a given downwash $w(x)$, we solve (1) with $K = 0$ for $N = 1$ to $N = 20$ and determine the rate of convergence empirically as in Tables 1 and 2. We then observe the period of oscillation of the errors and select a *monotone* subsequence suitable for extrapolation. For the results given in Tables 5 and 6, the subsequence corresponding to $N = 2, 4, 8, 16$ was used, and for those in Table 7, $N = 3, 6, 12$ was employed.

Table 5. Extrapolation of C_{L_δ} and norm for a midchord flap ($M = 0$, $k = 0$, $\eta = \infty$).

	Norm	C_{L_δ}	% Error (norm)	% Error (C_{L_δ})
Exact	1.60337	5.14159		
Z_8^1	1.60241	5.13466	−0.0599	−0.1348
Z_4^2	1.60327	5.14069	−0.0062	−0.0175
Z_2^3	1.60335	5.14139	−0.0012	−0.0039

Table 6. Extrapolation of C_{L_δ} for an oscillating flap hinged at the 50% chord
$(M = 0, k = 0.1, \eta = \infty)$.

	Real	Imaginary	Magnitude	Phase	% Error (magnitude)
Exact	4.33227	−0.43177	4.35373	−5.69	
Z_8^1	4.32647	−0.43122	4.34791	−5.69	−0.1337
Z_4^2	4.33153	−0.43171	4.35299	−5.69	−0.0171
Z_2^3	4.33210	−0.43176	4.35359	−5.69	−0.0033

We have chosen a simple form of extrapolation based on the formulas
(Refs. 5, 10)

$$Z_N^1 = -\alpha_N + 2\alpha_{2N}, \tag{39a}$$

$$Z_N^2 = \alpha_N/3 - 2\alpha_{2N} + 8\alpha_{4N}/3, \tag{39b}$$

$$Z_N^3 = -\alpha_N/21 + 2\alpha_{2N} - 8\alpha_{4N}/3 + 64\alpha_{8N}/21, \tag{39c}$$

where $\{\alpha_N\}_{N=1}^\infty$ is a given convergent sequence of complex numbers. If
$\{\alpha_N\}_{N=1}^\infty$ has the asymptotic form

$$\alpha - \alpha_N = a_1/N + a_2/N^2 + a_3/N^3 + \cdots,$$

where $\alpha = \lim_{N\to\infty} \alpha_N$, then it is easily shown that

$$\alpha - Z_N^1 = O(1/N^2),$$

$$\alpha - Z_N^2 = O(1/N^3),$$

and

$$\alpha - Z_N^3 = O(1/N^4).$$

Due to our present limitation of $N = 20$, we have been unable to use higher
than the third-order extrapolation given by Eq. (39c).

Tables 5–7 show the effect of extrapolating the results in Tables 1–4.
For a midchord flap $(x_1 = 0)$, 4-figure accuracy is achieved using third-
order extrapolation, and 3-figure accuracy is obtained for a three-quarter

Table 7. Extrapolation of C_{L_δ} for an oscillating flap hinged at the 75% chord
$(M=0, k=0.1, \eta=\infty)$.

	Real	Imaginary	Magnitude	Phase	% Error (magnitude)
Exact	3.20444	−0.49892	3.24166	−8.69	
Z_6^1	3.19235	−0.48792	3.22942	−8.69	−0.3776
Z_3^2	3.20191	−0.48944	3.23910	−8.69	−0.0790

Table 8.   Effect of using incorrect subsequences for extrapolation.

	Real	Imaginary	Magnitude	Phase	% Error (magnitude)
Midchord flap					
Z_6^1	4.32230	−0.43083	4.34372	−5.69	−0.2300
Z_3^2	4.57927	−0.45301	4.60163	−5.65	+5.6939
Three-quarter chord flap					
Z_8^1	3.51114	−0.53452	3.55159	−8.66	9.5608
Z_4^2	3.77727	−0.57288	3.82352	−8.92	17.9494
Z_2^3	4.00144	−0.60507	4.04693	−8.60	24.8413

chord flap $(x_1 = 1/2)$. The results of using incorrect subsequences for extrapolation are shown in Table 8. As is to be expected, the successive extrapolants appear to diverge. Additional calculations are reported in Ref. 10, where similar phenomena are observed.

At present the above procedure appears to produce sufficiently accurate solutions for engineering purposes (Ref. 10) with little additional computational effort over that needed for problems with smooth downwashes.

5. Conclusions

We have shown how the use of a simple extrapolation method can be effective in solving Bland's integral equation with discontinuous right-hand sides. Even so, many problems remain, the most pressing being that of dealing with the oscillatory nature of the convergence of the primary collocation method. At present, it is possible to handle this problem on a case-by-case basis with little difficulty. However, extrapolation would become more efficient if it were possible to either predict the period in advance or to modify the method so that monotone convergence could be achieved.

Both of these possibilities are under investigation and will be the subject of future work.

References

1. BLAND, S. R., *The Two-Dimensional Oscillating Airfoil in a Wind Tunnel in Subsonic Compressible Flow*, Society for Industrial and Applied Mathematics Journal on Applied Mathematics, Vol. 18, pp. 830–848, 1970.

2. FROMME, J. A., and GOLBERG, M. A., *Unsteady Two-Dimensional Airloads Acting on Oscillating Thin Airfoils in Subsonic Ventilated Wind Tunnels*, National Aeronautics and Space Administration, Contractor Report No. 2967, 1978.

3. EDWARDS, J. W., BREAKWELL, J. V., and BRYSON, A. E., *Active Flutter Control Using Generalized Unsteady Aerodynamic Theory*, Journal of Guidance and Control, Vol. 1, pp. 32–40, 1978.

4. FROMME, J. A., and GOLBERG, M. A., *Numerical Solution of a Class of Integral Equations Arising in Two-Dimensional Aerodynamics*, this volume, Chapter 4.

5. FROMME, J. A., and GOLBERG, M. A., *On the L_2 Convergence of Collocation for the Generalized Airfoil Equation* (to appear).

6. NISSIM, E., and LOTTATI, I., *Oscillatory Subsonic Piecewise Continuous Kernel Function Method*, Journal of Aircraft, Vol. 14, pp. 515–516, 1977.

7. JAGER, E. M., *Tables of the Aerodynamic Aileron Coefficients for an Oscillating Wind-Aileron System in Subsonic Compressible Flow*, National Luchtvaartlaboratorium, Report F-155, 1954.

8. ROWE, W., SEBASTIAN, J., and REDMAN, M., *Some Recent Developments in Predicting Unsteady Loadings Caused by Control Surface Motions*, American Institute of Aeronautics and Astronautics, Paper No. 75-101, 1975.

9. ROWE, W., REDMAN, M., EHLERS, F., and SEBASTIAN, J., *Prediction of Unsteady Loadings Caused by Leading and Trailing Edge Control Surface Motions in Subsonic Compressible Flow—Analysis and Results*, National Aeronautics and Space Administration, Contractor Report No. 2543, 1975.

10. FROMME, J. A., GOLBERG, M. A., and WERTH, J., *Computation of Aerodynamic Effects on Oscillating Airfoils with Flaps in Ventilated Subsonic Wind Tunnels* (to appear).

11. SLOAN, I., BURN, B., and DATYNER, N., *A New Approach to the Numerical Solution of Integral Equations*, Journal of Computational Physics, Vol. 18, pp. 92–105, 1975.

12. FLAX, A., *Reverse Flow and Variational Theorems for Lifting Surfaces in Non-Stationary Compressible Flow*, Journal of the Aeronautical Sciences, Vol. 20, pp. 120–126, 1953.

13. MILNE, R., *Application of Integral Equations for Fluid Flows in Unbounded Domains*, Finite Elements in Fluids, Vol. 2, Edited by R. H. Gallagher, J. T. Oden, and O. C. Zienkiewicz, John Wiley and Sons, New York, New York, 1975.

14. WILLIAMS, M., *The Resolvent of Singular Integral Equations*, Quarterly Journal of Applied Mathematics, Vol. 28, pp. 99–110, 1977.

15. WILLIAMS, M., *Exact Solutions in Thin Oscillating Airfoil Theory*, American Institute of Aeronautics and Astronautics Journal, Vol. 15, pp. 875–876, 1977.

16. WILLIAMS, M., *The Solution of Singular Integral Equations by Jacobi Polynomials* (to appear).

17. BAKER, C. T. H., *The Numerical Treatment of Integral Equations*, Cambridge University Press, London, England, 1977.

18. FROMME, J. A., *Extension of the Subsonic Kernel Function Analysis to Incorporate Partial Span Trailing Edge Control Surfaces*, North American Aviation, Technical Report NA 64H-209, 1964.
19. ATKINSON, K., *A Survey of Numerical Methods for the Solution of Fredholm Integral Equations of the Second Kind*, Society for Industrial and Applied Mathematics, Philadelphia, Pennsylvania, 1976.
20. FROMME, J. A., and GOLBERG, M. A., *Projection Methods for the Generalized Airfoil Equation*, Society for Industrial and Applied Mathematics Fall Meeting, Albuquerque, New Mexico, 1977.
21. SÖHNGEN, H., *Zur Theorie der Endlichen Hilbert-Transformation*, Mathematische Zeitschrift, Vol. 60, pp. 31–51, 1954.
22. BLAND, S. R., *The Two Dimensional Oscillating Airfoil in a Wind Tunnel in Subsonic Compressible Flow*, North Carolina State University, PhD Thesis, 1968.
23. VON KARMAN, T., *Supersonic Aerodynamics—Principles and Applications*, Journal of the Aeronautical Sciences, Vol. 14, pp. 373–409, 1947.
24. KELLER, H. B., *Accurate Difference Methods for Nonlinear Two Point Boundary Value Problems*, Society for Industrial and Applied Mathematics Journal on Numerical Analysis, Vol. 11, pp. 305–320, 1974.
25. STRANG, W. G., and FIX, G., *An Analysis of the Finite Element Method*, Prentice-Hall, Englewood Cliffs, New Jersey, 1973.

Additional Bibliography[5]

26. MUNK, M., *General Theory of Thin Wing Sections*, National Advisory Committee for Aeronautics, Report 142, 1922.
27. BIRNBAUM, W., *Die Tragende Wirbei Fläche als Hiffsmittle zur Behandlung des Ebenen Problems der Tragflügeltheorie*, Zeitschrift fur Angewandte Mathematik und Mechanik, Vol. 4, pp. 227–292, 1924.
28. MUNK, M., *Elements of the Wing Sections Theory and of the Wing Theory*, National Advisory Committee for Aeronautics, Report 191, 1924.
29. KÜSSNER, H. G., *Schwingungen von Flugzeugflügeln*, Luftfahrtforschung, Vol. 4, pp. 41–62, 1929.
30. THEODORSEN, T., *General Theory of Aerodynamic Instability and the Mechanism of Flutter*, National Advisory Committee for Aeronautics, Report 496, 1935.
31. KÜSSNER, H. G., *Zusammenfossender Bericht über den Instationären Auftreib von Flügeln*, Luftfahrtforschung, Vol. 13, pp. 410–424, 1936.
32. MULTHOPP, H., *Die Berechung der Auftriebsverteilung von Tragflügeln*, Luftfahrtforschung, Vol. 15, pp. 153–169, 1938.
33. POSSIO, C., *L'Azione Aerodinamica sul Profilo Oscillante in un Fluido Compressible a Velocita Iposonora*, L'Aerotecnica, Vol. 18, pp. 421–458, 1938.

---

[5] This bibliography is based on the files of Professor J. A. Fromme.

34. VON KARMAN, T., and SEARS, W. R., *Airfoil Theory for Nonuniform Motion*, Journal of Aeronautical Sciences, Vol. 5, pp. 370–378, 1938.

35. KÜSSNER, H. G., *General Airfoil Theory*, National Advisory Committee for Aeronautics, TM 979, 1940.

36. KÜSSNER, H. G., *Allgemeine Tragflachentheorie*, Luftfahrtforschung, Vol. 17, pp. 370–378, 1940.

37. SCHWARZ, L., *Berechung der Druchnerteilung einer Harmonisch sich Verformenden Tragflache in Eliener Stromung*, Luftfahrtforschung, Vol. 17, pp. 379–386, 1940.

38. SÖHNGEN, H., *Bestimmung der Auftriebsverteilung fur Beliebige Instationäre Begungen (Ebenes Problem)*, Luftfahrtforschung, Vol. 17, pp. 401–420, 1940.

39. KÜSSNER, H. G., and SCHWARZ, L., *The Oscillating Wing with Aerodynamically Balanced Elevator*, National Advisory Committee on Aeronautics, TM 991, 1941.

40. SCHADE, T., *Numerical Solution of Possio's Integral Equation for Oscillating Airfoils in Plane Subsonic Flow*, British Aeronautical Research Council, Research Memorandum 9506, 1944.

41. BIOT, M. A., *The Oscillating Deformable Airfoil of Infinite Span in Compressible Flow*, Proceedings of the Sixth International Congress of Applied Mechanics, Paris, France, 1946.

42. DIETZE, F., *The Air Forces of the Harmonically Vibrating Wing in a Compressible Medium at Subsonic Velocity, Part II, Numerical Tables and Curves*, American Air Force, Report No. F-TS-948-RE, 1947.

43. HASKIND, M. D., *Oscillations of a Wing in Subsonic Flow*, Brown University, Report No. A9-T-22, 1947.

44. KARP, S. N., SHU, S. S., and WEIL, H., *Aerodynamics of the Oscillating Airfoil in Compressible Flow*, Brown University, Technical Report No. F-TR-1167-ND, 1947.

45. KARP, S. N., and WEIL, H., *The Oscillating Airfoil in Compressible Flow, Part II, A Review of Graphical and Numerical Data*, Brown University, 1948.

46. MULTHOPP, H., *Methods for Calculating the Lift Distribution of Wings*, Great Britain Aeronautical Research Council, Reports and Memoranda No. 2884, 1950.

47. OSWATITSCH, K., *Die Geschwindigkeitsverteilung Bei Lokalen Überschallgabienten an Flachen Profilen*, Acta Physica Austriaca, Vol. 4, pp. 228–271, 1950.

48. REISSNER, E., *On the Application of Mathieu Functions in the Theory of Subsonic Compressible Flow Past Oscillating Airfoils*, National Advisory Committee on Aeronautics, TN 2363, 1951.

49. FETTIS, H. E., *An Approximate Method for the Calculation of Nonstationary Air Forces at Subsonic Speeds*, United States Air Force, WADCTR 52-56, 1952.

50. KÜSSNER, H. G., *A Review of the Two-Dimensional Problem of Unsteady Lifting Surface Theory During the Last Thirty Years*, Max Planck Institut für Strommungforschung, Göttingen, Germany, 1953.

51. RUNYAN, H. L., and WATKINS, C. E., *Considerations of the Effect of Tunnel Walls on the Forces on an Oscillating Airfoil in Two Dimensional Subsonic Compressible Flow*, National Advisory Committee on Aeronautics, Report 1150, 1953.

52. BALDWIN, B. TURNER, J., and KNECHTEL, E., *Wall Interference in Wind Tunnels with Slotted and Porous Boundaries at Subsonic Speeds*, National Advisory Committee on Aeronautics, TN 3176, 1954.

53. SPREITER, J. R., and ALKSNE, R., *Theoretical Prediction of Pressure Distributions on Non-lifting Airfoils at High Subsonic Speeds*, National Advisory Committee on Aeronautics, TN 2096, 1954.

54. BISPLINGHOFF, R. L., ASHLEY, H., and HALFMAN, R. A., *Aeroelasticity*, Addison-Wesley Publishing Company, Cambridge, Massachusetts, 1955.

55. WATKINS, C. E., RUNYAN, H. L., and WOOLSTON, D. S., *On the Kernel Function of the Integral Equation Relating the Lift and Downwash Distributions of Oscillating Finite Wings at Subsonic Speeds*, National Advisory Committee on Aeronautics, Report No. 1234, 1955.

56. WILLIAMS, D. E., *On the Integral Equations on Two-Dimensional Flutter Derivative Theory*, British Aeronautical Research Council, Reports and Memoranda No. 3057, 1955.

57. RUNYAN, A. L., WOOLSTON, D. S., and RAINEY, A. G., *Theoretical and Experimental Investigation of the Effects of Tunnel Walls on the Forces on an Oscillating Airfoil in Two-Dimensional Subsonic Compressible Flow*, National Advisory Committee on Aeronautics, Report 1262, 1956.

58. DRAKE, D., *The Oscillating Two-Dimensional Airfoil Between Porous Walls*, The Aeronautical Quarterly, Vol. 8, pp. 226–239, 1957.

59. WOODS, L., *On the Theory of Two-Dimensional Wind Tunnels with Porous Walls*, Proceedings of the Royal Society, Series A, Vol. 242, pp. 341–354, 1957.

60. DRAKE, D., *Quasi-steady Derivatives for the Subsonic Flow Past an Oscillating Airfoil in a Porous Wind Tunnel*, The Aeronautical Quarterly, Vol. 10, pp. 211–229, 1959.

61. WILLIAMS, D. E., *Some Mathematical Methods in Three-Dimensional Subsonic Flutter-Derivative Theory*, British Aeronautical Research Council, Reports and Memoranda No. 3302, 1961.

62. FROMME, J. A., *Least Squares Approach to Unsteady Kernel Function Theory*, Journal of the American Institute of Aeronautics and Astronautics, Vol. 7, pp. 1349–1350, 1964.

63. GARNER, H., ROGERS, E. W. E., ACUM, W. E. A., and MASKELL, E. C., *Subsonic Wind Tunnel Wall Corrections*, NATO Advisory Group for Aerospace Research and Development, AGARDograph 109, 1966.

64. BLAND, S. R., RHYNE, R. H., and PIERCE, H. B., *Study of Flow Induced Vibrations of a Plate in Narrow Channels*, Transactions of the American Society of Mechanical Engineers, Series B, Vol. 89, pp. 824–830, 1967.

65. HESS, J. L., and SMITH, A. M. O., *Calculation of Potential Flow About Arbitrary Bodies*, Progress in Aeronautical Sciences, Vol. 8, Edited by D. Kuchemann, Pergamon Press, New York, New York, 1967.

66. LANDAHL, M. T., and STARK, V. J., *Numerical Lifting Surface Theory—Problems and Progress*, Journal of the American Institute of Aeronautics and Astronautics, Vol. 6, pp. 2049–2060, 1968. (See this paper for an additional bibliography on three-dimensional problems.)

67. WHITE, R., and LANDAHL, M., *Effects of Gaps on the Loading Distributions of Planar Lifting Surfaces*, Journal of the American Institute of Aeronautics and Astronautics, Vol. 6, pp. 626–631, 1968.

68. FUNG, Y. C., *An Introduction to the Theory of Aeroelasticity*, Dover Publications, New York, New York, 1969.

69. NIAZ, S. A., *The Application of Projection Methods to Linearized Unsteady Lifting Surface Theory*, London University, PhD Thesis, 1970.

70. NØRSTRUD, H., *Numerische Lösungen von Schallnahen Strömungen um Ebene Profile*, Zeitschrift fur Flugwissenschaften, Vol. 18, pp. 149–157, 1970.

71. GRAHM, J. M. R., *A Lifting Surface Theory for the Rectangular Wing in Non-Stationary Flow*, Aeronautical Quarterly, Vol. 22, pp. 83–100, 1971.

72. EBIHARA, M., *A Study of Subsonic Two Dimensional Wall Interference Effects in a Perforated Wind Tunnel*, National Aerospace Laboratory, Report No. TR-252T, 1972.

73. DOWELL, E. H., and VENTRES, C. S., *Derivation of Aerodynamic Kernel Functions*, Journal of the American Institute of Aeronautics and Astronautics, Vol. 11, pp. 1586–1588, 1973.

74. NØRSTRUD, H., *Transonic Flow Past Lifting Wings*, Journal of the American Institute of Aeronautics and Astronautics, Vol. 11, pp. 754–757, 1973.

75. NØRSTRUD, H., *The Transonic Airfoil Problem with Embedded Shocks*, The Aeronautical Quarterly, Vol. 24, pp. 129–138, 1973.

76. OSBORNE, C., *Unsteady Thin Airfoil Theory for Subsonic Flow*, Journal of the American Institute of Aeronautics and Astronautics, Vol. 11, pp. 205–209, 1973.

77. FROHN, A., *Problems and Results of the Integral Equation Methods for Transonic Flows*, Symposium Transonicum II, Edited by K. Oswatitsch and D. Rues, Springer-Verlag, Berlin, Germany, 1974.

78. NIXON, D., and HANCOCK, G. J., *High Subsonic Flow Past a Steady Two Dimensional Airfoil*, Aeronautical Research Council, Paper No. 1280, 1974.

79. YATES, J. E., *Linearized Integral Theory of Three Dimensional Unsteady Flow in a Shear Layer*, Journal of the American Institute of Aeronautics and Astronautics, Vol. 12, pp. 596–602, 1974.

80. KRAFT, E., *An Integral Equation Method for Boundary Interference in Perforated Wind Tunnels at Transonic Speeds*, University of Tennessee, PhD Thesis, 1975.

81. MOKRY, M., *Integral Equation Method for Subsonic Flow Past Airfoils in Ventilated Wind Tunnels*, Journal of the American Institute of Aeronautics and Astronautics, Vol. 15, pp. 47–53, 1975.

82. NIXON, D., *A Comparison of Two Integral Equation Methods for High Subsonic Flows*, The Aeronautical Quarterly, Vol. 26, pp. 56–58, 1975.

83. SEARS, W. R., and TELIONIS, D. P., *Boundary-Layer Separation in Unsteady Flow*, Society for Industrial and Applied Mathematics Journal on Applied Mathematics, Vol. 28, pp. 215–235, 1975.

84. NIXON, D., *Calculation of Transonic Flows Using Integral Equation Methods*, London University, PhD Thesis, 1976.

85. NØRSTRUD, H., *Comment on an Extended Integral Equation Method for Transonic Flows*, Journal of the American Institute of Aeronautics and Astronautics, Vol. 14, pp. 820–826, 1976.

86. ROWE, W., SEBASTIAN, J., and REDMAN, M., *Some Recent Developments in Predicting Unsteady Airloads Caused by Control Surface Motions*, Journal of Aircraft, Vol. 13, pp. 955–961, 1976.

87. CHAKRABORTY, S. K., and NIYOGI, P., *Integral Equation Formulation for Transonic Lifting Profiles*, Journal of the American Institute for Aeronautics and Astronautics, Vol. 15, pp. 1816–1817, 1977.

88. JASWON, M. A., and SYMM, G. T., *Integral Equation Methods in Potential Theory and Elastostatics*, Academic Press, London, England, 1977.

89. KRAFT, E., and LO, C., *Analytical Determination of Blockage Effects in a Perforated Wall Transonic Wind Tunnel*, Journal of the American Institute of Aeronautics and Astronautics, Vol. 15, pp. 511–517, 1977.

90. NIXON, D., *Calculation of Transonic Flows Using an Extended Integral Equation Method*, Journal of the American Institute of Aeronautics and Astronautics, Vol. 15, pp. 295–296, 1977.

91. WILLIAMS, M. H., CHI, M. R., and DOWELL, E. H., *Effects of Inviscid Parallel Shear Flow on Unsteady Aerodynamics and Flutter*, Paper No. 77–158, 15th Aerospace Sciences Meeting, Los Angeles, California, 1977.

92. NIXON, D., *An Alternative Treatment of Boundary Conditions for the Flow over Thick Wings*, Aeronautical Quarterly, Vol. 15, pp. 295–296, 1977.

93. CIELAK, Z. M., and KINNEY, R. B., *Analysis of Viscous Flow Past an Airfoil, Part II—Numerical Formulation and Results*, Journal of the American Institute of Aeronautics and Astronautics, Vol. 16, pp. 105–110, 1978.

94. JORDAN, P. F., *Reliable Lifting Surface Solutions for Unsteady Flow*, Paper No. 78-228, 16th Aerospace Sciences Meeting, Huntsville, Alabama, 1978.

95. NIXON, D., *Calculation of Unsteady Transonic Flows Using the Integral Equation Method*, Paper No. 78-13, 16th Aerospace Sciences Meeting, Huntsville, Alabama, 1978.

96. NIYOGI, P., *Transonic Integral Equation Formulation for Lifting Profiles and Wings*, Journal of the American Institute of Aeronautics and Astronautics, Vol. 16, pp. 92–94, 1978.

97. YATES, J. E., *Viscous Thin Airfoil Theory and the Kutta Condition*, Paper No. 78-152, 16th Aerospace Sciences Meeting, Huntsville, Alabama, 1978.

98. WILLIAMS, M. H., *Generalized Theodorsen Solution for Singular Integral Equations of the Airfoil Class* (to appear).

6

Applications of Integral Equations in Particle-Size Statistics

A. Goldman[1] and W. Visscher[2]

Abstract. We discuss the application of integral equations techniques to two broad areas of particle statistics, namely, stereology and packing. Problems in stereology lead to the inversion of Abel-type integral equations; and we present a brief survey of existing methods, analytical and numerical, for doing this. Packing problems lead to Volterra equations which, in simple cases, can be solved exactly and, in other cases, need to be solved numerically. Methods for doing this are presented along with some numerical results.

1. Introduction

There are two distinct problems found in the field of particle-size statistics whose solutions utilize integral equations. The subject matters of concern are stereology and packing. Both of these topics have wide applications, and numerous examples can be found in the literature of many technological and scientific areas. The problem in stereology will be formulated as a prescription of a procedure for finding a three-dimensional (3D) statistical characterization of particles with data obtained from a two-dimensional (2D) analysis. The problem in packing is concerned with setting up a model to determine the so-called *packing fraction* and other properties of a random array.

2. Application to Stereology

We quote the following definition, due to Elias (Ref. 1): "Stereology, sensu stricto, deals with a body of methods for the exploration of three-dimensional space, when only two-dimensional sections through solid

---

[1] Professor, Department of Mathematics, University of Nevada at Las Vegas, Las Vegas, Nevada.
[2] Staff Member, T-11, Theoretical Division, Los Alamos Scientific Laboratory, Los Alamos, New Mexico.

bodies or their projections on a surface are available. Thus, stereology could also be called extrapolation from two-to-three-dimensional space."

The basic problem from stereology, which involves the solution of an integral equation, was introduced in a classic paper by Wicksell (Ref. 2). In order to extend his work, we wish to obtain the distribution of *sizes* of arbitrarily shaped particles that are dispersed in an opaque material from measurements that are obtained by figures fashioned by their intersections with a random plane. There are two sets of measures describing the array of particles: the real measures of the distribution of 3D particles and the apparent measures describing the distribution of 2D shapes intersected by the sectioning plane. A measure could be a size (such as diameter) or a shape (such as a sphere). The problem is to find the distribution of the real measures from the distribution of the apparent measures. Wicksell (Ref. 2), Kendall and Moran (Ref. 3), Scheil (Ref. 4), Reid (Ref. 5), and Santalo (Ref. 6) are among the many authors in varied fields who have considered the problem for the special case of spherical particles and obtained a solution that requires solving an integral equation of Abel's type. We will present a derivation of the equation as well as a discussion of the difficulties arising in its solution.

A start at the general problem of stereology may be made by considering the following problem: given an arbitrarily-shaped 3D object imbedded in an opaque medium, and given a randomly oriented plane which passes through the object, what probabilistic statements can be made about the 3D object from a study of the loci of sections generated by many such random planes? The geometry is illustrated in Fig. 1.

It is clear that the statement of the problem of stereology is not precisely the problem encountered in practice. In most applied situations, we have not one but many 3D objects imbedded in such a way that the proximity of neighboring particles (objects) causes correlations between the loci of sections and makes the problem vastly more difficult than the problem under consideration. We define a single 3D object or, equivalently, many 3D objects which are explicitly assumed to be so

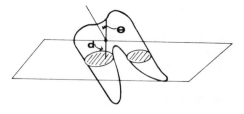

Fig. 1.   An irregularly shaped 3D object intersected by a random plane. The problem is to describe the 3D object by studying the curves of section.

dilutely distributed that their positions and orientations are independent of one another.

In addition to the fact that our model overlooks correlations among positions and orientations, we will necessarily restrict the shapes to families described by a few variables (measures). Let q denote the random variables determined by size, shape, and other physical properties. It is obvious that we have simplified the assumptions of arbitrary shapes.

Let $A^{(l)}(q, d, \theta)$ denote a sample of M observable quantities $l = 1, 2, \ldots, M$, which may be obtained from the surface of section (e.g., areas, perimeters, colors, densities of intersected objects). The quantities $A^{(l)}$ are determined by q, whose probability density function will be denoted by $g(q)$; by d, the measure of impact, i.e., the distance of the center of gravity of the object from the plane; and by $\theta = (\sigma_1, \sigma_2)$, a vector consisting of two angles describing the orientation of the object relative to d (see Fig. 1).

The characteristic function of $A^{(l)}$ is

$$E[\exp[ikA^{(l)}]] = \int (1/4\pi) \, d\theta \int (1/L) \, dd \int g(q) \exp[ikA^{(l)}(q, d, \theta)] \, dq, \tag{1}$$

which is an integral equation for $g(q)$, since $A^{(l)}$ and $E[\exp[ikA^{(l)}]]$ are observables. Equation (1) can in principle be solved for $g(q)$, with q consisting of an arbitrarily large number of variables. This involves making measurements of $E[\exp[ikA^{(l)}]]$ for sufficiently large M. In addition, we have assumed the center of gravity of our single 3D object to be confined within a cube of volume L^3. From its expansion in powers of k, we may write equations for all of the moments of $A^{(l)}$, and the problem of determining $g(q)$, which up to now is a function of an arbitrarily large set of measures, becomes the formidable task of solving a set of simultaneous linear equations.

An integral equation in which the density function of $A^{(l)}$ occurs on the left is easily obtained from (1) as follows: Multiplying by $(2\pi)^{-1} \exp[-ika^{(l)}]$ and integrating over k gives the result:

$$f(a^{(l)}) = \int (1/4\pi) \, d\theta \int g(q) \, dq \int \delta[a^{(l)} - A^{(l)}(q, d, \theta)](1/L) \, dd. \tag{2}$$

Equation (2) can be verified to be a density function and can be transformed, by integration over d, into

$$f(a^{(l)}) = \int (1/4\pi L) \, d\theta \int g(q)[\partial A^{(l)}(q, d, \theta)/\partial d]_{A^{(l)}}^{-1} \, dq. \tag{3}$$

Note that d has been eliminated from $\partial A^{(l)}/\partial d$ by solving

$$A^{(l)}(q, d, \theta) = a^{(l)}$$

for d. In general, (3) has multiple roots, two for the objects of high symmetry which we consider below; however, for simplicity, we omit the sum over these roots.

For illustrative purposes, we shall drop pretenses of generality. Let $A^{(l)}$ be the area of the intercepted curve, and reduce the set of size and shape variables q to a single *size*. Two detailed examples will be presented: the sphere and the thin disc. The radius is used as the size variable (q).

We assume that a collection of randomly sized spheres are embedded in a solid. In this case, A is the intercepted area, which is independent of θ (see Fig. 2), and is given by

$$A(R, d, \theta) = \pi (R^2 - d^2).\tag{4}$$

We obtain

$$\partial A/\partial d = -2\pi d,$$

and Eq. (2) can be written as

$$f(a) = (1/\pi L) \int_{\sqrt{(a/\pi)}}^{\infty} g(R)(R^2 - a/\pi)^{-1/2}\, dR.\tag{5}$$

Note that, because of the double roots which would arise from Eq. (4), we have included the factor 2 in the calculation of Eq. (5).

A formal solution for (5) is accomplished by using a simple trick. Let $h(u, a)$ be some function so that

$$\int_0^{\pi R^2} h(u, a)(R^2 - a/\pi)^{-1/2}\, da = \begin{cases} 1, & u < R, \\ 0, & u \geq R. \end{cases}\tag{6}$$

Then, from (5) and (6), we find that

$$\int_0^{\infty} f(a)h(a)\, da = (1/\pi L) \int_u^{\infty} g(R)\, dR.\tag{7}$$

Methods found in standard calculus texts can be used to justify the interchange of order in the R and a integrations, which was used in

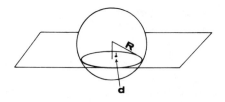

Fig. 2.   Surface of section for a sphere with diameter R.

obtaining (7). A convenient choice of $h(u, a)$ turns out to be

$$h(u, a) = \begin{cases} (1/\pi^2)(a/\pi - u^2)^{-1/2}, & a > \pi u^2, \\ 0, & a \geq \pi u. \end{cases} \tag{8}$$

This satisfies (6); consequently,

$$(L/\pi) \int_{\pi u^2}^{\infty} f(a)(a/\pi - u^2)^{-1/2}\, da = \int_{u}^{\infty} g(R)\, dR. \tag{9}$$

Equation (9) is an exact formal expression for the distribution function which we seek; i.e., by definition,

$$G(u) = \int_{u}^{\infty} R g(R).$$

Nevertheless, its evaluation involves considerable numerical difficulties, because of the singular integral on the left-hand side. Coping with (9) is one of the principal tasks of stereological data analysis. We shall defer a review of the literature dealing with this problem until later. In concluding this example, we note that, iff

$$g(R) = \delta(R - R_0),$$

then

$$f(a) = (1/\pi L)(R_0^2 - a/\pi)^{-1/2}. \tag{10}$$

A second example, slightly more complicated, but still a simple solvable problem in stereology, is that posed by a random suspension of thin circular discs. Here, the intersection of sections are line segments in the plane; and, from the apparent distributions of length, we will derive an equation for the distribution of diameters of the discs. The geometry is illustrated in Fig. 3.

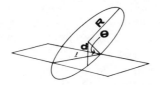

Fig. 3. Interception of a thin circular disc by a random plane. l is the length of line-segment intercepted, d is the impact parameter, and θ is the angle between normals to the disc and to the plane.

For this situation, Eq. (2), with l substituted for A, becomes

$$f(l) = (2l/L) \int_0^{\pi/2} d(\cos\theta) \int_{l/2}^\infty g(R)[\sin\theta/\sqrt{(4R^2 - l^2)}]\, dR, \qquad (11)$$

because

$$R^2 = (l/2)^2 + (d/\sin\theta)^2.$$

After the integration over θ is performed, this equation is nearly identical with (5), and the same method can be used to solve it. The result is

$$(4L/\pi^2) \int_{2u}^\infty f(l)(l^2 - 4u^2)^{-1/2}\, dl = \int_u^\infty g(R)\, dR. \qquad (12)$$

If

$$g(R) = \delta(R - R_0),$$

then we have, from (11),

$$f_0(l) = (\pi/2)(l/L)(4R_0^2 - l^2)^{-1/2},$$

which, when integrated over l, yields

$$\int_0^{2R_0} f_0(l)\, dl = (\pi/4)(2R_0/L).$$

This says that the probability that the plane will intersect the disc at all is $\pi/4$ times the ratio of the disc diameter to the side of the box. This is exactly Buffon's needle problem arrived at by a roundabout route.

We shall now examine some of the numerical procedures available for obtaining a solution to the integral equation given by (9). Kendall and Moran (Ref. 3) write Eq. (9) in a more convenient manner by using a formula given in Ref. 7; they suggest integrating by parts first and then numerically computing $g(r)$ from $\varphi(x)$. This procedure has the inherent difficulty of dealing with random data as exact numerical quantities, a practice which usually leads to unsatisfactory results.

Anderssen and Jakeman (Ref. 3) have categorized and appraised various approaches to reach the conclusion that a formula based on product integration and spectral differentiation is the best of the recently published inversion procedures (Refs. 9–11). Considerable labor is involved in their rather complicated procedure, and a computer program is a must in order to obtain a solution. One basic requirement is that the data be of the form of a sample cumulative frequency function. Product integration is performed first because of limitations (described in Ref. 8) that preclude the initial use of spectral differentiation. They prove some advantageous statistical and numerical properties of their method; however, it should be noted that, although the method is valid, there appears to be a lack of simplicity and flexibility in its use. For more details, interested readers are referred to Refs. 12–14.

The two-step procedure used in Ref. 8 was compared with a single method used by Saltikov in Refs. 15–16. Saltikov's method was poor in the comparison using exact data; however, its simplicity and ease of application might justify its use. We will briefly review the procedure.

Saltikov's method is a way to solve (5) or (9) by discretization. It exploits the fact that circles of section of a given size are produced only by spheres of equal or larger size. Therefore, the discretization of (6) is a triangular set of linear equations which can be readily solved sequentially. Saltykov's method uses a discretization in which the logarithmic decrement between successive radii is uniform. It leads to a simple expression for g_i:

$$g_i = R_i^{-1}(1.646f_i - 0.4561f_{i-1} - 0.1162f_{i-2} - 0.0415f_{i-3} + \cdots), \quad (13)$$

where f_i denotes the number of circles in the ith bin (area $= \pi R_i^2$) and g_i denotes the number of spheres (volume $= 4\pi/3R_i^3$). For details the reader is referred to Ref. 17.

Some methods, not as demanding in application as the two-step method, but more exact in nature than Saltikov's methods include the log normal model given by Dehoff (Ref. 18) and the chi-square model offered by Kaiding et al. (Ref. 19). Although these parametric approaches appear to be reasonable, large errors occur in the tails, unless a tremendous sample size is chosen. A function used by Wiggins (Ref. 20) to fit cumulative data y obtained from measurements taken from lung tissues x of dogs provided a reasonably good fit and is given by

$$y = [1 - \exp(-\alpha x)]^{\beta}. \quad (14)$$

Probably, other curve-fitting practitioners have arrived at various functions that would represent the data in a tolerable way; however, as stated earlier, the problem of avoiding errors in extremity points is hard to surmount, because of the large sample required. In addition, any parametric approach should use a *simple* curve to be beneficial, and this is usually unobtainable.

Equation (5) may be used to obtain an interesting relation between the moments of real and apparent radii. If ρ is the radius of the circle of sector,

$$a = \pi\rho^2;$$

and

$$f(\rho) = (da/d\rho)f(a) = 2\pi\rho f(a)$$

is the apparent radius density function. It then follows immediately that

$$m_{n-1} = (\sqrt{\pi}/L)[\Gamma((n+1)/2)/\Gamma((n+2)/2)]M_n, \qquad (15)$$

where m_n is the nth moment of apparent radii and M_n is the nth moment of real radii. For $n = 0$, Eq. (15) yields the superficially surprising result that the harmonic mean of apparent radii is independent of the distribution of real radii! The reason, of course, is that the larger contribution of a small radius to the harmonic mean is exactly compensated by the reduced probability that the smaller sphere will be intersected at all.

Watson (Ref. 21) has demonstrated that any experimental method involving a planar or linear probe used to estimate linear functionals or ratios will produce results which have bad statistical properties. Errors of estimation will be either incalculable or so large that the results are meaningless. Goldsmith (Ref. 22) has worked out the integral equation for consideration when a *thick slice* is taken through the medium. Hilliard (Ref. 23) has amassed a listing of relationships for obtaining numbers of spheres per unit volume. Moran (Ref. 24) presents an excellent discussion of the state of the stereological art, including the spherical model which we have alluded to in this paper. Underwood (Ref. 25) deals with the mathematical aspects which relate to our study.

3. Application to Packing

An important topic which has widespread application and appeal relates to packing geometrical objects in a container. An unsolved problem is to determine how much space is occupied by a large number of random-sized or random-shaped objects, or both, relative to the total volume available. The ratio is called the *packing fraction*. An adequate solution (theoretical, analytical, or simulated) has not been found. Investigators have utilized a tractable, but specialized, one-dimensional approach in an effort to solve a simpler case, namely the *parking fraction*. The following development, which makes use of integral equation theory, was first published by Renyi (Ref. 26).

Cars of unit length are parked on a street $[0, x]$ of length $x \geq 1$. The first car is parked so that the center is uniformly distributed on $[\frac{1}{2}, x - \frac{1}{2}]$. Assuming that there is still space for another car, a second car is parked according to a similar uniform distribution for the random variable, such that the two cars do not overlap. Cars continue to park until all the remaining gaps are shorter than a car. Let $\mu(x)$ denote the expected value of the total number of cars $N(x)$ parked on a street of length x. We are interested in $N(x)$ as $x \to \infty$.

Let $[t, t+1]$ be the random position of the first car parked on a street $[0, x+1]$ of length $x+1$. The available space remaining consists of the two intervals $[0, t]$ on the left and $[t+1, x+1]$ on the right. Let $N(y)$ denote the random variable designating the number of cars parked on a street of length y. Note that the expected or average value of $N(y)$ is $\mu(y)$. The distribution of $N(t)$ and $N(x-t)$ are independent. It is also true that the conditional distribution of $N(x+1)$, given that $[t, t+1]$ is occupied by the first car, is the same as the distribution of $N(t)+N(x-t)+1$. Consequently,

$$\mu(x+1) = 1 + (1/x) \int_0^t \mu(t)\, dt + (1/x) \int_{t+1}^x \mu(x-t)\, dt$$

$$= 1 + (1/x) \left[\int_0^x \mu(t)\, dt + \int_0^x \mu(x-t)\, dt \right]$$

$$= 1 + (2/x) \left[\int_0^x \mu(t)\, dt \right]. \tag{16}$$

For illustrative purposes, we shall examine the integral equation (16) for various values of x. We can slightly simplify (16) by letting

$$f(x) = \mu(x) + 1.$$

Then,

$$f(x+1) = (2/x) \int_0^x f(t)\, dt. \tag{17}$$

The initial conditions stipulate that

$$f(x) = 1, \qquad f(1) = 2, \qquad \text{when } 0 \le x < 1. \tag{18}$$

We can obtain

$$f(x) = 2, \qquad 1 < x \le 2, \tag{19}$$

$$f(x) = 4 - 2/(x-1), \qquad 2 < x \le 3, \tag{20}$$

$$f(x) = 8 - 10/(x-1) - [4/(x-1)]\log(x-2), \qquad 3 < x \le 4. \tag{21}$$

At this point, the integration becomes difficult, and we must look elsewhere for a solution. It is interesting to evaluate $f(x)/x$ for a few values to be compared with the solution obtained later:

$$f(1)/1 = 1.00,$$

$$f(2)/2 = 2/2 = 1.00,$$

$$f(3)/3 = 3/3 = 1.00,$$

$$f(4)/4 = 3.7424705/4 = 0.9356. \tag{22}$$

A procedure for finding a solution to (16) is given by Hua (Ref. 27) and is described in detail by Renyi (Ref. 26). A *retarded differential equation* is obtained by multiplying (16) by x and differentiating both sides with respect to x:

$$x\mu'(x+1)+\mu(x+1)=2\mu(x)+1, \qquad x>0. \qquad (23)$$

Renyi finally obtained the relation

$$\lim_{x\to\infty}[\mu(x)/x]=2\int_0^\infty \exp\left\{-t-2\int_0^t[(1-\exp(-u))/u]\,du\right\}dt=C$$

and calculated

$$C\approx 0.748.$$

Blaisdell and Solomon (Ref. 28) computed

$$C=0.7475979202\,53398\ldots,$$

using a method given by Dvoretzky and Robbins (Ref. 29). The latter paper, as well as those of Mannion (Ref. 30) and Ney (Ref. 31), solve a similar integral equation to obtain higher moments.

A more general integral equation was obtained by Goldman, Lewis, and Visscher (Ref. 32). They assume that car lengths have a given density function $g(r)$ and are parked according to rules involving a termination probability. One gap is considered at a time. If the gap has length less than the smallest car, the space remains vacant. If a car selected from $g(r)$ is shorter than the gap, it is parked. If the car selected is longer than the gap, then, with a specified probability ρ, no cars will be allowed to park in that particular space. Alternatively, with probability $1-\rho$, another car is selected, and the process is repeated. The integral equation is found to be

$$\mu(x)=\{G(x)/[1-(1-G(x))(1-\rho)]\}$$
$$\times\left\{1+[2/G(x)]\int_0^x[g(r)/(x-r)]\,dr\int_0^{x-r}\mu(t)\,dt\right\}, \qquad (24)$$

where $G(r)$ denotes the distribution function of car lengths, ρ is a specified probability, and $g(r)$ is the p.d.f. of car lengths, i.e.,

$$G=\int_0^r g(r)\,dr.$$

If cars are of unit length, then (24) reduces to (16). Some results are given in Ref. 32 for the binary, uniform, and triangular distributions.

For illustrative purposes, we shall give a solution for the binary distribution (i.e., two different-sized cars), which is defined as

$$G(r) = \begin{cases} 0, & \text{for } r < r_0, \\ f, & \text{for } r_0 \le r < r_1, \\ 1, & \text{for } r \ge r_1, \end{cases}$$

where f is the fraction of short cars. Note that r_1 is the length of the larger-sized cars and r_0 is the length of the smaller-sized cars. Let

$$c = r_1/r_0,$$

and $G(r)$ can be determined by the parameters c and f. A method for solving (24) without using an exorbitant amount of computer time is to approximate $G(r)$ with a step function and discretize the t-integral. A system of equations can then be solved to obtain the parking fraction (PF). Some results of PF as a function of c and f are shown in the 3D picture displayed in Fig. 4 for $p = 0.5$. Other pictures and related tables can be found in Ref. 33.

The integral equation can be written so as to describe any one of the many variants of the parking problem. For example, one might allow parking only in spaces b units longer than the cars. This has the effect of changing the upper limit in the r-integral of (24) from x to $x + b$. Solomon (Ref. 34) allowed cars of unit length to park anywhere along the street; when overlap (double parking) occurred, the car was allowed to move up half its length, so as to park bumper-to-bumper in the nearest vacant slot.

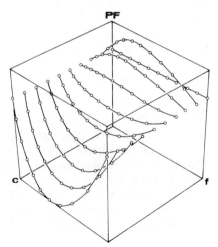

Fig. 4.   Parking fraction versus c and f, $1 \le c \le 10$ and $0 \le f \le 1$, for a binary distribution with $p = 0.5$. Note that PF = 0.74759 if $f = 0$ or 1, or if $c = 1$.

This procedure increased PF to 0.80865 . . .. Solomon's equation is a special case of an integral equation given in Ref. 29.

$$(x+c-1)N(x)=(x+c-1)+cN(x-1)+2\int_0^{x-1} N(y)\,dy, \qquad (25)$$

where c is the allowed overlap ($c/2$ at each end of a gap).
Note that, if

$$c = 0,$$

we get Eq. (16); and, if

$$c = 1,$$

we get

$$N(x)=1+(1/x)N(x-1)+(2/x)\int_0^{x-1} N(y)\,dy,$$

which is the equation used by Solomon.

References

1. ELIAS, H., *Steriologia*, Vol. 1, Editorial, p. 1, 1962.
2. WICKSELL, S. D., *The Corpuscle Problem*, Biometrika, Vol. 17, pp. 84–89, 1925.
3. KENDALL, M. G., and MORAN, P. A. P., *Geometrical Probability*, Hafner Publishing Company, New York, New York, 1963.
4. SCHEIL, E., *Die Berechnung der Anzahl and Grossenverteilung Kugelformiger Korpern mit Hilfe der Durch Abenen Schnitt Erhaltenen Schnitt Kreise*, Zeitschrift für Anorganische und Allgemeine Chemie, Vol. 201, pp. 259–264, 1931.
5. REID, W. P., *Distribution of Sizes of Spheres in a Solid from a Study of Slices of the Solid*, Journal of Mathematics and Physics, Vol. 34, pp. 95–102, 1955.
6. SANTALO, L. A., *Sobre la Distribucion de los Tamaños de Corpúsculos Contenides en un Cuerpo a Partir de la Distribucion en Sus Secciones o Proyecciones*, Trabajos de Estadistica, Vol. 6, pp. 181–196, 1955.
7. COURANT, R., and HILBERT, D., *Methods of Mathematical Physics*, Vol. 1, John Wiley and Sons (Interscience Publishers), New York, New York, 1953.
8. ANDERSSEN, R. S., and JAKEMAN, A. J., *Abel-Type Integral Equations in Stereology*, Journal of Microscopy, Vol. 105, pp. 121–153, 1975.
9. MINERBO, G. N., and LEVY, M. E., *Inversion of Abel's Integral Equation by Means of Orthogonal Polynomials*, SIAM Journal on Numerical Analysis, Vol. 6, pp. 598–616, 1969.
10. EINARRSON, B., *Numerical Solution of Abel's Integral Equation with Spline Functions*, Forsvarets Forskning Sanstalt, Report No. 1-C2455-11(25), 1971.

11. PIESSENS, R., and VERBAETEN, P., *Numerical Solution of the Abel Integral Equation*, Bit, Vol. 13, pp. 451–457, 1973.
12. NICHOLSON, W. L., *Estimation of Linear Properties of Particle Size Distributions*, Biometrika, Vol. 57, pp. 273–297, 1970.
13. JAKEMAN, A. J., *The Numerical Inversion of Abel-Type Integral Equations*, Australian National University, PhD Thesis, 1975.
14. ANDERSSEN, R. S., and JAKEMAN, A. J., *Product Integration for Functionals of Particle Size Distributions*, Utilitas Mathematica, Vol. 8, pp. 111–126, 1975.
15. SALTIKOV, S. A., *The Determination of the Size Distribution of Particles in an Opaque Material from a Measurement of the Size Distribution of Their Sections*, Stereology, Edited by H. Elias, Springer-Verlag, New York, New York, 1967.
16. ANDERSSEN, R. S., and JAKEMAN, A. J., *On Computational Stereology*, Proceedings of the 6th Australian Computer Conference, Sydney, Australia, 1974.
17. LEWIS, H. D., WALTERS, K. L., and JOHNSON, K. A., *Particle Size Distributions by Area Analysis: Modifications and Extensions of the Saltikov Method*, Metallography, Vol. 6, pp. 93–101, 1973.
18. DEHOFF, R. T., *The Estimation of Particle Distribution from Simple Counting Measurements Made on Random Plane Sections*, Transactions of the Metallurgical Society, AIME, Vol. 233, pp. 25–29, 1965.
19. KEIDING, N., JENSEN, S. T., and RANEK, L., *Maximum Likelihood Estimation of the Size Distribution of Linear Cell Nuclei From the Observed Distribution in a Plane Section*, Biometrics, Vol. 28, pp. 813–830, 1972.
20. WIGGINS, A., Unpublished Report, 1976.
21. WATSON, G. S., *Estimating Functions of Particle Size Distributions*, Biometrika, Vol. 58, pp. 483–490, 1971.
22. GOLDSMITH, P. L., *Calculation of True Particle Size Distributions from Sizes Observed in a Thin Slice*, British Journal of Applied Physics, Vol. 18, pp. 813–830, 1967.
23. HILLIARD, J. E., *The Counting and Sizing of Particles in Transmission Microscopy*, Transactions of the Metallurgical Society of AIME, Vol. 224, pp. 906–917, 1962.
24. MORAN, P. A. P., *The Probabilistic Basis of Stereology*, Advances in Applied Probability, Supplement, pp. 69–91, 1972.
25. UNDERWOOD, E. E., *The Mathematical Foundations of Quantitative Stereology*; *Stereology and Quantitative Metallography*, Special Technical Publication No. 504, ASTM, Philadelphia, Pennsylvania, 1972.
26. RENYI, A., *On a One-Dimensional Problem Concerning Space Filling*, Publications of the Mathematical Institute of the Hungarian Academy of Sciences, Vol. 3, pp. 109–127, 1958.
27. HUA, L. K., *Estimation of an Integral*, Chinese Mathematical Society, Vol. 11, pp. 393–402, 1951.
28. BLAISDELL, E., and SOLOMON, H., *On Random Sequential Packing in the Plane and a Conjecture by Palasti*, Journal of Applied Probability, Vol. 7, pp. 667–698, 1970.

29. DVORATZKY, A., and ROBBINS, H., *On the Parking Problem*, Publications of the Mathematical Institute of The Hungarian Academy of Sciences, Vol. 9, pp. 209–224, 1964.
30. MANNION, D., *Random Space Filling in One Dimension*, Publications of the Mathematical Institute of the Hungarian Academy of Sciences, Vol. 9, pp. 143–154, 1964.
31. NEY, P. E., *A Random Integral Filling Problem*, Annals of Mathematical Statistics, Vol. 33, pp. 702–718, 1962.
32. GOLDMAN, A., LEWIS, H. D., and VISSCHER, W. M., *Random Packing Model*, Technometrics, Vol. 16, pp. 301–309, 1974.
33. GOLDMAN, A., LEWIS, H. D., and VISSCHER, W. M., *Random Packing of Particles: Simulation with a One-Dimensional Parking Model*, Los Alamos Scientific Laboratory, Report No. LA-DC-72-1537, 1974.
34. SOLOMON, H., *Random Packing Density*, Proceedings of the Fifth Berkeley Symposium on Mathematical Statistics and Probability, University of California Press, Berkeley, California, 1967.

7

Smoothing and Ill-Posed Problems[1]

G. WAHBA[2,3]

Abstract. The method of weighted cross-validation is applied to the problem of solving linear integral equations of the first kind with noisy data. Numerical results illustrating its efficacy are given for estimating derivatives and for solving Fujita's equation.

1. Introduction

Consider the regression problem

$$y = X\beta + \epsilon, \tag{1}$$

where $X : E_p \to E_n$ is a known $n \times p$ (design) matrix; ϵ is an n-dimensional noise vector with $E\epsilon = 0$, $E(\epsilon\epsilon^T) = \sigma^2 I$; β is a p-dimensional vector of regression coefficients to be estimated; and y is the n-dimensional data vector. The problem of recovering β from y is ill-conditioned if the ratio of the largest to the smallest singular value of X is large. In this case the Gauss–Markov or minimum variance unbiased estimate of β may be bad from the point of view of mean square error. One can frequently reduce the mean square error by using a biased estimate for β.

A Hilbert space analogue of this problem is the following: Let

$$y(t) = \int_0^1 K(t, s)f(s)\, ds + \epsilon(t), \qquad t \in [0, 1], \tag{2}$$

where the linear operator $\mathcal{K}$ defined by

$$(\mathcal{K}f)(t) = \int_0^1 K(t, s)f(s)\, ds$$

---

[1] This work was supported by the United States Air Force under Grant No. AF-AFOSR-77-3272 and the Office of Naval Research under Grant No. N00014-77-C-0675.

[2] Professor of Statistics, University of Wisconsin, Madison, Wisconsin.

[3] This paper contains some aspects of joint work with Peter Craven, Gene Golub, Mike Smith, and Svante Wold.

183

is viewed as a map from one reproducing kernel Hilbert space to another. The noise model is $E(\epsilon(t)) = 0$, $E(\epsilon(s)\epsilon(t)) = \sigma^2 \delta(s-t)$. It can be shown that if the domain of $\mathcal{K}$ is a reproducing kernel Hilbert space (r.k.h.s.) $\mathcal{H}_R$ with reproducing kernel (r.k.) R, then $\mathcal{K}(\mathcal{H}_R) = \mathcal{H}_Q$, the r.k.h.s. with r.k. Q given by

$$Q(s, t) = \int_0^1 \int_0^1 K(s, u)R(u, v)K(t, v) \, du \, dv. \qquad (3)$$

It is also useful to consider the map $\mathcal{K}_n$ defined by $\mathcal{K}_n f = [(\mathcal{K}f)(t_1), \ldots, (\mathcal{K}f)(t_n)]^T$, $f \in \mathcal{H}_R$. One observes $(y(t_1), \ldots, y(t_n))^T = \mathcal{K}_n f + \epsilon_n$, $\epsilon_n = (\epsilon(t_1), \ldots, \epsilon(t_n))$, and wishes to recover f.

This problem has all the features of the ill-conditioned regression problem and more if n is large, since, under general conditions, the singular values of $\mathcal{K}$ tend toward 0. The estimate of f, which is the analogue of the Gauss–Markov estimate of β, is $\mathcal{K}_n^+ y$, where $\mathcal{K}_n^+$ is the generalized inverse of $\mathcal{K}_n: \mathcal{H}_R \to E_n$; that is, $\mathcal{K}_n^+ y$ is that element $\hat{f}_n$ in $\mathcal{H}_R$ of minimal $\mathcal{H}_R$ norm that satisfies $\mathcal{K}_n \hat{f}_n = y$. Generally, as n becomes large, $\hat{f}_n$ becomes unduly sensitive to noise, and a better estimate must be found. There is a variety of interesting ways to consider for doing this and many interesting and elegant problems for study in the Hilbert space setting.

We hasten to point out, at this time, that the Hilbert space analogue of regression is of far more than merely "academic" interest; equations of the form of (2) are pervasive in many branches of science and engineering. One useful example is Fujita's equation (Ref. 1)

$$(\mathcal{K}f)(t) = \int_0^{s_{\max}} \{\theta s \exp(-\theta st)/[1 - \exp(-\theta s)]\}f(s) \, ds = g(t), \qquad (4)$$

relating molecular weight distribution f to the steady-state concentration or optical density in a centrifuged sample. Here

$$s = \text{molecular weight},$$

$$t = (r_b^2 - r^2)/(r_b^2 - r_a^2),$$

where r is the radial distance along the centrifuge column and r_b and r_a are the maximum and minimum radii,

$$f(s) = \text{molecular weight distribution},$$

$$g(t) = \text{concentration},$$

$$\theta = (1 - V\rho)\omega^2(r_b^2 - r_a^2)/2RT,$$

$$\omega = \text{angular velocity},$$

$$\rho = \text{density},$$

$$V = \text{partial specific volume},$$

$$T = \text{absolute temperature},$$

and

$$R = \text{universal gas constant}.$$

Values of $y(t) = g(t) + \epsilon(t)$, $t = t_1, \ldots, t_n$, are observed experimentally, and it is desired to recover f.

The problem of numerical differentiation falls in this class; if, for example,

$$g(t) = \int_0^1 [(t - s)_+^{m-1}/(m - 1)!] f(s) \, ds, \tag{5}$$

then $f = g^{(m)}$.

2. Estimation

In the remainder of this chapter we shall consider the problem of finding biased estimates of β and f that have good mean square error properties. We shall consider three types of methods which are known in the regression literature as ridge, subset selection, and principal components methods and in the integral equations context as regularization, regression methods, and singular value truncation.

The ridge estimate β_λ for β is the solution to the following minimization problem: Find $\beta \in E_p$ to minimize

$$(\|y - X\beta\|^2/n) + \lambda\|\beta\|^2.$$

Then

$$\beta_\lambda = (X^T X + n\lambda I)^{-1} X^T y. \tag{6}$$

The regularized estimate $f_{n,\lambda}$ for f is the solution to the following minimization problem: Find $f \in \mathcal{H}_R$ to minimize

$$\sum_{i=1}^n \{[y(t_i) - (\mathcal{H}f)(t_i)]^2/n\} + \lambda\|f\|_R^2,$$

where $\|\cdot\|_R$ is the norm in $\mathcal{H}_R$ or, sometimes, a seminorm, for example, $\|f\|_R^2 = \int (f''(u))^2 \, du$. A Bayesian could derive β_λ and $f_{n,\lambda}$ from the following priors: if $\beta \sim \mathcal{N}(0, bI)$, then

$$\beta_\lambda = E\{\beta | y\}$$

where $\lambda = \sigma^2/nb$. If f is a normally distributed zero mean stochastic process with $E(f(s)f(t)) = bR(s, t)$, then

$$f_{n,\lambda}(t) = E[f(t)|y]$$

with $\lambda = \sigma^2/nb$. We prefer to view β and f as fixed but unknown members of E_p and $\mathcal{H}_R$, respectively. Then λ controls the squared bias–variance or infidelity–roughness trade-off. Large λ gives $\|\beta_\lambda\|$ and $\|f_\lambda\|_R$ small and gives $\|y - X\beta_\lambda\|^2$ and $\sum_{i=1}^{n} [y(t_i) - (\mathcal{H}f_{n,\lambda})(t_i)]^2/n$ large. Small λ gives the reverse. As a criterion for choosing λ, we suppose it is desired to minimize the "predictive mean square error" $T(\lambda)$ defined by

$$T(\lambda) = \|X\beta - X\beta_\lambda\|^2/n, \tag{7}$$

$$T(\lambda) = \sum_{i=1}^{n} \{[(\mathcal{H}f_{n,\lambda})(t_i) - (\mathcal{H}f)(t_i)]^2/n\}. \tag{8}$$

Other "goodness" criteria, for example, $\|\beta - \beta_\lambda\|^2$ and $\sum_{j=1}^{N} \{[f(j/n) - f_{n,\lambda}(j/n)]^2/n\}$ come to mind. We choose for the moment to attempt to minimize $T(\lambda)$ because it is the easiest. In principle these other goodness criteria can be handled in similar ways, but their computation may itself be an ill-conditioned problem. If σ^2 is known or can be estimated unbiasedly, then an unbiased estimate $\hat{T}(\lambda)$ of $ET(\lambda)$ is available and can be written compactly as follows. Let $A(\lambda)$ be the "hat" matrix when λ is used; that is, $A(\lambda)$ is the matrix satisfying

$$A(\lambda)y = X\beta, \tag{9}$$

$$A(\lambda)y = \mathcal{H}_n f_{n,\lambda}. \tag{10}$$

In the regression case

$$A(\lambda) = X(X^T X + n\lambda I)^{-1} X^T,$$

and in the Hilbert space case it can be shown (Ref. 2) that

$$A(\lambda) = Q_n(Q_n + n\lambda I)^{-1},$$

where Q_n is the $n \times n$ matrix with jkth entry $Q(t_j, t_k)$. Then it is not hard to verify that $\hat{T}(\lambda)$ defined by

$$\hat{T}(\lambda) = [y^T(I - A(\lambda))^2 y - \sigma^2 \text{ Trace}(I - A(\lambda))^2 + \sigma^2 \text{ Trace } A^2(\lambda)]/n \tag{11}$$

is an unbiased estimate for $E(T(\lambda))$. The minimizer $\hat{\lambda}$ of $\hat{T}(\lambda)$ may be taken as a good value of λ. In the ridge regression case, this suggestion is due to Mallows (Ref. 3) [see also Hudson (Ref. 4)]; its use in the Hilbert space case seems to have been recognized only recently (Refs. 5–6). We note that this approach may be used in principle to choose λ to attempt to minimize $\|\beta - \beta_\lambda\|$ or $(1/n) \sum_{j=1}^{n} [\mathcal{H}_n^+ \mathcal{H}_n (f - f_\lambda)(j/n)]^2$; however, the calculation of $\hat{\lambda}$ will in this case itself be an ill-conditioned problem, and so the variance of the analogue of $\hat{T}$ will be large.

Provided σ^2 is known or estimable, this approach can also be used in subset selection [this is Mallows' C_p (Ref. 3)] or in a principal components method. (See also Ref. 7.) The subset selection method selects all but k components of β to be *a priori* equal to 0. The principal components method replaces the design matrix X by a new design matrix—call it $X^{(k)}$—which is obtained from X by setting all but $p - k$ of the singular values of X to 0. Each subset candidate of components of β or of singular values has a hat matrix $A(k)$ associated with it which satisfies $A(k)y = X\hat{\beta}$, $\hat{\beta}$ being the estimate of β where the particular subset is chosen. One chooses that subset for which $\hat{T}(k)$ is minimized.

The analog of the subset selection method in the Hilbert space problem might be considered to be a method which begins with a sequence of convenient basis functions for $\mathcal{H}_R$ (B-splines, for example) and then proceeds by least squares, pretending that the solution f is in some k-dimensional subspace spanned by k of the basis functions. The singular value truncation method consists of replacing $\mathcal{H}$ by a rank k operator obtained by setting all but k of the singular values of $\mathcal{H}$ equal to zero and using the generalized inverse of the resulting operator. Usually the k largest singular values are retained. In both cases there is a hat matrix $A(k)$ which satisfies $A(k)y = \mathcal{H}_n \hat{f}$, $\hat{f}$ being the estimate of f. Again, one chooses the subset which minimizes $\hat{T}$.

The problem with using $\hat{T}$ arises when σ^2 is unknown and no estimate of it is available. Typically, in the Hilbert space problem an unbiased estimate of σ^2 is not available from the data.

How, then, can one choose λ (or a subset of parameters, basis functions, or singular values)? The answer we propose is the method of generalized cross-validation (GCV). The GCV estimate for ridge regression is a coordinate-free or rotation-invariant version of Allen's PRESS (see Refs. 8–9), or "ordinary" cross-validation.

PRESS goes as follows: Let $\beta_\lambda^{(k)}$ be the solution to the following minimization problem: Find $\beta \in E_p$ to minimize

$$(\|y^{(k)} - X^{(k)}\beta\|^2 / n) + \|\beta\|^2,$$

where $y^{(k)}$ and $X^{(k)}$ are obtained from y and X by leaving out the kth entry and the kth row, respectively. The idea is that if λ is a good choice, then the kth component of $X\beta_\lambda^{(k)}$ should be a good predictor of the missing data point

y_k (the kth component of y). This is measured by $V_0(\lambda)$, defined by

$$V_0(\lambda) = \sum_{k=1}^{n} \{[X\beta_\lambda^{(k)}]_k - y_k\}^2/n \qquad (12)$$

(where $[\cdot]_k$ denotes kth component), and the minimizer of $V_0(\lambda)$ is the PRESS estimate of λ. It can be shown (see Ref. 6) that

$$V_0(\lambda) \equiv \sum_{k=1}^{n} \{[Ay]_k - y_k\}^2/(1-a_{kk})^2 n, \qquad (13)$$

where a_{kk} is the kkth entry of A. The GCV estimate of λ is the minimizer of $V(\lambda)$, defined by

$$V(\lambda) = \sum_{i=1}^{n} (\{[Ay]_k - y_k\}^2/n) \Big/ \Big(1 - \sum_{j=1}^{n} a_{jj}/n\Big)^2$$
$$= (\|(I-A)y\|^2/n)/[\text{Trace}(I-A)/n]^2. \qquad (14)$$

Note that $V(\lambda)$ and $V_0(\lambda)$ coincide if A is constant down the diagonal, which happens if, e.g., XX^T is circulant. In general the GCV estimate can be obtained by rotating the E_n-coordinate system so that the new design matrix $\tilde{X}$, say, has the property that $\tilde{X}\tilde{X}^T$ is circulant and then doing PRESS in the new system. The advantage of using V over V_0 is that in general, for large n,

$$V \approx T + \sigma^2$$

in the neighborhood of the minimizer of T.

Specifically, it can be shown that for ridge regression, subset selection, or singular value truncation

$$|(EV - ET + \sigma^2)/ET| \le p(1 + O(p/n))/n.$$

In the Hilbert space setting, we define A as the matrix satisfying

$$\mathcal{K}_n\hat{f} = Ay$$

for any of the three methods and the optimum choice of λ or the subset as the minimizer of

$$V(\lambda) = (\|(I-A)y\|^2/n)/\{[\text{Tr}(I-A)]/n\}^2. \qquad (15)$$

Letting $\mu_1 = \text{Tr } A/n$ and $\mu_2 = \text{Tr}(A^2)/n$, it can be shown (Ref. 5) that if $f \in \mathcal{H}_R$,

$$|(EV - ET + \sigma^2)/ET| = O(\mu_1) + O(\mu_1^2/\mu_2).$$

For the regularization method, it can be shown (Ref. 2) that the sequence $\lambda^* = \lambda^*(n)$ minimizing $E(T(\lambda))$ has the property that their associated A's result in $\mu_1 \to 0$, $\mu_1^2/\mu_2 \to 0$. In general, an intuitive argument shows that

$\mu_1 \to 0$, $\mu_1^2/\mu_2 \to 0$ are exactly the conditions one needs in order for the "signal" to be separable from the "noise."

The method has been implemented experimentally for ridge regression (see Ref. 6). Experiments have also been conducted to demonstrate its efficacy in smoothing, in estimating first and second derivatives, and in solving Fujita's equation numerically via regularization. Some examples of these latter results are presented here. The smoothing example is taken from Ref. 5, and details of the computation can be found there. See Ref. 10 for earlier work.

3. Numerical Results[4]

To smooth data, the solution $g_{n,\lambda}$ to the following minimization problem is obtained: Find $g \in \mathcal{H}_Q$: $\mathcal{H}_Q = \{g: g, g'$ abs. cont., $g'' \in \mathcal{L}_2[0, 1]\}$, to minimize

$$\sum_{j=1}^{n} \{[g_{n,\lambda}(t_j) - y(t_j)]^2/n\} + \lambda \int_0^1 [g''(u)]^2 \, du. \tag{16}$$

The function $g_{n,\lambda}$ is a cubic smoothing spline. (See Ref. 11.) The parameter λ is chosen as $\hat{\lambda}$, the minimizer of $V(\lambda)$, where the hat matrix $A(\lambda)$ satisfies $(g_{n,\lambda}(t_1), \ldots, g_{n,\lambda}(t_n))^T = A(\lambda)y$. The estimate of g' [in the model $y(t) = g(t) + \epsilon(t)$] is taken as $g'_{n,\hat{\lambda}}$. The data $y(t_i)$, $t_i = i/n$ were generated by Monte Carlo methods from the model $y(i/n) = g(i/n) + \epsilon_i$, $\epsilon_i \sim$ i.i.d. (independent identically distributed), $\mathcal{N}(0, \sigma^2)$. The data, the "true" function $g(\cdot)$, and the estimated function $g_{n,\hat{\lambda}}(\cdot)$ are given in Fig. 1.

Figure 2 demonstrates how closely the minimizer of $V(\lambda)$ comes to the minimizer of $T(\lambda)$. In this example, $T(\hat{\lambda})/\min_\lambda T(\lambda) = 1.0$. $\hat{T}(\lambda)$ is also plotted. Similar results have been obtained with other examples. The derivative $g'_{n,\hat{\lambda}}$ of $g_{n,\hat{\lambda}}$ is a good estimate of g', particularly when σ^2 is not too large; see Ref. 5 for details.

Figure 3 demonstrates the accuracy obtainable in estimating the second derivative. The model is

$$y(t) = \int_0^1 K(t, s)f(s) \, ds + \epsilon(t), \tag{17}$$

where

$$K(t, s) = (|t - s|^2 - |t - s| + 1/6)/2. \tag{18}$$

---

[4] The computer program behind Figs. 1 and 2 were written by Michael Akritas, and the program behind Figs. 3 and 4 was written by Michael Smith.

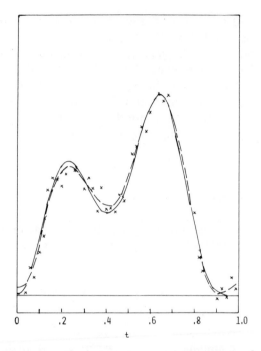

Fig. 1.   True function $g(\cdot)$ (solid curve) and $g_{n,\hat{\lambda}}(t)$ (dashed curve). $\sigma^2 = 0.01$, $N = 50$.

Here $K(t, s)$ is a Green's function for the second-derivative operator, so that if $g = \mathcal{K}f$, g is the solution to $g'' = f$, $\int_0^1 g(u)\,du = 0$, $g(1) = g(0) = 0$. The function $g = \mathcal{K}f$ is plotted in Fig. 3(a) as well as the data $y(t_i) = g(t_i) + \epsilon_i$ and $g_{n,\hat{\lambda}} = \mathcal{K}f_{n,\hat{\lambda}}$, where $\hat{\lambda}$ is the minimizer of $V(\lambda)$. Formulas for $g_{n,\lambda}$ and $\mathcal{K}f_{n,\lambda}$ appear in Ref. 2.

Figure 3(b) plots $T(\lambda)$, $V(\lambda)$, and the mean-square error of the solution $T_D(\lambda)$, defined by

$$T_D(\lambda) = \sum_{j=1}^{n} [f_{n,\lambda}(j/n) - f(j/n)]^2/n.$$

It is seen that the minimizer of $T(\lambda)$ is also close to the minimizer of $T_D(\lambda)$ in this example. Figure 3(c) plots f and $f_{n,\hat{\lambda}}$ ($f_{n,\hat{\lambda}} \equiv g''_{n,\hat{\lambda}}$).

The three pictures in Fig. 4 parallel those of Fig. 3 except that the equation being solved is the Fujita equation, $K(t, s) = [\theta s \exp(-\theta st)]/[1 - \exp(-\theta s)]$, $\theta = 4.25$. The recovery of the solution f is impressive in this example. However, while Figs. 1, 2, and 3 represent results

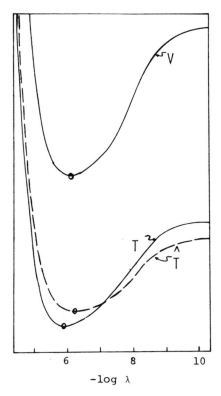

Fig. 2.   The cross-validation function V, the predictive mean-square error T, and $\hat{T}$. $\sigma^2 = 0.01$, $N = 50$.

that are *typical* over a range of f, σ^2, and n, the accuracy indicated in Fig. 4(c) was *not* typical of our ability to solve Fujita's equation. Some other n, σ^2, and f gave very poor results. The reason behind this was not hard to determine from examination of the eigenvalues of Q_n, which appeared in our printout. For the second-derivative problem of Fig. 3 all but two of these eigenvalues are essentially nonzero, but for the Fujita equation we used, only four to five of them were nonzero to double-precision machine accuracy (only slightly dependent on n). Basically this means that $[(\mathcal{K}_n^+\mathcal{K}_n f)(1/n), \ldots, (\mathcal{K}_n^+\mathcal{K}_n f)(n/n)]$ can be recovered very well if it is in a certain five-dimensional subspace of $\mathcal{H}_R$. However, any component of this vector not in that subspace will be estimated as 0. Further work is needed to develop appropriate methods for entering the missing information and/or understanding and quantifying the implications of its absence. See also Varah (Ref. 12) for a discussion of this problem.

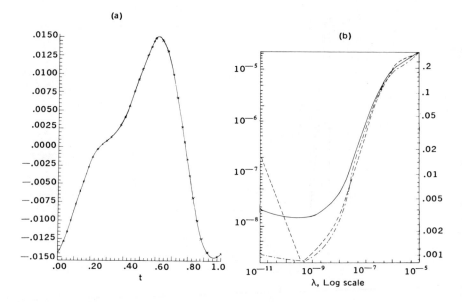

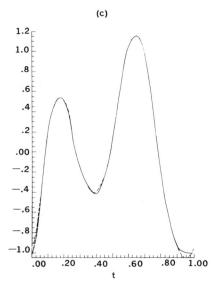

Fig. 3. Estimation of second derivative: (a)
$g_{n,\lambda} = \mathcal{K} f_{n,\lambda}$ (dashed curve, which
coincides with solid curve); $g = \mathcal{K} f$
(solid curve); data ($\times$); $\sigma = 0.1(-3)$;
$N = 41$; (b) GCV (lambda), $V(\lambda)$
(solid curve); range MSE, $T(\lambda)$ (dot-
dashed curve); domain MSE, $T_D(\lambda)$
(dashed curve); (c) True solution f
(solid curve); approximate solution
$f_{n,\lambda}$ (dashed curve).

(a)

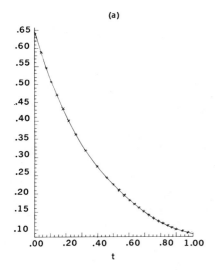

(b)

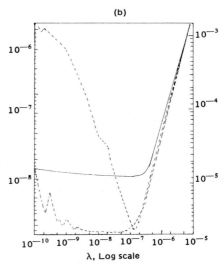

(c)

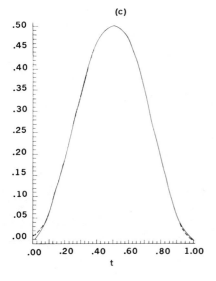

Fig. 4. Estimation of solution to Fujita's equation: (a) $g = \mathcal{H}f$ (solid curve); $g_{n,\lambda} = \mathcal{H}f_{n,\lambda}$ (dashed curve, which coincides with solid curve); $n = 21$; $\sigma^2 = 0.0001$; (b) $V(\lambda)$ (solid curve); $T_D(\lambda)$ (dashed curve); $T(\lambda)$ (dot-dashed curve); (c) f (solid curve); $f_{n,\lambda}$ (dashed curve).

References

1. GEHATIA, M., and WIFF, D. R., *Solution of Fujita's Equation for Equilibrium Sedimentation by Applying Tikhonov's Regularizing Functions*, Journal of Polymer Science Series A2, Vol. 8, pp. 2039–2050, 1970.
2. WAHBA, G., *Practical Approximate Solutions to Linear Operator Equations when the Data Are Noisy*, Society for Industrial and Applied Mathematics Journal on Numerical Analysis, Vol. 14, pp. 651–667, 1977.
3. MALLOWS, C. L., *Some Comments on Cp*, Technometrics, Vol. 15, pp. 661–675, 1973.
4. HUDSON, H. M., *Empirical Bayes Estimation*, Department of Statistics, Stanford University, Technical Report No. 58, 1974.
5. CRAVEN, P., and WAHBA, G., *Smoothing Noisy Data with Spline Functions: Estimating the Correct Degree of Smoothing by the Method of Generalized Cross-Validation*, Numerische Mathematik, Vol. 31, pp. 377–403, 1979. Report No. 445, 1977.
6. GOLUB, G., HEATH, M., and WAHBA, G., *Generalized Cross-Validation as a Method for Choosing a Good Ridge Parameter*, Department of Statistics, University of Wisconsin at Madison, Technical Report No. 491, 1977.
7. ALLEN, D. M., *Mean Square Error of Prediction as a Criterion for Selecting Variables*, Technometrics, Vol. 13, pp. 469–475, 1971.
8. ALLEN, D. M., *The Relationship Between Variable Selection and Data Augmentation and a Method for Prediction*, Technometrics, Vol. 16, pp. 125–127, 1974.
9. STONE, M., *Cross-Validatory Choice and Assessment of Statistical Prediction*, Journal of the Royal Statistical Society Series B, Vol. 36, pp. 111–147, 1974.
10. WAHBA, G., and WOLD, S., *A Completely Automatic French Curve: Fitting Spline Functions by Cross-Validation*, Communications in Statistics, Vol. 4, pp. 1–17, 1975.
11. REINISCH, C. M., *Smoothing by Spline Functions*, Numerische Mathematik, Vol. 10, pp. 177–183, 1967.
12. VARAH, J. E., *A Practical Examination of Numerical Methods for Linear Discrete Ill-Posed Problems*, Department of Computer Sciences, University of British Columbia, Technical Report No. 76-08, 1976.

8

Imbedding Methods for Integral Equations with Applications

H. Kagiwada[1] and R. Kalaba[2]

Abstract. During the last decade or two, significant progress has been made in the development of imbedding methods for the analytical and computational treatment of integral equations. These methods are now well known in radiative transfer, neutron transport, optimal filtering, and other fields. In this review paper, we describe the current status of imbedding methods for integral equations. The paper emphasizes new analytical and computational developments in control and filtering, multiple scattering, inverse problems of wave propagation, and solid and fluid mechanics. Efficient computer programs for the determination of complex eigenvalues of integral operators, analytical investigations of stability for significant underlying Riccati integrodifferential equations, and comparisons against other methods are described.

1. Introduction

The importance of integral equations is well known in such applications as multiple scattering, filtering, mechanics, potential theory, rarefied gas dynamics, wave propagation, lateral inhibition, econometrics, and so on. Much analytical and computational progress has been made, as can be seen in the books by Courant and Hilbert (Ref. 1), Noble (Ref. 2), and Cochran (Ref. 3). There has also been a high level of development in imbedding methods for integral equations, as described in the book by the authors (Ref. 4), in the bibliography by Scott (Ref. 5), and in other books and papers. Analytical and computational treatments have been developed for both inhomogeneous and homogeneous linear integral equations and for nonlinear ones as well. Interesting new functional equations have

---

[1] President, HFS Associates, Los Angeles, California.
[2] Professor, Departments of Economics and Biomedical Engineering, University of Southern California, Los Angeles, California.

arisen. Imbedding approaches characteristically lead to stable *initial-value problems*. New and efficient variable stepsize integration procedures enhance the computational utility of such formulations (Shampine and Gordon, Ref. 6). The journal *Applied Mathematics and Computation* plays a major role in the dissemination of information about imbedding methods.

The general notion that the solution of an integral equation be regarded as a function of a parameter is not new. Poincaré and Fredholm espoused such an idea, although they were led to their beautiful results by thinking of expansions in the parameter. The concept is found in Bellman's principle of optimality in dynamic programming, as well as in Ficken's continuation method (Ref. 7). The prime distinctive feature of invariant imbedding is *the exact transformation of various classes of functional equations into Cauchy problems*. Such Cauchy systems are generally *easier* to solve than the original problems. It is conceptually interesting that this transformation can be made for wide classes of integral equations, boundary-value problems, variational problems, and differential–difference equations.

Various types of imbedding are possible. Consider the linear Fredholm integral equation

$$u(t) = g(t) + \int_0^x k(t, y)u(y)\, dy, \qquad 0 \le t \le x.$$

We imbed this integral equation within a class of such equations by considering the interval length x to be a variable,

$$0 \le t \le x \le x_1.$$

Then, we write

$$u = u(t) = u(t, x),$$

and we assume that there is a unique solution when x_1 is sufficiently small. We can derive a complete Cauchy system for the function u and for certain auxiliary functions in which the interval length x plays the role of the independent variable. For nonlinear integral equations, similar reductions can be made. Thus, integral equations can be conceptually related to the theory of semigroups (Ref. 8).

We have studied invariant imbedding for various types of functional equations. Historically, our work in the imbedding of integral equations grew out of our interest in the integral equation for the source function of radiative transfer theory (Sobolev, Ref. 9). That equation has a kernel which is the first exponential integral function. We knew from the paper of Ambarzumian (Ref. 10) that a certain functional on its solution could be determined as a function of the thickness of the medium, without first determining the entire solution inside the layer. This quantity is known as the reflection function, and it satisfies a Riccati equation and a known initial condition. In 1964, we saw how to physically imbed a given problem for the source function in a class of problems for source functions for layers of varying thickness. This resulted in a complete initial-value problem for the source function, in which the reflection function played a basic role. The next step was to derive the imbedding equations analytically, making full use of the analytical form of the kernel. Then, the imbedding theory was developed for general displacement kernels of exponential form. Later developments include special consideration for degenerate kernels, semi-degenerate kernels, infinite intervals, homogeneous integral equations, nonlinear integral equations, parameter imbedding, Volterra equations, and dual integral equations.

Imbedding equations are very stable and also very attractive from the computational standpoint. Though generally nonlinear, they can readily be solved by methods available for the numerical integration of systems of ordinary differential equations. The new highly efficient integration codes enhance the practicality of this method.

Many analytical matters remain in the study of imbedding methods. Some have been discussed by Golberg (Ref. 11). There are questions regarding the stable nature of the equations (Nelson, Ref. 12). Such stability is always seen computationally, though it is not well understood theoretically. Other topics concern the existence of functions of fewer variables which are equivalent to the solutions of the basic Riccati equations. These and other fascinating questions are described in the following pages.

2. Parameter Imbedding of Integral Equations

Functional equations of analytic interest result from the imbedding of integral equations in a class of similar equations. They are also of computational utility for they can be accurately solved by the method of numerical integration of ordinary differential equations. Let us discuss the parameter imbedding of the linear inhomogeneous equation

$$u(t) = g(t) + \lambda \int_0^1 k(t, y) u(y) \, dy, \qquad 0 \le t \le 1,$$

where λ is the Poincaré parameter, and the nonlinear inhomogeneous equation

$$u(t) = g(t, \lambda) + \lambda \int_0^1 k(t, y, \lambda, u(y)) \, dy, \qquad 0 \le t \le 1.$$

In addition, we consider the homogeneous equation

$$u(t) = \lambda \int_0^1 k(t, y)u(y)\, dy, \qquad 0 \le t \le 1.$$

The aim is to study how the solution depends upon the parameter λ. The computational method is described, and selected numerical results presented for a number of applications. We shall discuss a new technique for computing the complex eigenvalues and eigenfunctions of general kernels. The technique combines some classical results of integral equations and complex variable theory with an imbedding method for converting Fredholm integral equations into a system of initial-valued differential equations. The method provides a flexible and accurate method for computing the eigenvalues and eigenfunctions.

3. Initial-Value Problems for Eigenvalues and Eigenfunctions of Integral Operators

We discuss the application of the imbedding method to the computation of the eigenvalues and eigenfunctions of complex-valued symmetric kernels which occur in the theory of lasers. These equations are usually of the form

$$u(t) = \lambda \int_a^b k(t, y)u(y)\, dy, \qquad a \le t \le b, \tag{1}$$

where $k(t, y)$ is a complex-valued symmetric kernel which is not Hermitian, that is,

$$k(x, y) = k(y, x) \quad \text{but} \quad k(x, y) \ne \overline{k(y, x)}.$$

These kernels are particularly interesting in that many of the classical techniques (such as Rayleigh–Ritz) for estimating the eigenvalues may give erroneous results. The technique discussed here is applicable to any kernel for which Fredholm's original theory applies. In an excellent paper, Cochran (Ref. 13) has shown that Fredholm's theory does apply and has proven the existence of the eigenvalues for such kernels.

We wish to solve the homogeneous equation

$$u(t) = \lambda \int_a^b k(t, y)u(y)\, dy. \tag{2}$$

In his collected works, Fredholm (Ref. 14) showed that the resolvent $K(t, y, \lambda)$ of the kernel $k(t, y)$ could be expressed as

$$K(t, y; \lambda) = D(t, y; \lambda)/D(\lambda), \tag{3}$$

where $D(t, y; \lambda)$ and $D(\lambda)$ are *entire* functions of λ. He also showed that $D(t, y; \lambda)$ and $D(\lambda)$ are related by the differential equation

$$D_\lambda(\lambda) = -\int_a^b D(y', y'; \lambda) \, dy' \qquad (4)$$

and that

$$D(0) = 1. \qquad (5)$$

The subscript in Eq. (4) denotes differentiation with respect to λ. The eigenvalues of Eq. (2) are the roots of the equation

$$D(\lambda) = 0. \qquad (6)$$

At a simple root $\lambda = \lambda^*$, an eigenfunction is given by $D(t, y; \lambda^*)$ for any $y \in [a, b]$ for which $D \neq 0$.

It will be shown by an imbedding approach that the resolvent kernel $K(t, y; \lambda)$ satisfies the Cauchy system

$$K_\lambda(t, y; \lambda) = \int_a^b K(t, y'; \lambda) K(y', y; \lambda) \, dy', \qquad a \leq t, y \leq b, \qquad (7)$$

subject to the initial condition

$$K(t, y; 0) = k(t, y). \qquad (8)$$

If desired, this system can be easily solved numerically to actually produce the resolvent K.

It is then a simple matter to show [using Eqs. (3), (4), and (7)] that the function $D(t, y; \lambda)$ satisfies the differential equation

$$D_\lambda(t, y; \lambda) = [1/D(\lambda)][\textstyle\int_a^b D(t, y'; \lambda) D(y', y; \lambda) \, dy'$$
$$- D(t, y; \lambda) \textstyle\int_a^b D(y', y'; \lambda) \, dy']; \qquad (9)$$

and, from Eqs. (3), (5), and (8), we find that the initial condition for Eq. (9) is

$$D(t, y; 0) = k(t, y). \qquad (10)$$

One of the interesting features of this approach is that we avoid the need to integrate (9) through or even near the zeros of $D(\lambda)$ in order to calculate the eigenvalues and eigenfunctions.

From the theory of complex variables, the number of zeros N of $D(\lambda)$ (each counted with proper multiplicity) in a closed contour C, with none on C, is given by

$$N = (1/2\pi i) \oint_c [D_\lambda(\lambda)/D(\lambda)] \, d\lambda. \qquad (11)$$

We shall propose a numerical scheme for evaluating Eq. (11).

Assume that there are no roots *on* the circle of radius R. To find the number of eigenvalues enclosed by the circle of radius R in the λ-plane, we numerically integrate the initial-value problem for the functions $D(t, y; \lambda)$ and $D(\lambda)$ in Eqs. (4), (5), (9), and (10) from the origin, along a curve having no root of $D(\lambda)$, to $\lambda = P$, a point on the circle. In particular, this gives the values of $D(P)$ and $D_\lambda(P)$. Then, we adjoin the differential equation

$$d\Psi/d\lambda = (1/2\pi i)[D_\lambda(\lambda)/D(\lambda)] = (1/2\pi i)[-1/D(\lambda)] \int_a^b D(y', y'; \lambda)\, dy'$$

(12)

and the initial condition

$$\Psi(P) = 0$$

(13)

to the Cauchy system for $D(t, y; \lambda)$ and $D(\lambda)$ described above. We now integrate this new system around the circle of radius R. It follows from Eq. (11) that the increment of Ψ, which must be zero or a positive integer, is the number of roots enclosed by the circle of radius R.

In an obvious manner, we can further localize the position of a root to various sectors, as when finding roots of a polynomial numerically. The formula

$$(1/2\pi i) \oint_c \lambda^k [D_\lambda(\lambda)/D(\lambda)]\, d\lambda = \sum_{i=1}^M \lambda_i^k, \qquad k = 1, 2, \ldots,$$

(14)

where $\lambda_1, \lambda_2, \ldots, \lambda_M$ are the roots enclosed by the circle of radius R, can be used to aid in the determination of both locations and multiplicities. The above formulas represent M nonlinear equations in the M unknowns, $\lambda_1, \ldots, \lambda_M$. A Newton–Raphson technique is then used to solve for $\lambda_1, \ldots, \lambda_M$. Many obvious possibilities exist for finding the roots outside the circle of radius R.

An important tool in our calculation of the eigenfunctions is Cauchy's integral formula which states: for a function $f(z)$ which is analytic inside and on a simple closed curve C, and for any point a inside C, we have

$$f(a) = (1/2\pi i) \oint_c [f(z)/(z - a)]\, dz.$$

(15)

Since $D(t, y; \lambda)$ is analytic (in fact entire) in λ, then for $a = \lambda^*$, a simple eigenvalue, the relation

$$D(t, y; \lambda^*) = (1/2\pi i) \oint_c [D(t, y; \lambda)/(\lambda - \lambda^*)]\, d\lambda,$$

(16)

yields, for any $\lambda \in [a, b]$, an eigenfunction corresponding to the eigenvalue λ^*, assuming $D \neq 0$.

At the point P described in the previous section, we adjoin the differential equations

$$d\Phi_j/d\lambda = (1/2\pi i)[D(t_j, y_k; \lambda^*)/(\lambda - \lambda^*)], \qquad j = 1, 2, \ldots, N,$$
$$k \in \{1, 2, \ldots, N\}, \quad (17)$$

and the initial conditions

$$\Phi_j(P) = 0, \tag{18}$$

where N is the number of quadrature points used to evaluate the integrals in Eqs. (4) and (9) numerically and y_k corresponds to one of the N points. We now integrate this new system around the curve C. It follows from Eq. (16) that the increment in Φ_j is the eigenfunction for the eigenvalue λ^*.

If the eigenfunction is desired at a value of t which does not correspond to one of the above quadrature points, we may produce the desired value by using Eq. (1) in the following manner. Suppose that we wish the eigenfunction at

$$t = t^* \in [a, b].$$

Then, using the computed $D(t_j, y_k; \lambda^*)$, $j = 1, \ldots, N$, we have

$$u(t^*) = \lambda^* \int_a^b k(t^*, z)D(z, y_k; \lambda^*)\, dz$$

$$\cong \lambda^* \sum_{j=1}^{N} k(t^*, z_j)D(z_j, y_k; \lambda^*)w_j. \tag{19}$$

In addition to producing the eigenfunction at points not corresponding to quadrature points, it also provides a check on our computation by comparing the computed eigenfunction with Eq. (19) at the quadrature points.

We shall present an example which has been extensively investigated to illustrate the efficacy of the method. A Gaussian quadrature scheme of order N is used to approximate the integrals in Eqs. (4) and (9), and a standard fourth-order Runge–Kutta scheme employing complex arithmetic is used to numerically integrate the resulting system of differential equations. For a quadrature scheme of order N, we get N^2 differential equations from Eq. (9) and one equation from each of Eqs. (4) and (12). Additional equations may be added from Eq. (14), by letting

$$d\Omega_k/d\lambda = (1/2\pi i)\lambda^k[D_\lambda(\lambda)/D(\lambda)], \qquad k = 1, 2, \ldots. \tag{20}$$

If the eigenfunctions are also desired, we get N additional equations from Eq. (17).

Table 1. Power loss per transit for dominant mode between parallel strip mirrors.

| | Loss $= 1 - |1/\lambda_1|^2$ | | |
M	Variational technique	Iterative technique	Initial-value method
$\frac{1}{2}$	0.1639	0.1740	0.1694
1	0.0607	0.0800	0.0746
2	0.0217	0.0320	0.0304
3	0.0108	0.0194	0.0165
4	0.0080	0.0129	0.0127

Consider the integral equation

$$u(t) = \lambda \int_{-1}^{1} k(t, y)u(y)\, dy, \qquad -1 \le t \le 1,$$

where the kernel $k(t, y)$ is given by

$$k(t, y) = M^{1/2} \exp(i\pi/4) \exp[-iM\pi(t-y)^2].$$

This particular kernel is the nondimensional form of the kernel which relates to laser plane parallel reflectors. There are a number of difficulties associated with kernels of this type. For example, although the kernel is symmetric, it is not Hermitian. Hence, many classical techniques are not necessarily applicable. Our technique is directly applicable.

We have computed the eigenvalues for

$$M = \tfrac{1}{2}, 1, 2, 3, \text{ and } 4.$$

Table 2. Comparison of computed eigenfunctions for first eigenvalue for $M = 1$.

| | $u(t)$ from (16) | | $u(t)$ from (19) | |
t	Re (u)	Im (u)	Re (u)	Im (u)
-0.949	$-0.8388 - 1$	-0.1124	$-0.8397 - 1$	-0.1114
-0.742	-0.1202	-0.1962	-0.1206	-0.1946
-0.406	$-0.5697 - 1$	-0.3530	$-0.5853 - 1$	-0.3510
0.0	$-0.7652 - 1$	-0.4007	$-0.7817 - 1$	-0.3984
0.406	$-0.5715 - 1$	-0.3531	$-0.5863 - 1$	-0.3511
0.742	-0.1204	-0.1962	-0.1208	-0.1946
0.949	$-0.8395 - 1$	-0.1123	$-0.8413 - 1$	-0.1114

The quantity of greatest physical interest is not λ, but the relative power loss per transit, which is $1-|\mu|^2$, where

$$\mu = 1/\lambda.$$

Table 1 shows the power loss as computed for the lowest even mode by a variational technique, an iterative technique, and the above initial-value method. The number of quadrature points varied from $N = 7$ for $M = \frac{1}{2}$ and 1 to $N = 20$ for $M = 4$. The radii of the contours varied from $R = 1.03$ for $M = 4$ to $R = 1.5$ for $M = \frac{1}{2}$. The integration stepsize was chosen to be $\Delta\theta = \frac{1}{8}$ degrees.

As a check on our computed eigenfunctions, we compared the results of Eqs. (16) by using (19). Table 2 illustrates this for the first eigenfunction for $M = 1$.

We have described a technique for calculating the eigenvalues and eigenfunctions for integral equations with complex-valued symmetric kernels of laser theory. The method involves the combination of classical results of integral equations and complex variable theory with a recent technique for converting Fredholm integral equations into initial-valued differential equation systems. The method is both flexible and accurate, whereas some classical techniques yield results of doubtful accuracy for complex-valued symmetric kernels.

4. Initial-Value Problem for Linear Integral Equations

We shall derive an initial-value problem for the solution of a class of linear Fredholm integral equations. To call attention to the dependence of the solution u upon the parameter λ, as well as upon t, we write

$$u = u(t) = u(t, \lambda).$$

Consider the family of Fredholm integral equations

$$u(t, \lambda) = g(t) + \lambda \int_0^1 k(t, y)u(y, \lambda)\, dy, \qquad 0 \le t \le 1, \qquad 0 \le \lambda \le \Delta, \qquad (21)$$

for λ having a range of positive values near zero. For λ sufficiently near zero, it is assumed that there is a unique solution of Eq. (21) for all inhomogeneous functions g. The function $u = u(t, \lambda)$ and the resolvent function $K = K(t, y, \lambda)$ satisfy the Cauchy system

$$u_\lambda(t, \lambda) = \int_0^1 K(t, y, \lambda)u(y, \lambda)\,dy,$$

$$K_\lambda(t, y, \lambda) = \int_0^1 K(t, y', \lambda)K(y', y, \lambda)\,dy', \qquad 0 \le \lambda \le \Delta,$$

$$u(t, 0) = g(t),$$

$$K(t, y, 0) = k(t, y), \qquad 0 \le t, y \le 1.$$

The subscript λ denotes differentiation with respect to λ.

Let us show that, if the function u satisfies Eq. (21), then it also satisfies the Cauchy system. In terms of the resolvent, the solution of Eq. (21) may be represented in the form

$$u(t, \lambda) = g(t) + \lambda \int_0^1 K(t, y, \lambda)g(y)\,dy, \qquad 0 \le \lambda \le \Delta, \qquad 0 \le t \le 1. \quad (22)$$

The resolvent satisfies the integral equation

$$K(t, y, \lambda) = k(t, y) + \lambda \int_0^1 k(t, y')K(y', y, \lambda)\,dy',$$

$$0 \le \lambda \le \Delta, \qquad 0 \le t, y \le 1. \quad (23)$$

Differentiate both sides of Eq. (23) with respect to λ. The result is

$$K_\lambda(t, y, \lambda) = \int_0^1 k(t, y')K(y', y, \lambda)\,dy' + \lambda \int_0^1 k(t, y')K_\lambda(y', y, \lambda)\,dy'.$$

Regard this as an integral equation for the function K_λ; its solution can be expressed in terms of the resolvent. This leads to the desired integrodifferential equation [using superposition with Eq. (23)]

$$K_\lambda(t, y, \lambda) = \int_0^1 K(t, y', \lambda)K(y', y, \lambda)\,dy', \quad 0 \le \lambda \le \Delta, \quad 0 \le t, y \le 1. \quad (24)$$

The initial condition,

$$K(t, y, 0) = k(t, y) \quad (25)$$

clearly follows by setting $\lambda = 0$ in Eq. (23). In a similar manner, we derive the equation

$$u_\lambda(t, \lambda) = \int_0^1 K(t, y, \lambda)u(y, \lambda)\,dy, \qquad 0 \le \lambda \le \Delta, \qquad 0 \le t \le 1. \quad (26)$$

This is the desired integrodifferential equation for the function u. The initial condition at $\lambda = 0$,

$$u(t, 0) = g(t), \qquad 0 \le t \le 1, \tag{27}$$

follows from putting $\lambda = 0$ in Eq. (21).

5. Numerical Method

The numerical solution of the Cauchy system is direct using the method of lines. Consider Eqs. (24) and (25) for the resolvent. Use a quadrature formula with N points, with $z_1, z_2, \ldots, z_N$ being the points and $w_1, w_2, \ldots, w_N$ being the weights:

$$\int_0^1 f(z) \, dz \cong \sum_{m=1}^N f(z_m) w_m.$$

Write

$$K(z_i, z_j, \lambda) = K_{ij}(\lambda), \qquad i, j = 1, 2, \ldots, N.$$

Then, the functions $K_{ij}(\lambda)$ satisfy the approximate system of N^2 ordinary differential equations

$$(d/d\lambda) K_{ij}(\lambda) = \sum_{m=1}^N K_{im}(\lambda) K_{mj}(\lambda) w_m, \tag{28}$$

and the complete set of initial conditions

$$K_{ij}(0) = k(z_i, z_j), \tag{29}$$

for $i, j = 1, 2, \ldots, N$. For $N \cong 50$, this is a reasonable computational problem, and the exact system can be approximated as closely as desired. The function u is determined either through the Cauchy system of Eqs. (6) and (7) or through the representation formula (22).

6. Application to the Dirichlet Problem: Boundary Integral Method

The imbedding approach was applied to the Dirichlet problem (Courant and Hilbert, Ref. 15). Let R be a plane region bounded by a boundary curve Γ which will be assumed to be represented parametrically by two functions $x(t)$ and $y(t)$ which are differentiable up to the fourth order. The problem is to find a function $u(x, y)$ such that

$$\nabla^2 u(x, y) = 0 \qquad \text{in } R,$$

$$u(x(t), y(t)) = f(t) \qquad \text{on } \Gamma,$$

where $f(t)$ is a known function. If the function $u(x, y)$ is represented as

$$u(x, y) = \int_\Gamma \sigma(t)(\partial\gamma/\partial\nu)\, dt,$$

where

$$\gamma = \log(1/r),$$
$$r = [(x - x(t))^2 + (y - y(t))^2]^{1/2},$$

and if ν is the exterior unit normal to Γ at $(x(t), y(t))$, then $\sigma(t)$ satisfies a Fredholm integral equation having the form

$$\sigma(t, \lambda) = (1/\pi)f(t) + \lambda \int_\Gamma k(t, s)\sigma(s, \lambda)\, ds.$$

The desired solution is that for

$$\lambda = 1.$$

This integral equation was solved by the imbedding method. Integration of the system of ordinary differential equations was performed on an IBM 360/91 using a fourth-order Runge–Kutta scheme with a stepsize $\Delta\lambda = 0.02$. In the first example, the region was a unit circle discretized into 50 equal intervals. The boundary condition was

$$f(t) = \cos^2 t, \qquad 0 \le t \le 2\pi,$$

and the exact solution is

$$\sigma(t, 1) = (1/\pi)(\cos^2 t - 0.25).$$

The imbedding calculations were accurate to five or more significant figures, which is a remarkable accuracy for partial differential equations.

The second example was the computation of the potential inside the rectangular region

$$0 \le x \le 12, \qquad 0 \le y \le 8,$$

with the boundary conditions

$$f = x(12 - x),$$

on the sides parallel to the x-axis, and

$$f = -y(8 - y),$$

on the other sides. The boundary was discretized in 40 intervals. The computed solution presented in Kalaba and Ruspini (Ref. 16) shows an accuracy of up to five significant figures. Computing times were 99 seconds for the first example and 58 seconds for the second, and could have been reduced by the use of coarser grids and other quadrature formulas.

Besides the high accuracy of the procedure, several other advantages are worth mentioning. The integration procedure depends only on the particular region and boundary used. This reduces sensitivity of the computation time to the particular problem. The integration time is dependent only on the integration method used and on the number of discrete partitions used in the boundary. The generality of the method with respect to region shapes is obvious. Although (depending on the integration method used) considerable storage is needed during the integration, storage of the unidimensional function σ is sufficient to compute the potential at any point inside the region. This compares favorably with the storage of a two-dimensional function, as required by conventional difference methods. The approach could be generalized to higher dimensions, and other types of potential problems can be subjected to a similar treatment.

7. Application to the Lebedev and Ufliand Problem in Elasticity

The method described has been applied to various physical problems. The linear integral equation of Lebedev and Ufliand is well known in elasticity. A naturally occurring parameter is λ, the ratio of the radius of contact of a punch to the thickness of the layer. Some numerical results are reported in the original paper, having been obtained through a quadrature formula and the solution of linear algebraic systems. In our initial-value method, we use $N = 7$ points with a Gaussian quadrature formula, and an integration stepsize $\Delta\lambda = 0.01$ with a fourth-order Adams–Moulton integration method. Since it is desired to compare results at $t = 0, 0.1, 0.2, \ldots, 1.0$, rather than at Gaussian quadrature points, we adjoin additional differential equations for those 11 values of t. Thus, we simultaneously integrate $7^2 + 7 + (7+1) \cdot 11 = 144$ ordinary differential equations. Our solutions are in excellent agreement with those of Lebedev and Ufliand for $\lambda \le 2$, the discrepancy being no greater than 0.25%. For $2 < \lambda \le 4$, our results are new and no comparisons could be made. It must be remembered that we solve no linear algebraic equations, and we make a complete parameter study for $0 \le \lambda \le 4$ in steps of 0.01 (see Ref. 4).

8. Initial-Value Problem for Nonlinear Integral Equations

The imbedding of nonlinear integral equations offers very little added complexity. Consider the equation

$$u(t, \lambda) = g(t, \lambda) + \lambda \int_0^1 k(t, y, \lambda, u(y, \lambda)) \, dy, \quad 0 \le t \le 1, \quad 0 < \lambda \le \Delta. \qquad (30)$$

Assume that the solution is differentiable in λ, and differentiate both sides of Eq. (30) to obtain the relation

$$u_\lambda(t, \lambda) = \Psi(t, \lambda) + \lambda \int_0^1 k_u(t, y, \lambda, u(y, \lambda)) u_\lambda(y, \lambda) \, dy, \qquad (31)$$

where the forcing function is

$$\Psi(t, \lambda) = g_\lambda(t, \lambda) + \int_0^1 k(t, y, \lambda, u(y, \lambda)) \, dy + \lambda \int_0^1 k_\lambda(t, y, \lambda, u(y, \lambda)) \, dy. \qquad (32)$$

Equation (31) is a linear Fredholm integral with kernel $k_u(t, y, \lambda, u(y, \lambda))$. Assume that the linear integral equation

$$w(t, \lambda) = F(t, \lambda) + \int_0^1 k_u(t, y, \lambda, u(y, \lambda)) w(y, \lambda) \, dy,$$

$$0 \le t \le 1, \qquad 0 < \lambda \le \Delta, \qquad (33)$$

possesses a unique solution for u being a solution of Eq. (30) and F being arbitrary. Then, represent the solution in terms of the resolvent K for the integral operator with kernel k_u:

$$w(t, \lambda) = F(t, \lambda) + \lambda \int_0^1 K(t, y, \lambda) F(y, \lambda) \, dy, \qquad (34)$$

where K satisfies the integral equation

$$K(t, y, \lambda) = k_u(t, y, \lambda, u(t, \lambda)) + \lambda \int_0^1 k_u(t, y', \lambda, u(y', \lambda)) K(y', y, \lambda) \, dy'. \qquad (35)$$

Then, the function u_λ may be represented in the form

$$u_\lambda(t, \lambda) = \Psi(t, \lambda) + \lambda \int_0^1 K(t, y', \lambda) \Psi(y', \lambda) \, dy', \qquad 0 < \lambda \le \Delta, \qquad 0 \le t \le 1. \qquad (36)$$

Equation (36) may also be regarded as an integrodifferential equation for the function u, in view of Eq. (32). The initial condition at $\lambda = 0$ is

$$u(t, 0) = g(t, 0), \qquad 0 \le t \le 1. \qquad (37)$$

In a similar manner, we obtain the integrodifferential equation for the function K:

$$K_\lambda(t, y, \lambda) = Q(t, y, \lambda) + \lambda \int_0^1 K(t, y', \lambda) Q(y', y, \lambda) \, dy',$$

$$0 < \lambda \le \Delta, \qquad 0 \le t, y \le 1, \qquad (38)$$

where the forcing function Q is

$$Q(t, y, \lambda) = k_{u\lambda}(t, y, \lambda, u(y, \lambda))$$

$$+ k_{uu}(t, y, \lambda, u(t, \lambda)) \left\{ \Psi(y, \lambda) + \lambda \int_0^1 K(y, y', \lambda) \Psi(y', \lambda) \, dy' \right\}$$

$$+ \int_0^1 k_u(t, y', \lambda, u(y', \lambda)) K(y', y, \lambda) \, dy'$$

$$+ \lambda \int_0^1 k_{u\lambda}(t, y', \lambda, u(y', \lambda)) K(y', y, \lambda) \, dy'$$

$$+ \lambda \int_0^1 k_{uu}(t, y', \lambda, u(y', \lambda))$$

$$\cdot \left\{ \Psi(y', \lambda) + \lambda \int_0^1 K(y', y'', \lambda) \Psi(y'', \lambda) \, dy'' \right\} \cdot K(y', y, \lambda) \, dy',$$

$$(39)$$

and we have assumed that the required derivatives exist. The initial condition is

$$K(t, y, 0) = k_u(t, y, 0, g(t, 0)), \qquad 0 \le t, y \le 1. \tag{40}$$

9. Application to the Ambarzumian Integral Equation

The nonlinear integral equation of Ambarzumian,

$$\Phi(\eta) = 1 + \tfrac{1}{2}\lambda \eta \Phi(\eta) \int_0^1 [\Phi(\xi)/(\eta + \xi)] \, d\xi, \qquad 0 \le \eta \le 1, \qquad 0 \le \lambda \le 1,$$

has been studied by the imbedding method. It is especially interesting because of the bifurcation point at $\lambda = 1.0$, where the solution becomes imaginary. Tables presented by Ivanov, and by Abhyankar and Fymat are used for checking. The imbedding calculations are done on an IBM 360/44. For the most part, we use $N = 7$ points with a Gaussian quadrature formula, an integration stepsize $\Delta\lambda = 0.01$, and a fourth-order Adams–Moulton integration scheme. We are in perfect agreement with Ivanov for $\lambda \le 0.95$. At $\lambda = 1.0$, the greatest discrepancy is 5%. With a reduced stepsize 0.005 and an increase in N to $N = 9$, we find $\Phi = 2.5835$, compared with $\Phi = 2.5873$ given by Ivanov for $\eta = 1.0$ and $\lambda = 0.995$, a considerable improvement. These experiments show that the solution can be determined up to, and including, the bifurcation point. There is reason to believe that the computation may routinely be extended past this point by a method such as that suggested by Willers (Ref. 17) and the use of complex integration. The investigation of bifurcation phenomena is extremely interesting, and certainly should be carried further.

10. Integral Equations with Displacement Kernels

Imbedding theory is especially rich for integral equations with displacement kernels having the form

$$u(t, x) = g(t) + \lambda \int_0^x k(|t - y|)u(y, x)\, dy, \qquad 0 \le t \le x.$$

We can study how the solution varies when the upper limit x is varied. Expressing the kernel in exponential form,

$$k(s) = \int_0^1 \exp(-s/z)\, d\mu(z), \qquad s \ge 0,$$

we can derive the initial-value problem for the function u and the auxiliary functions $J(t, x, z)$, $X(x, z) = J(x, x, z)$, and $Y(x, z) = J(0, x, z)$ (see Ref. 4):

$$X_x(x, z) = \lambda Y(x, z) \int_0^1 Y(x, z')\, d\mu(z'),$$

$$Y_x(x, z) = -(1/z)Y(x, z) + \lambda X(x, z) \int_0^1 Y(x, z')\, d\mu(z'),$$

$$J_x(t, x, z) = -(1/z)J(t, x, z) + \lambda X(x, z) \int_0^1 J(t, x, z')\, d\mu(z'),$$

$$e_x(x, z) = -(1/z)e(x, z) + \left[g(x) + \lambda \int_0^1 e(x, z')\, d\mu(z') \right] X(x, z),$$

$$u_x(t, x) = \lambda \left[g(x) + \lambda \int_0^1 e(x, z')\, d\mu(z') \right]$$

$$\times \int_0^1 J(t, x, z')\, d\mu(z'), \qquad 0 \le t \le x, \qquad 0 \le x \le x_{\max}.$$

The initial conditions are

$$X(0, z) = 1,$$

$$Y(0, z) = 1,$$

$$J(t, t, z) = X(t, z),$$

$$e(0, z) = 0,$$

$$u(t, t) = g(t) + \lambda \int_0^1 e(t, z') \, d\mu(z').$$

The Sobolev function Φ plays an important role. It is the solution of the integral equation (Ref. 9)

$$\Phi(t, x) = k(x - t) + \int_0^x k(|t - y|)\Phi(y, x) \, dy.$$

It is related to the function J through the formula

$$\Phi(t, x) = \int_0^1 J(t, x, z)w(z) \, dz,$$

and it is readily available by solving the imbedding equations. The function Φ is important, because it gives the solution u through the new representation formula

$$u(t, x) = U(t) + \int_t^x \Phi(t, x)U(s) \, ds,$$

where

$$U(x) = g(x) + \int_0^x \Phi(y, x)g(y) \, dy.$$

This equation is to be compared with the usual formula for expressing a solution in terms of the Fredholm resolvent $K(t, y, x)$. The function Φ can be identified as the resolvent evaluated at $y = x$:

$$\Phi(t, x) = K(t, x, x).$$

It is noteworthy that the solution can be expressed not only in terms of the resolvent (a function of three variables), but also in terms of the function Φ (a function of two variables), which is merely a part of the resolvent. The solution u itself may be directly produced through the solution of its initial-value problem, if desired.

Much numerical computation has already been performed and attests to the high accuracy and complete feasibility of this imbedding approach (see the following section). It would now be desirable to write *efficient* computer programs utilizing the best of currently available numerical integration procedures. It should become a routine matter for a user to generate the solution for arbitrary forcing functions at desired internal points and for desired interval lengths. This should be done for both inhomogeneous and homogeneous Fredholm integral equations.

11. Computational Solution of Integral Equations: Comparisons

The effectiveness of obtaining a computational solution via the imbedding method is shown in a recent paper by Cali, Casti, and Juncosa (Ref. 18). Their abstract reads as follows:

"This paper compares the relative efficiencies of the invariant imbedding method with the traditional solution techniques of successive approximations (Picard method), linear algebraic equations, and Sokolov's method of averaging functional corrections in solving numerically two representatives of a class of Fredholm integral equations. The criterion of efficiency is the amount of computing time necessary to obtain the solution to a specified degree of accuracy. The results of this computational investigation indicate that invariant imbedding has definite numerical advantages; more information was obtained in the same length of time as with the other methods, or even less time. The conclusion emphasized is that a routine application of invariant imbedding may be expected to be computationally competitive with, if not superior to, a routine application of other methods for the solution of some classes of Fredholm integral equations."

Two examples were studied in Ref. 18:

(i) $k(|t-y|) = \exp(-|t-y|)$,

(ii) $k(|t-y|) = \lambda \int_0^1 \exp(-|t-y|/z) z^{-1} \, dz.$

Briefly, the results were as follows: (a) linear algebraic equations required long computing times for the first example; (b) linear algebraic equations were totally unsuccessful in the second example; (c) Picard's successive approximation method required longer computing times than Sokolov's method; and (d) Sokolov's method required longer computing times than the imbedding method. Computing times for the first example and $x = 1.56$ are given in Table 3. Of course, generalizations should not be made on the basis of these limited experiments. But additional experiments should be carried out for a better assessment of general problems for which imbedding would seem to be superior.

Table 3.   Computing times.

Imbedding technique	35.2 sec
Sokolov's method	176.0 sec
Successive approximation method	250.7 sec
Linear algebraic equations	(unsuccessful)

12. Special Functions b and h

The functions b and h, first introduced by Kagiwada and Kalaba, play especially important roles in the theory of integral equations with displacement kernels. The functions X and Y and various other functionals of solutions of such integral equations can be expressed algebraically in terms of these functions. The functionals depend on as many as four variables. The functions b and h have only three variables. Therefore, they are important new functions for new types of decomposition formulas. They provide alternatives to Fourier and other expansion formulas. These functions should be studied, as Legendre and other special functions have been.

Let $J(t, x, u)$ be the solution of the integral equation

$$J(t, x, u) = \exp[-(x-t)/u] + \int_0^x k(|t-y|)J(y, x, u)\, dy, \qquad 0 \le t \le x. \qquad (41)$$

Equations having this form play significant roles in the theory of multiple scattering. Let $\Phi(t, x)$ be the solution of the related integral equation

$$\Phi(t, x) = k(x-t) + \int_0^x k(|t-y|)\Phi(y, x)\, dy. \qquad (42)$$

Introduce the basic functions b and h:

$$b(t, v, x) = \begin{cases} (1/v) \displaystyle\int_0^t \exp[-(t-y)/v]\, \Phi(y, x)\, dy, & v > 0, \\[2ex] (1/v)\left[\exp[-(t-x)/v] + \displaystyle\int_t^x \exp[-(t-y)/v]\, \Phi(y, x)\, dy\right], \\[2ex] \hfill v < 0, \end{cases} \qquad (43)$$

$$h(t, v, x) = \begin{cases} (1/v)\left[\exp(-t/v) + \displaystyle\int_0^t \exp[-(t-y)/v]\, \Phi(x-y, x)\, dy\right], & v > 0, \\[2ex] (1/v) \displaystyle\int_t^x \exp[-(t-y)/v]\, \Phi(x-y), x)\, dy, & v < 0, \\[2ex] \hfill 0 \le t \le x. \end{cases} \qquad (44)$$

Also introduce the following functionals of J:

$$I(t, v, x, u) = \begin{cases} (1/v) \int_0^t \exp[-(t-y)/v] J(y, x, u)\, dy, & v > 0, \\[2mm] (1/v) \int_t^x \exp[-(t-y)/v] J(y, x, u)\, dy, & v < 0, \end{cases} \tag{45}$$

$$r(v, u, x) = I(x, v, x, u) = (1/v) \int_0^x \exp[-(x-y)/v] J(y, x, u)\, dy, \qquad v > 0, \tag{46}$$

$$\tau(v, u, x) = I(0, -v, x, u) = (1/v) \int_0^x \exp(-y/v) J(y, x, u)\, dy, \qquad v > 0, \tag{47}$$

$$X(x, u) = J(x, x, u), \qquad Y(x, u) = J(0, x, u). \tag{48}$$

Note that the b and h functions are really one, since

$$b(t, -v, x) = h(x - t, v, x), \qquad h(t, -v, x) = b(x - t, v, x), \qquad v > 0. \tag{49}$$

Then, the functions $J, X, Y, I, r, \tau, \Phi$ may all be expressed algebraically in terms of the b and h functions by means of the formulas

$$J(t, x, u) = u\{[1 + ub(x, u, x)]b(t, -r, x) - uh(x, u, x)h(t, -r, x)\}, \tag{50}$$

$$X(x, u) = 1 + ub(x, u, x), \qquad Y(x, u) = uh(x, u, x), \tag{51}$$

$$[(u + v)/u]I(t, v, x, u) = [1 + ub(x, u, x)][ub(t, -u, x) + vb(t, v, x)]$$
$$- uh(x, u, x)[uh(t, -u, x) + vh(t, v, x)], \tag{52}$$

$$[(u + v)/u]r(v, u, x) = [1 + ub(x, u, x)][1 + vb(x, v, x)]$$
$$- uh(x, u, x)vh(x, v, x), \tag{53}$$

$$[(u - v)/u]\tau(v, u, x) = uh(x, u, x)[1 + vb(x, v, x)]$$
$$- [1 + ub(x, u, x)]vh(x, v, x), \tag{54}$$

$$\Phi(t, x) = b(t, 0, x). \tag{55}$$

We saw earlier that Φ can be expressed as an integral of J. A representation for the solution of the integral equation (41) in terms of the solution of Eq. (42) is given in Eq. (50):

$$J(t, x, u) = \left[1 + \int_0^x \exp[-(x-y)/u] \, \Phi(y, x) \, dy \right]$$

$$\times \left[\exp[-(x-t)/u] + \int_t^x \exp[-(y-t)/u] \, \Phi(y, x) \, dy \right]$$

$$- \left[\exp(-x/u) + \int_0^x \exp[-(x-y)/u] \, \Phi(x-y, x) \, dy \right]$$

$$\times \left[\int_t^x \exp[-(y-t)/4] \, \Phi(x-y, x) \, dy \right]. \tag{56}$$

Formulas for the solution of Eq. (41) are useful for more general forcing functions. The solution of the integral equation

$$\varphi(t, x) = g(t) + \int_0^x k(|t-y|)\varphi(y, x) \, dy, \tag{57}$$

where

$$g(t) = \int_c \exp(t/u) \, d\rho(u), \tag{58}$$

may be represented as

$$\varphi(t, x) = \int_c \exp(x/u) J(t, x, u) \, d\rho(u). \tag{59}$$

Formulas (50)–(55) make the numerical determination of J, X, Y, I, r, τ, Φ a practical matter, once b and h are known. The b and h functions satisfy an initial-value problem for the differential equations:

$$b'(t, v, x) = \Phi(0, x)h(t, v, x) + v^{-1}b(t, v, x) - v^{-1}\Phi(t, x), \tag{60}$$

$$h'(t, v, x) = \Phi(0, x)b(t, v, x), \qquad x > t. \tag{61}$$

The prime denotes differentiation with respect to x, and initial conditions are given at $x = t$. The equations needed for the determination of Φ have been supplied above. A tabulation of b and h functions when the kernel is a constant multiplied by the first exponential integral is found in Ref. 19.

13. Application to Multiple Scattering

In the theory of isotropic multiple scattering, the kernel is a constant multiplied by the first exponential integral function. Physically, the function J is the source function, and I is the internal intensity function for the case of monodirectional illumination of the upper boundary of a plane-parallel medium. In the case of isotropic illumination of the upper boundary, the source function is Φ, and the internal intensity function is b. The equivalence of the monodirectional and isotropic source problems is expressed by Eqs. (50) and (52), in the sense that knowledge of b and h leads to the determination of J and I.

While the analytical derivation of results is readily followed, a firm understanding of the principles involved is lacking. This is necessary if we are to be able to extend the theory to other integral equations and other functionals. In a work as yet unpublished, we have extended the theory to Sobolev's integral equations for anisotropic scattering (Ref. 20).

14. Computation of the Functions b and h

The b and h functions which occur in multiple scattering theory satisfy unstable two-point boundary-value problems as well as systems of singular integral equations. These functions have been computed, for the case of isotropic scattering, as solutions of stable Cauchy systems via the imbedding method of Eqs. (60) and (61). Representative tables of these functions are available in Kagiwada and Kalaba (Ref. 19). More recently, computations have been performed by these authors and their colleagues Scott, Fymat, and Garfinkle. The new computational method is based on Gram–Schmidt orthogonalization, and the method of complementary solutions for the two-point boundary-value problem. The orthogonalization process is crucial to successful computation by this method. As an example, several such orthogonalizations were necessary for a medium of optical thickness of twenty. This is because the independent solutions have a strong tendency to become almost dependent. All results are in excellent agreement with the earlier results.

Since the feasibility and high accuracy of the new method have been proven, it is now important to write an efficient computer program for the solution of multiple scattering problems. It would produce not only the functions b and h, but the other functionals of physical interest through the decomposition formulas. This should be done for both the isotropic and anisotropic scattering cases. The computer programs must be written, checked out, documented, and made available to all interested persons.

15. Direct and Inverse Problems in Radiative Transfer

Basic functions in radiative transfer theory (the internal and external radiation fields, the source function, the X and Y functions) all have accurately been calculated by solving the initial-value problem of the previous section. A survey of the results is presented in Ref. 19.

An important class of problems is that of *inverse problems*, in which physical parameters of a scattering medium, a reflector, or sources are unknown and are to be estimated on the basis of radiation measurements, possibly remote ones. The reduction of the basic problem to an initial-value problem as discussed above is significant, for then the inverse problem becomes a *system identification problem*. In such a problem, there are differential equations containing unknown constants which are to be estimated on the basis of observations. The method of quasilinearization is effective in solving such problems, as shown in Refs. 19 and 21.

16. Integrodifferential Equations and Inverse Problems in Wave Propagation

Differential–integral equations of the Riccati type arise in many different settings. Three such instances are contact problems of elasticity theory, inverse problems of wave propagation, and anisotropic multiple scattering. In many cases, the method of lines leads to effective numerical solution of the resultant initial-value problems. On the other hand, these problems possess the common difficulty of the evaluation of definite integrals over intervals in which large and rapid fluctuations of the integrand occur.

It has been shown by Kay and Balanis that certain inverse problems in wave propagation reduce to the solution of the Fredholm integral equation

$$0 = R(x+t) + U(x, t) + \int_{-x}^{+x} R(t+y)U(x, y)\, dy, \qquad -x \le t \le x, \qquad 0 \le x,$$

where the function R is an observed reflected wave, and where the function U is to be determined. More precisely, it is the function

$$V(x) = (d/dx)U(x, x), \qquad 0 \le x,$$

that we wish to find, for it is directly related to the density of the medium.

Kalaba and Zagustin have shown that the function V may be determined directly, i.e., without solving the integral equation for all values of t in the interval $[-x, +x]$. The problem can be reduced to the solution of the

Riccati differential–integral equation

$$\rho_x(v, z, x) = -\left[\exp(-zx) + (1/2\pi)\int_{-\infty}^{\infty} \exp(-iz'x)\rho(z', z, x)r(z')\,dz'\right]$$

$$\times \left[\exp(-vz) + (1/2\pi)\int_{-\infty}^{\infty} \rho(v, u, x)\exp(-iux)r(u)\,du\right]$$

$$+ \left[\exp(ixz) + (1/2\pi)\int_{-\infty}^{\infty} \exp(iz'x)\rho(z', z, x)r(z')\,dz'\right]$$

$$\times \left[\exp(ivx) + (1/2\pi)\int_{-\infty}^{\infty} \rho(v, u, x)\exp(iux)r(u)\,du\right],$$

$$-\infty < v, z < \infty, \qquad 0 \le x.$$

The function $\rho = \rho(v, z, x)$ also satisfies the condition at $x = 0$:

$$\rho(v, z, 0) = 0, \qquad -\infty < v, z < \infty.$$

Solution of this Cauchy system leads to the determination of the desired function $V(x)$ as a double integral of ρ. There is a similar Cauchy problem associated with dual integral equations (see Kagiwada and Kalaba, Ref. 4) as well as with the integral equation of multiple scattering (Ref. 19).

Several important tasks remain for problems of this type. One of them has to do with the computation of the function ρ. Difficulties will arise in trying to evaluate the integrals which appear in the differential–integral equation. What type of quadrature formula would be appropriate? Should new quadrature formulas along the lines of Filon's be developed? Should the method of lines be used?

Other tasks are of an analytical nature. Are there functions of fewer variables which are equivalent to the function ρ which satisfies a Riccati equation? What equations do they satisfy? What are their physical interpretations? Such questions have been addressed by Kailath (Ref. 22), Sidhu and Casti (Ref. 23), Kagiwada and Kalaba (Ref. 4), and others. However, much more remains to be done.

17. Milne–Thomson Integral Equation of Hydrodynamics

The study of free streamlines for flow under gravity is an old and difficult problem. It is a mixed Dirichlet-Neumann boundary-value problem where part of the boundary is unknown. For a large class of free streamline problems, the basic equation is an integral equation of the form (Ref. 24)

$$\theta(t) = g(t) + c\mu \int_a^b K(s, t) H(s) \sin \theta(s) \left[1 + \mu \int_\alpha^s H(u) \sin \theta(u) \, du \right]^{-1} ds,$$

where $g(t)$ is zero in some cases. Although in 1915 Villat derived an integrodifferential equation for a special case, he did not solve it. Equivalent initial-value problems should be derived and validated. Computational procedures must be developed and tested, and applications made for various situations.

18. Integral Equations for the Optimal Control of Differential–Difference Systems

The problem of the optimal control of systems containing time lags occurs in various engineering and social science applications. The solution is extremely difficult to produce, in most cases. The integral equation for the feedback gain matrix has been discussed by Koivo (Ref. 25) and others. It is

$$R(t, y, \tau) = M(t, y) - \int_\tau^T M(t, y') R(y', y, \tau) \, dy', \qquad \tau \le t, y, y' \le T,$$

where

$$M(t, y) = \int_{\max[t, y]}^T \bar{X}(t', t) \bar{X}(t', y) \, dt,$$

$$\bar{X}_t(t, y) = a\bar{X}(t, y) + b\bar{X}(t - 1, z), \qquad 0 \le y \le t,$$

$$\bar{X}(t, t) = I,$$

$$\bar{X}(t, y) = 0, \qquad y > t.$$

An initial-value problem of possible computational utility may be derived using the general imbedding method of Kagiwada and Kalaba (see Ref. 4, Chapter 8). However, it would seem that special equations for this particular kernel would be desirable. There are many more questions, both theoretical and computational. Only a start has been made.

19. Inversion of the Integral Equation of Atmospheric Temperature Sensing

An area of active study is the estimation of the temperature profile in the Earth's atmosphere on the basis of satellite measurements of upwelling radiation (Fleming and Smith, Ref. 26; Duncan, Ref. 27). Traditionally, the mathematical formulation takes the form of the integral equation of the first kind:

$$g(\nu) = \int_0^x k(\nu, t)u(t)\, dt, \qquad \nu_a \le \nu \le \nu_b.$$

The unknown function $u(t)$ represents the temperature as a function of the generalized altitude t, and the function $g(\nu)$ represents the measured radiance as a function of frequency. It is well known that this equation is extremely difficult to solve. Approximation by a system of linear algebraic equations is a standard approach, but these equations are usually ill-conditioned.

To avoid these difficulties, we have formulated a new approach which completely obviates the need for linear algebraic equations. We first formulate this problem as an optimization problem in the calculus of variations. This leads to a two-point boundary-value problem. The imbedding method then transforms this problem to an initial-value problem. The optimal temperature function is finally represented in terms of some basic shape functions. This procedure has been tested and is described in Kagiwada and Kalaba (Ref. 29). Refinements will be made with regard to the mathematical model of the physical world, the mathematical formulation of the problem, and the computational solution. There are many theoretical and computational problems which should be pursued. This may be the first new significant development in this highly important problem.

20. Concluding Remarks

The imbedding approach plays two important roles in the study of integral equations. On the theoretical side, it shows how to effect exact transformations of integral equations into Cauchy systems, and it provides many interesting new functional equations and relationships for further analysis. Computationally, the imbedding method provides practical procedures for obtaining accurate numerical solutions. Where analysis is difficult, computational studies may shed light on properties of solutions and provide clues to additional theoretical investigation.

The successes of the past indicate that there are even more exciting discoveries to be made in the years to come.

References

1. COURANT, R., and HILBERT, D., *Methods of Mathematical Physics*, Vol. 1, John Wiley and Sons (Interscience Publishers), New York, New York, 1953.

2. NOBLE, B., *A Bibliography of Methods for Solving Integral Equations*, University of Wisconsin, Mathematics Research Center, Report No. 73, 1971.
3. COCHRAN, J., *Analysis of Linear Integral Equations*, McGraw-Hill Publishing Company, New York, New York, 1972.
4. KAGIWADA, H., and KALABA, R., *Integral Equations Via Imbedding Methods*, Addison-Wesley Publishing Company, Reading, Massachusetts, 1974.
5. SCOTT, M. R., *A Bibliography on Invariant Imbedding and Related Topics*, Sandia Laboratories, Report No. SC-71-0886, 1971.
6. SHAMPINE, L., and GORDON, M. K., *Computer Solution of Ordinary Differential Equations: The Initial Value Problem*, Freeman Press, San Francisco, California, 1975.
7. FICKEN, F. A., *The Continuation Method for Functional Equations*, Communications on Pure and Applied Mathematics, Vol. 4, pp. 435–455, 1951.
8. HILLE, E., and PHILLIPS, R., *Functional Analysis and Semi-Groups*, American Mathematical Society, Providence, Rhode Island, 1957.
9. SOBOLEV, V. V., *A Treatise on Radiative Transfer*, D. Van Nostrand Company, Princeton, New Jersey, 1963.
10. AMBARZUMIAN, V. A., *Diffuse Reflection of Light by a Foggy Medium*, Doklady Akademiia Nauk SSSR, Vol. 38, pp. 229, 1943.
11. GOLBERG, M., *Initial-Value Methods in the Theory of Fredholm Integral Equations*, Journal of Optimization Theory and Applications, Vol. 9, pp. 112–119, 1972.
12. NELSON, W., *Existence, Uniqueness, and Stability of Solutions to Chandrasekhar's Integrodifferential Equation for X and Y Functions*, Journal of Mathematical Analysis and Applications, Vol. 37, pp. 580–606, 1972.
13. COCHRAN, J. A., *The Existence of Eigenvalues for the Integral Equations of Laser Theory*, Bell System Technical Journal, Vol. 44, pp. 77–88, 1965.
14. FREDHOLM, I., *Oevres Completes de Ivar Fredholm*, Malmo, Lund, Sweden, 1955.
15. COURANT, R., and HILBERT, D., *Methods of Mathematical Physics*, Vol. 2, John Wiley and Sons (Interscience Publishers), New York, New York, 1965.
16. KALABA, R., and RUSPINI, E. H., *Invariant Imbedding and Potential Theory*, International Journal of Engineering Science, Vol. 7, pp. 1091–1101, 1969.
17. WILLERS, I. M., *A New Integration Algorithm for Ordinary Differential Equations Based on Continued Fraction Approximations*, Communications of the ACM, Vol. 17, pp. 504–510, 1974.
18. CALI, M., CASTI, J., and JUNCOSA, M., *Invariant Imbedding and the Solution of Fredholm Integral Equations with Displacement Kernels—Comparative Numerical Experiments*, Applied Mathematics and Computation, Vol. 1, pp. 287–393, 1975.
19. KAGIWADA, H., KALABA, R., and UENO, S., *Multiple Scattering Processes: Inverse and Direct*, Addison-Wesley Publishing Company, Reading, Massachusetts, 1975.
20. SOBOLEV, V. V., *Light Scattering in Planetary Atmospheres*, Pergamon Press, Oxford, England, 1975.

21. KAGIWADA, H., *System Identification*, Addison-Wesley Publishing Company, Reading, Massachusetts, 1974.
22. KAILATH, T., *Some New Algorithms for Recursive Estimation In Constant Linear Systems*, IEEE Transactions on Information Theory, IT-19, pp. 750–760, 1973.
23. SIDHU, G., and CASTI, J., *A Rapprochement of the Theories of Radiative Transfer and Linear Stochastic Estimation*, Applied Mathematics and Computation, Vol. 1, pp. 295–323, 1975.
24. CONWAY, W., and THOMAS, J., *Free Streamline Problems and the Milne–Thompson Integral Equation*, Journal of Mathematical and Physical Science, Vol. 8, pp. 67–92, 1975.
25. KOIVO, H., *On the Equivalence of Maximum Principle Open-Loop Controllers and the Caratheodory Feedback Controllers for Time-Delay Systems*, Journal of Optimization Theory and Applications, Vol. 14, pp. 163–178, 1974.
26. FLEMING, H. E., and SMITH, W. L., *Inversion Techniques For Remote Sensing of Atmospheric Temperature Profiles*, Paper Presented at the Fifth Symposium on Temperature, Washington, DC, 1971.
27. DUNCAN, L. D., *An Improved Algorithm for the Iterated Minimal Information Solution for Remote Sounding of Temperature*, United States Army Electronics Command, Report No. ECOM-5571, 1975.
28. KAGIWADA, H., and KALABA, R., *Imbedding Methods for Temperature Retrieval*, Nonlinear Analysis, Vol. 1, pp. 65–74, 1976.

Additional Bibliography

29. KAGIWADA, H., and KALABA, R., *Direct and Inverse Problems for Integral Equations via Initial Value Methods*, SIAM–AMS Symposium on Transport Theory, American Mathematical Society, Providence, Rhode Island, 1967.
30. KAGIWADA, H., and KALABA, R., *An Initial Value Method Suitable for the Computation of Certain Fredholm Resolvents*, Journal of Mathematical and Physical Sciences, Vol. 1, pp. 109–122, 1967.
31. KAGIWADA, H., and KALABA, R., *Initial Value Methods for the Basic Boundary Value Problem and Integral Equations of Radiative Transfer*, Journal of Computational Physics, Vol. 1, pp. 322–329, 1967.
32. KAGIWADA, H., and KALABA, R., *A New Initial Value Method for Internal Intensities in Radiative Transfer*, Astrophysical Journal, Vol. 147, pp. 301–319, 1967.
33. KAGIWADA, H., and KALABA, R., *An Initial Value Method for Fredholm Integral Equations of Convolution Type*, International Journal of Computer Mathematics, Vol. 2, pp. 143–155, 1968.
34. KAGIWADA, H., and KALABA, R., *An Initial Value Method for Nonlinear Integral Equations*, Journal of Optimization Theory and Applications, Vol. 12, pp. 329–337, 1973.

35. KAGIWADA, H., KALABA, R., and SHUMITZKY, A., *A Representation for the Solution of Fredholm Integral Equations*, Proceedings of the American Mathematical Society, Vol. 23, pp. 37–40, 1969.

36. GOLBERG, M., *An Initial Value Method for the Computation of the Characteristic Values and Functions of an Integral Operator-II: Convergence*, Journal of Mathematical Analysis and Applications, Vol. 49, pp. 773–781, 1975.

37. GOLBERG, M., *Convergence of an Initial Value Method for Solving Fredholm Integral Equations*, Journal of Optimization Theory and Applications, Vol. 12, pp. 344–356, 1973.

9

On an Initial-Value Method for Quickly Solving Volterra Integral Equations: A Review

J. M. BOWNDS[1]

Abstract. A method of converting nonlinear Volterra equations to systems of ordinary differential equations is compared with a standard technique, the *method of moments*, for linear Fredholm equations. The method amounts to constructing a Galerkin approximation when the kernel is either finitely decomposable or approximated by a certain Fourier sum. Numerical experiments from recent work by Bownds and Wood serve to compare several standard approximation methods as they apply to smooth kernels. It is shown that, if the original kernel decomposes exactly, then the method produces a numerical solution which is as accurate as the method used to solve the corresponding differential system. If the kernel requires an approximation, the error is greater, but in examples seems to be around 0.5% for a reasonably small number of approximating terms. In any case, the problem of excessive kernel evaluations is circumvented by the conversion to the system of ordinary differential equations.

1. Introduction

The problem considered here is that of quickly, numerically solving the Volterra equation of the form

$$u(x) = f(x) + \int_a^x K(x, t, u(t))\, dt, \qquad a \le x \le b, \tag{1}$$

where, in addition to other assumptions below, K and f are assumed to be scalar-valued and have sufficient structure to guarantee the existence of a unique solution of (1) on $[a, b]$. The entire discussion generalizes to vector-valued functions in an obvious way. The sources for equations of this type

---

[1] Associate Professor, Department of Mathematics, University of Arizona, Tucson, Arizona.

probably require no discussion here; these equations obviously arise from the integration of basic initial-value problems but, more significantly, they generally appear in the formulation of problems where a rate of change depends not only on an instantaneous state but on the history of the state as well (Refs. 1–4).

The problem of numerically solving (1) has been effectively described in several papers (Refs. 5–8), where the idea of treating this equation (and generalizations of same) as a generalized initial-value problem has evidently been quite satisfactory from the point of view of generating algorithms which resemble some of those for differential equations.

The principal difficulty encountered with these methods seems to be that, since the kernel K depends on the additional variable x, the usual step-by-step method for computing a solution loses efficiency, due to the fact that the integrand in (1) changes from interval to interval. Of course, this is an inherent property of problems which require a cumulative or hereditary effect. To be more specific, it is not difficult to see that, if a single-step method is used, then the total number of kernel evaluations needed to tabulate a solution at N points is at least a constant times N^2 (Refs. 6–8). This can account for the main computational effort in the problem, if the solution is to be tabulated at many points; see Ref. 9 for further details, and see Ref. 10 for a more economical method which still, however, involves N^2 kernel evaluations. This kernel evaluation difficulty is not unique to step-by-step methods; the so-called block-by-block methods described in Refs. 11–14 require the same order of evaluations as does the modified initial-value method described in Ref. 15. However, this is not to say that certain other methods require so many evaluations. At the expense of constructing a spline approximation to the solution, it is shown in Ref. 16 that only $O(N)$ kernel evaluations are required.

2. An Initial-Value Method

The purpose here is to describe the application of an initial-value method for quickly solving (1) via an approximation of the original equation by a hopefully small system of ordinary differential equations. The order of kernel evaluations with this approach is $O(N)$ if N is the number of tabulation points. The basic idea of this particular conversion dates back to a result of Goursat (Ref. 17) and is further motivated by analogous considerations for linear Fredholm equations with kernels of finite rank (see below). A special case of this conversion has been previously used to study stability for linear Volterra equations (Refs. 18–19); a similar independent result for linear equations is found in Ref. 20.

Lemma 2.1. *Basic Conversion Lemma.* If Eq. (1) has a unique solution on $[a, b]$ and K has the form

$$K(x, y, z) = \sum_{i=1}^{s} a_i(x)b_i(y, z), \qquad a \le y \le x \le b, \qquad |z| < +\infty, \qquad (2)$$

where each a_i is continuous and each $b_i(y, g(y))$ is integrable on $[a, b]$ if g is continuous, then the solution of (1) is given by

$$u(x) = f(x) + \sum_{j=1}^{s} a_j(x)y_j(x), \qquad a \le x \le b, \qquad (3)$$

where

$$y_j'(x) = b_j(x, f(x) + \sum_{i=1}^{s} a_i(x)y_i(x)), \qquad a < x \le b, \qquad (4\text{-}1)$$

$$y_j(a) = 0, \qquad j = 1, 2, \dots, S. \qquad (4\text{-}2)$$

Thus, if the integral equation has a decomposable kernel, then it is not difficult to show that the integral equation is equivalent to the system of differential equations (4). This is, of course, reminiscent of the basic theorem of linear Fredholm theory which states that an equation with a finitely decomposable kernel is completely equivalent to a certain system of linear *algebraic* equations. The proof of the above lemma could be accomplished by direct verification; however, in order to draw as much analogy as possible with the Fredholm case, a simple, more constructive proof will be supplied.

Proof. Let $u(x)$ be the unique solution of (1) and set

$$y_j(x) = \int_a^x b_j(t, u(t))\, dt, \qquad j = 1, 2, \dots, S.$$

Then, since

$$u(x) = f(x) + \int_a^x \sum_{j=1}^{s} a_j(x)b_j(t, u(t))\, dt,$$

it is clear that

$$u(x) = f(x) + \sum_{j=1}^{s} a_j(x)y_j(x),$$

and

$$y_j'(x) = b_j(x, u(x)) = b_j(x, f(x) + \sum_{i=1}^{s} a_i(x)y_i(x)),$$

$$y_j(a) = 0, \qquad j = 1, 2, \dots, S. \qquad \square$$

We can obviously obtain the suggested result for Fredholm equations by replacing $y_i(x)$ by $y_i(b)$, a special case of which would be the classical theorem mentioned above. We also point out here that this lemma seems to generalize and possibly supply a simpler proof for a certain result used in an imbedding method for certain Hammerstein equations; see Ref. 2, Eqs. (9)–(18).

It is clear that, if the integral equation (1) has an obvious decomposition of the type (2), then the problem of numerical solution can be completely referred to the problem of numerically solving (4) for which there exists many fast, efficient methods. Probably more important, however, is the fact that even though the original equation (1) involves an additional variable (x) which is usually responsible for $O(N^2)$ kernel evaluations, the conversion to (4) effectively replaces this difficulty with that associated with the problem of numerically solving a *system* which is as large as the number of terms S in the right-hand side of (2). From the point of view of using this technique to develop a rapid method for solving (1), it is certainly reasonable to require S, the system size, to be as small as possible. In this regard, given a Volterra equation (1) which has no exact decomposition (2), one major aspect of using this particular method effectively is that of finding appropriate approximation methods which resemble, in some sense, the right-hand side of (2). Before mentioning some techniques and experiments which have actually been tried, some further motivation for considering the differential system (4) will be supplied in the following section for an important class of equations.

3. Volterra Equations of Hammerstein Type

Galerkin Approximations. By drawing further analogy with some of the basic linear Fredholm theory, it is possible to outline a *best approximation* of sorts to which the conversion in Section 2 applies for a special case of the integral equation. If Eq. (1) is specialized to the form

$$u(x) = f(x) + \int_a^x K(x, t)g(u(t))\, dt, \qquad a \le x \le b, \tag{5}$$

then, referring to the above lemma, it is not difficult to see that the decomposition in (2) need only involve $K(x, t)$. More specifically, it will be assumed here that this kernel is written as

$$K(x, t) = \sum_{i=1}^n \phi_i(x)\psi_i(t) + \epsilon_n(x, t), \qquad a \le t \le x \le b, \tag{6}$$

where the ϕ_i and ψ_i are continuous. In general, it is not clear how to make efficient approximations of this type; however, some cases have been treated in Ref. 9; further remarks on this and examples are found below.

The problem of constructing approximations of the form (6) naturally could be considered without specific regard to the integral equation. However, it is surely more appropriate to take the original problem into consideration; consequently, it is useful to consider approximate solutions of (5) of the form

$$u_n(x) = f(x) + \sum_{i=1}^{n} \phi_i(x) y_i(x), \qquad (7)$$

where $\{\phi_i\}_{i=1}^{\infty}$ is a complete, continuous, orthonormal system in $L^2[a, b]$, and the functions y_i are to be selected in such a way that the error, given by

$$\delta_n(x) = u_n(x) - f(x) - \int_a^x K(x, t) g(u_n(t)) \, dt, \qquad (8)$$

is orthogonal to the first n basis functions. The existence of such y_i is given in the next theorem where it is shown that a sufficient condition for this indicated projection to vanish is that the y_i should solve the appropriate initial-value problem.

Let $\langle,\rangle$ denote the usual inner product for $L^2[a, b]$.

Theorem 3.1. Suppose that K satisfies (6), with $\phi_1, \phi_2, \ldots$, as above, and let $\psi_1(t), \psi_2(t), \ldots, \psi_n(t)$ solve the system

$$\sum_{j=1}^{n} \left(\int_t^1 \phi_j(x) \phi_l(x) \, dx \right) \psi_j(t) = \int_t^1 K(x_1 t) \phi_l(x) \, dx, \qquad l = 1, 2, \ldots, n. \qquad (9)$$

If $\mathrm{col}(y_1, y_2, \ldots, y_n)$ satisfies the initial-value problem

$$y_i'(x) = \psi_i(x) g(u_n(x)), \qquad a < x \le b, \qquad (10\text{-}1)$$

$$y_i(a) = 0, \qquad i = 1, 2, \ldots, n, \qquad (10\text{-}2)$$

with $u_n(x)$ given by (7), then

$$\langle \delta_n, \phi_l \rangle = 0, \qquad 1 \le l \le n,$$

where δ_n is given by (8).

For linear Fredholm equations of the second kind, the analog of the above method for determining $u_n(x)$ has been referred to as the *method of moments* (Ref. 22). In that case, it turns out that (9) is much simpler, due to the fact that the functions $\psi_j(t)$ are easily determined; it is clear that they are, in fact, the Fourier coefficients

$$\psi_j(t) = \int_0^1 K(x, t) \phi_j(x) \, dx.$$

The above theorem suggests a way to choose a kernel approximation method (Fourier approximation) which is directly related to a measure of how close the function $u_n(x)$ approximates the actual solution. Unfortunately, for the purposes of supplying a rapid, reasonably accurate technique for solving Eq. (5), the computation of the quantities $\psi_j(t)$ is not necessarily desirable and so does not necessarily provide an efficient computing method. On the other hand, it seems reasonable perhaps to consider approximations for which $\langle \delta_n, \phi_l \rangle$ is small; one could take the point of view here that the above approximation, for which

$$\langle \delta_n, \phi_l \rangle = 0,$$

is *best* for solving (5).

In view of the above discussion, Theorem 3.1 will be considered as a corollary of the following theorem, where the actual approximation is not of such specific form.

Theorem 3.2. If $K(x, t)$ satisfies (6), $\{\phi_i\}_{i=1}^{\infty}$ is as above, $\delta_n(x)$ is given by (8), and $\mathrm{col}(y_1, y_2, \ldots, y_n)$ satisfies (10), then

$$\langle \delta_n, \phi_l \rangle = \int_a^b E_l(t)g(u_n(t))\, dt, \qquad 1 \le l \le n, \tag{11}$$

where

$$E_l(t) = \sum_{j=1}^{n} \left(\int_t^b \phi_j(x)\phi_l(x)\, dx \right)\psi_j(t) - \int_t^b K(x, t)\phi_l(x)\, dx, \qquad a \le t \le b.$$

The practicality of this theorem would seem to be in choosing pairs $\{\phi_i, \psi_i\}_{i=1}^{\infty}$ in such a way that the functionals in (11) are as small as possible. To actually implement this, one of course needs the fact that, if $\epsilon_n(x, t)$ in (6) is uniformly small, then u_n is near the actual solution u and is therefore uniformly bounded in x and n, a fact which is not difficult to verify (Ref. 23). In other words, approximation methods should naturally have a small $\epsilon_n(x, t)$, but the *better* approximation methods will probably be characterized by a small projection $\langle \delta_n, \phi_l \rangle$ in (11), due to the fact that the consideration of this projection ties the approximation for the kernel to its use, which is in solving the original integral equation.

Proof. The proof amounts to the following computation:

$$\langle \delta_n, \phi_l \rangle = \int_a^b \left(u_n(x) - f(x) - \int_a^x K(x, t)g(u_n(t))\, dt \right)\phi_l(x)\, dx$$

$$= \int_a^b \sum_{i=1}^{n} \phi_i(x)\left(y_i(x) - \int_a^x \psi_i(t)g(u_n(t))\, dt \right)\phi_l(x)\, dx$$

$$- \int_a^b \int_a^x \left(K(x, t) - \sum_{i=1}^{n} \phi_i(x)\psi_i(t) \right)g(u_n(t))\phi_l(x)\, dt\, dx.$$

The first integral on the right immediately above is zero because of (10). The second integral can be rewritten as

$$-\int_a^b \int_t^b (K(x,t)\phi_l(x) - \sum_{i=1}^n \phi_i(x)\phi_l(x)\psi_i(t)) \, dx \, g(u_n(t)) \, dt,$$

which relates to $E_l(t)$ in the theorem.   ☐

Unfortunately, to this author's knowledge, there have been no exhaustive, definitive investigations of methods which use (11) as a measure of goodness of approximation for Volterra equations of this type. The approximation techniques which have been used are those which probably first come to mind from the point of view of approximating $K(x,t)$ without regard to the error projection (11). Some of these are mentioned below.

We mention here that, in the above proof, the inner product can, of course, easily be replaced by one involving a weight function. This simple observation will have relevance in future work on approximation methods.

A Priori Error Estimate. Using arguments very familiar in the analysis of Volterra equations of the second kind, it is not difficult to establish the following loose estimate.

Theorem 3.3. (*Ref. 9*).   Suppose that $\hat{K}(x,t)$ satisfies

$$\hat{K}(x,t) = \sum_{i=1}^s a_i(x)b_i(t), \qquad a \le t \le x < b, \qquad s < \infty,$$

such that

$$|\hat{K}(x,t) - K(x,t)| \le \epsilon_1 \qquad \text{for all } a \le t \le x \le b.$$

Further, suppose that g is Lipschitz and bounded, and let the system (10) be numerically solved at

$$a = x_0 < x_1 < x_2 < \cdots < x_n = b$$

in such a way that the maximum discretization error is ϵ_2; let $\hat{y}_{ik}$ denote the numerical approximation to $y_i(x_k)$, and let

$$\hat{u}_k = f(x_k) + \sum_{i=1}^s a_i(x_k)\hat{y}_{ik}, \qquad k = 1, 2, \ldots, n.$$

Then, for each k,

$$|u(x_k) - \hat{u}_k| = O(\epsilon_1) + O(\epsilon_2) \qquad \text{as } \epsilon_1, \epsilon_2 \to 0.$$

Not surprisingly, numerical experiments have shown the maximum error to be smaller than this prediction, and there appears to be a trade-off between the small ϵ_1 experienced by taking large numbers of terms in the approximation $\hat{K}(x, t)$ and increased error ϵ_2 existent because of a corresponding large number of equations in (10).

4. Linear Equations

If the integral equation is linear, then it can be shown that the approximate solution of the form (7) is actually just the first term in an infinite series which converges uniformly to the solution, no matter what kernel approximation is used. Of course, this first term will more closely represent the actual solution, the better the kernel approximation. This feature is reminiscent of the usual existence proof using successive approximations wherein the convergence is uniform, regardless of the choice for the initial approximation.

Consider the linear integral equation

$$u(x) = f(x) + \int_a^x K(x, t)u(t) \, dt, \qquad a \le x \le b. \tag{12}$$

The following theorem is proved in Ref. 9; however, a sketch of the proof is included here for completeness.

Theorem 4.1. Suppose that

$$\hat{K}(x, t) = \sum_{i=1}^s a_i(x)b_i(t),$$

and set

$$R(x, t) = K(x, t) - \hat{K}(x, t).$$

Let $v_0(x)$ denote the solution of the approximating equation

$$w(x) = f(x) + \int_a^x \hat{K}(x, t)w(t) \, dt, \qquad a \le x \le b, \tag{13}$$

and suppose that $u(x)$ solves (12). Then,

$$u(x) = v_0(x) + \sum_{j=1}^{m-1} v_j(x) + r_m(x) \tag{14}$$

for all $x \in [a, b]$, where

$$v_i(x) = f_i(x) + \sum_{k=1}^s a_k(x)y_{ki}(x), \tag{15-1}$$

$$y'_{kj}(x) = b_k(x)\left[f_j(x) + \sum_{m=1}^{S} a_m(x) y_{mj}(x) \right], \qquad (15\text{-}2)$$

$$y_{kj}(0) = 0, \qquad k = 1, 2, \ldots, S, \qquad j = 1, 2, \ldots, m-1, \qquad m \geq 2, \qquad (15\text{-}3)$$

$$f_j(x) = \int_0^x R(x, t) v_{j-1}(t)\, dt, \qquad j = 1, 2, \ldots. \qquad (15\text{-}4)$$

Furthermore, the remainder in (14) tends to zero uniformly in x and geometrically in m. A more descriptive estimate for $r_m(x)$ is possible using estimates which can be found in Ref. 25.

Theoretically, Eq. (14) describes how much error is involved in solving (12) with K replaced by $\hat{K}$. Practically, it may be appropriate only to compute several refinements to the basic result

$$u(x) \approx v_0(x).$$

The main point of the theorem is that the first approximation is actually the first term in the above series, and this series converges at the above rate, which is dependent on how closely $\hat{K}$ approximates K.

Proof. With v_0 and $\hat{K}$ as given, it follows that

$$u(x) - v_0(x) = \int_a^x K(x, t)[u(t) - v(t)]\, dt + \int_a^x R(x, t) v(t)\, dt.$$

Then, letting

$$r_1(x) = u(x) - v_0(x),$$

it is clear that r_1 solves the original integral equation, with $f(x)$ replaced by

$$f_1(x) = \int_a^x R(x, t) v_0(t)\, dt.$$

Using $v_1(x)$ as defined in the theorem, and letting

$$r_2(x) = r_1(x) - v_1(x),$$

it follows that

$$r_2(x) = \int_a^x R(x, t) v_1(t)\, dt + \int_a^x K(x, t) r_2(t)\, dt,$$

which, again, is the original linear integral equation, with f replaced by

$$f_2(x) = \int_a^x R(x, t) v_1(t)\, dt.$$

In general, it is clear that

$$r_m(x) = r_{m-1}(x) - v_{m-1}(x)$$

satisfies the equation

$$r_m(x) = \int_a^x R(x, t)v_{m-1}(t)\, dt + \int_a^x K(x, t)r_m(t)\, dt,$$

with v_j as in the theorem. In other words,

$$r_m(x) = u(x) - v_0(x) - \sum_{j=1}^{m-1} v_j(x),$$

from which (14) is obvious.

The fact that $r_m(x) \to 0$ as stated follows by first writing

$$f_m(x) = \int_a^x R(x, t)v_{m-1}(t)\, dt = \int_a^x R(x, t)\left[f_{m-1}(t) + \sum_{k=1}^{S} a_k(t)y_{k,m-1}(t)\right] dt.$$

Then, recalling the way in which $y_{kj}(t)$ is defined in the theorem, the usual variation-of-parameter formula implies that

$$y_{k,m-1}(t) = \int_a^t (Y(t, s)B(s))_k f_{m-1}(s)\, ds,$$

where

$$B(s) = \mathrm{col}(b_1(s), b_2(s), \ldots, b_S(s)),$$

$Y(t, s)$ is the fundamental matrix for (15) with $Y(t, t) = I$, and $(YB)_k$ is the kth component of the vector YB. When this is substituted into the above equation for f_m and the order of integration is changed, it follows that

$$f_m(x) = \int_a^x \left\{ R(x, s) + \int_s^x \sum_{k=1}^{S} a_k(t)(Y(t, s)B(s))_k\, dt \right\} f_{m-1}(s)\, ds, \quad (16)$$

$$m = 1, 2, 3, \ldots, \qquad f_0(x) \equiv f(x).$$

Let $\kappa(x, s)$ denote the kernel in the curly-bracketed expression in this integral. This kernel is bounded, and Eq. (16) is a recursive equation, which in fact defines a sequence of successive approximations to the solution of the linear homogeneous equation

$$\psi(x) = \int_a^x \kappa(x, s)\psi(s)\, ds,$$

which has

$$\psi \equiv 0$$

as its only solution. Hence, $f_m(x)$ tends to zero uniformly in x, and the rate of convergence is geometric in m, this following from the usual estimates for successive approximations; see Ref. 24 or, for more specific details, see Ref. 25. □

5. Numerical Examples and Discussion

The following examples represent a partial illustration and comparison with certain other methods. The numerical method used to solve the indicated system of differential equations was a standard fourth-order Runge–Kutta routine used with the extended precision feature on the IBM 1130 computer at the Department of Mathematics, University of Arizona. Although faster and larger computing equipment was available, the author took the point of view that a slower, smaller computer perhaps better serves the purpose of demonstrating methods which do not require more advanced features.

Example 5.1. *Kernels Which Decompose Exactly.* In Ref. 6, the following integrodifferential equation is solved with the methods of that paper:

$$u'(x) = 1 + u(x) - x \exp(-x^2) - 2 \int_0^x xt \exp(-u(t))^2 \, dt, \qquad 0 \le x \le 2,$$

$$u(0) = 0.$$

The exact solution is

$$u(x) = x.$$

The conversion described in Lemma 2.1 above is readily performed in this case, the exact details of which can be found in Ref. 9. The corresponding system of differential equations is only of dimension two, and the numerical results are as follows. A stepsize of 0.1 produced a maximum error on $[0, 2]$ of 1.1×10^{-6}; a stepsize of 0.05 produced an error of 1.3×10^{-7}. In this example, the error involved is, essentially, that created by the numerical integrator.

Example 5.2. *Kernels Which Require Approximate Expansions.* If the kernel $K(x, t)$ in (5) or (12) has no obvious decomposition, then approximate expansions must be used. This, of course, introduces additional error which may be too severe for some applications. The gain is fewer kernel evaluations, and so this method may be most applicable when a rapid, albeit rough, tabulation of the solution is required.

The integrodifferential equation (again taken from Ref. 6)

$$u'(x) = 1 + 2x - u(x) + \int_0^x x(1+2x) \exp[t(x-t)]u(t)\, dt, \qquad 0 \le x \le 1,$$

$$u(0) = 1,$$

has exact solution

$$u(x) = \exp(x^2).$$

A calculation similar to that required in the previous example produces the equivalent integral equation

$$u(x) = 1 + x + x^2 + \int_0^x K(x, t)u(t)\, dt, \qquad 0 \le x \le 1,$$

with

$$K(x, t) = -1 + \int_t^x s(1+2s) \exp(ts) \exp(-t^2)\, ds,$$

from which it follows that

$$K(x, 0) = -1 + \tfrac{1}{2}x^2 + \tfrac{2}{3}x^3,$$

and

$$K(x, t) = \exp(-x^2)\{(1-4/t)t^{-2}[\exp(tx)(tx-1)-(t^2-1)\exp(t^2)]$$
$$+ 2t^{-1}[x^2 \exp(tx)-t^2 \exp(t^2)]\} - 1 \qquad \text{for } x > t.$$

The approximation methods used for this kernel are described in detail in Ref. 9; to summarize, two-dimensional forms of interpolating polynomials and variation-diminishing splines were tried, with the interpolating polynomial (using Tchebycheff zeros) appearing to be superior. With a polynomial of degree seven, the corresponding system of differential equations consists of eight equations, and a stepsize of 0.02 produced a maximum error of 1.2×10^{-3}. It should be pointed out that the interpolating polynomial is simple to compute; in all of the other examples the approximation produced the best results, except where more sophisticated approximation methods applied, as in the case of convolution kernels (see below).

In another example, specifically

$$u(x) = \exp(-x) + \int_0^x \exp(-(x-t))(u(t)+\exp(-u(t)))\, dt, \qquad 0 \le x \le 1, \quad (17)$$

with exact solution

$$u(x) = \log(x + e),$$

the interpolating polynomial of degree 7 was superior to both the spline and the Taylor polynomial, which required systems of sizes 14 and 8, respectively. Note that this specific example does not actually require any approximation, since the exponential is trivially decomposable; also, the Taylor polynomial is simple to compute, a situation not always present for general kernels. The maximum computed error was 1.5×10^{-3}, when the stepsize was 0.05.

Example 5.3. *Kernels of Convolution Type.* If the kernel is of the form

$$K(x, t) = k(x - t),$$

as it is in many applications, then k, being a function of one variable, can be approximated using more refined techniques. Methods for these types of kernels have been explored in Ref. 24, and examples from that reference are reproduced here. On the basis of previous numerical experiments (listed in Section 5.2 above), comparisons were made for interpolating polynomials, Taylor polynomials, best uniform polynomials of a given degree, and finite Tchebycheff expansions. The error analysis for these methods is detailed by Wood in Ref. 24.

The kernel in the equation

$$u(x) = 1 + \int_0^x \exp[-(x - t)]u(t)\, dt, \qquad 0 \le x \le 1,$$

was approximated with all of the above polynomial approximations; not surprisingly, the uniform best and Tchebycheff polynomials produced the better numerical approximations to $u(x)$. A second-degree best uniform polynomial approximation (with corresponding differential system size of three) produced a maximum error of 4.26×10^{-3} with only slightly larger error using the Tchebycheff expansion of same degree (also, same system size). For essentially the same error, the respective use of the Taylor and interpolating polynomial required a system size of at least five.

The two best methods were tried on several other examples, one of which [Eq. (17)] was treated earlier by two-dimensional methods. It was found that the same order of error (5.7×10^{-3}) occurred using a second-degree Tchebycheff expansion on this particular example. Slightly lower error was experienced using the best uniform polynomial approximation of degree two. This amounts to a substantial improvement over the previous methods because of the much smaller system size.

In all of these examples, it was observed that taking more terms in the Tchebycheff expansion or higher degrees in the best uniform polynomial approximation did not appreciably improve the error. At first, there did not seem to be any obvious explanation for this, except possibly that the Runge–Kutta method used to solve the system of differential equations is sensitive to system size, and so there is a range of *best* system sizes which has to be balanced with a sufficiently high degree polynomial approximation for the kernel. This somewhat troublesome point has very recently been considered by Golberg (Ref. 26), who has shown that increased accuracy is evidently very dependent upon the properties of the particular numerical integrator used.

In summary, for smooth convolution kernels, at this writing it seems most efficient to use the finite Tchebycheff expansion as outlined in Ref. 24. On the numerical examples tried so far, such a polynomial of degree five or six produced an error in the numerical solution of the given integral equation of around 0.5%, and the number or type of computations involved are not nearly as inhibitive as those needed without using this conversion to a differential system. The uniform best polynomial approximation produces about the same results, but has the definite disadvantage of requiring a special algorithm to obtain the polynomial coefficients; again, Ref. 24 contains a description of this (Remez) algorithm. As mentioned previously, future work on constructing approximations should no doubt involve the integral equation more directly, principally by using Theorem 3.2 above.

6. Application to Other Equations

Equations of the First Kind. The above method will of course apply to any first-kind equation:

$$f(x) = \int_a^x K(x, t)u(t)\, dt, \qquad a \le x \le b, \tag{18}$$

which satisfies sufficient conditions to be equivalent to a second-kind equation considered above. Most methods for solving (18) in the current literature in fact assume this equivalence (Refs. 11–14). However, since these particular methods apply directly to (18), rather than to its differentiated form, they enjoy a special advantage in those applications where $f(x)$ is given only in tabulated form. The main computational chore, as usual, is the many computations required in evaluating $K(x, t)$; see Refs. 11–13. In this regard, for quicker, rougher solutions, when the equivalent second-kind equation for (18) is known explicitly, it may be appropriate to apply the above initial-value method to the second-kind equation.

In Ref. 23, the integral equation

$$f(x) = \int_0^x k(x, t)(x - t)^{-\alpha} u(t) \, dt, \qquad 0 \le x \le 1, \qquad 0 < \alpha < 1, \qquad (19)$$

is considered, and the usual assumptions that $f'(x)$, $k(x, t)$, and $\partial k_{(x,t)}/\partial x$ be continuous on their respective domains and $k(x, x)$ never vanishes are made. It turns out that this implies that (19) is equivalent to the second-kind equation

$$u(x) = [\sin(\pi\alpha)/\pi k(x, x)]\left[(d/dx) \int_0^x (x - t)^{\alpha - 1} f(t) \, dt + \int_0^x K(x, t) u(t) \, dt \right],$$
$$(20)$$

where

$$K(x, t) = \int_0^1 (1 - w)^{\alpha - 1} w^{1 - \alpha} k_1(t + [x - t]w, t) \, dw. \qquad (21)$$

Here, k_1 denotes differentiation of k with respect to the first argument. Using the approach outlined in Section 3, it is seen that the approximate solution takes the form, roughly, of a sum, the first term being the *closed-form* solution of the Abel equation ($k \equiv 1$), and the remaining terms being a combination of solution components for the corresponding system of equations [see (23) below]. To be more specific, the actual statement of the theorem is as follows. The proof is given in Ref. 23.

Theorem 6.1. Let $\{\phi_i\}_{i=1}^{\infty}$ be a complete, orthonormal basis of continuous L^2-functions on $[0, 1]$. If $\text{col}(y_1, y_2, y_3, \ldots, y_n)$ satisfies the initial-value problem

$$y_j(x) = \psi_j(x) u_n(x), \qquad 0 < x \le 1,$$
$$y_j(0) = 0, \qquad j = 1, 2, \ldots, n,$$

where $\psi_1(x), \psi_2(x), \ldots, \psi_n(x)$ solves the system

$$\sum_{j=1}^{n} \left[\int_{z=x}^1 \phi_j(z)\phi_l(z) \, dz \right] \psi_j(x) = [\sin(\pi\alpha)/\pi] \int_{z=x}^1 [\phi_l(x)K(z, x)/k(z, z)] \, dz,$$
$$l = 1, 2, \ldots, n, \quad (22)$$

and if u_n, the approximate solution of (19), is given by

$$u_n(x) = [\sin(\pi\alpha)/\pi k(x, x)](d/dx) \int_0^x (x - t)^{\alpha - 1} f(t) \, dt + \sum_{j=1}^{n} \phi_j(x) y_j(x), \qquad (23)$$

then

$$\langle \delta_n, \phi_l \rangle = 0 \qquad \text{for } 1 \le l \le n,$$

where the approximation error δ_n, is given by

$$\delta_n(x) = u_n(x) - [\sin(\pi \alpha)/\pi k(x, x)]$$

$$\cdot \left[(d/dx) \int_0^x (x-t)^{\alpha-1} f(t) \, dt + \int_0^x K(x, t) u_n(t) \, dt \right], \qquad 0 \le x \le 1,$$

with K given by (21).

In practice, as in Section 5, it may be useful to consider approximations for $K(x, t)$ which do not necessarily produce a zero-error projection on the first few test functions. If polynomial approximations are used, then some of the required computations are, in effect, at least *standardized*, because of the form of the kernel in (21). Specifically, if $k_1(x, t)$ is approximated in some fashion by

$$k_1(x, t) \approx \sum_{i+j=0}^{n} \alpha_{ij} x^i y^j,$$

then the corresponding approximation for K in (21) is

$$K(x, t) \approx \int_0^1 (1-w)^{\alpha-1} w^{1-\alpha} \sum_{i+j=0}^{n} \alpha_{ij} \sum_{l=0}^{i} \binom{i}{l} (wx)^l [t(1-w)]^{i-l} t^j \, dw,$$

which is just a polynomial with coefficients which are products of the α's and beta functions.

Multiple Volterra Integral Equations. The basic conversion lemma, Lemma 2.1 above, can be generalized for application to multiple Volterra equations of the form

$$u(x, y) = f(x, y) + \int_{a_1}^{x} \sum_{i=1}^{n_1} a_i(x, y) F_i(r, u(r, y)) \, dr$$

$$+ \int_{a_2}^{y} \sum_{j=1}^{n_2} b_j(x, y) G_j(s, u(x, s)) \, ds$$

$$+ \int_{a_1}^{x} \int_{a_2}^{y} \sum_{k=1}^{n_3} C_k(x, y) H_k(r, s, u(r, s)) \, dr \, ds. \qquad (24)$$

These equations are generally considered as generalized Goursat problems, but they also serve as generalized Cauchy problems for the wave equation (Ref. 27). Presumably, the added generality of the additional variables in the integrals (x and y) would allow for hereditary dependence as in the one-dimensional case.

This conversion has been described in more generality (for application to Fredholm equations) in Ref. 28; and, although more details are required, the basic idea of the proof follows that for Lemma 2.1. A special case of the following lemma was established in Ref. 29.

Lemma 6.1. If (24) has a unique, continuous solution $u(x, y)$, then this solution has the representation

$$u(x, y) = f(x, y) + \sum_{i=1}^{n_1} a_i(x, y)\alpha_i(x, y) + \sum_{j=1}^{n_2} B_j(x, y)\beta_j(x, y)$$

$$+ \sum_{k=1}^{n_3} c_k(x, y)\gamma_k(x, y), \tag{25}$$

where α_i, β_j, γ_i solve the characteristic problem

$$\partial\alpha_i/\partial x = F_i(x, f + S_1(x, y)), \qquad \alpha_i(a_1, y) = 0,$$
$$1 \le i \le n_1, \tag{26-1}$$

$$\partial\beta_j/\partial x = G_j(y, f + S_1(x, y)), \qquad \beta_j(x, a_2) = 0,$$
$$1 \le j \le n_2, \tag{26-2}$$

$$\partial^2\gamma_k/\partial x \partial y = H_k(x, y, f + S_1(x, y)), \qquad \gamma_k(a_1, y) = \gamma_k(x, a_2) = 0,$$
$$1 \le k \le n_3, \tag{26-3}$$

and where

$$S_1(x, y) = \sum_{i=1}^{n_1} a_i(x, y)\alpha_i(x, y) + \sum_{j=1}^{n_2} b_j(x, y)\beta_j(x, y) + \sum_{k=1}^{n_3} c_k(x, y)\gamma_k(x, y).$$

This of course means that, subject to the kernel decomposition depicted in (24), the numerical solution of (24) can be completed by solving (26) and using (25). Preliminary numerical work has shown that this approach seems feasible; and, to facilitate the solution of (26), simple, higher-order, *single-step* methods are currently being developed for solving this system (Ref. 30), although any method for solving the system could be used. Some error estimates analogous to those above are given in Ref. 28. The problem of approximating kernels which, for this problem, may depend on four variables has yet to be considered.

References

1. VOLTERRA, V., *Lecons sur la Theorie Mathematique de la Lutte pour la Vie*, Gauthier Villars, Paris, France, 1931.
2. VOLTERRA, E., *Vibrations of Elastic Systems Having Hereditary Characteristics*, Journal of Applied Mechanics, Vol. 14, pp. 363–371, 1950.
3. MILLER, R. K., *On Volterra's Population Equation*, SIAM Journal on Applied Mathematics, Vol. 14, pp. 446–452, 1966.
4. BOWNDS, J. M., and CUSHING, J. M., *On the Behavior of Predator–Prey Equations with Hereditary Terms*, Mathematical Biosciences, Vol. 26, pp. 47–54, 1975.
5. NOBLE, B., *The Numerical Solution of Nonlinear Integral Equations and Related Topics*, Nonlinear Integral Equations, Edited by P. M. Anselone, University of Wisconsin Press, Madison, Wisconsin, 1964.
6. LINZ, P., *Linear Multistep Methods for Volterra Integrodifferential Equations*, Journal of the Association for Computing Machinery, Vol. 16, pp. 295–301, 1969.
7. TAVERINI, L., *One-Step Methods for the Numerical Solution of Volterra Functional Differential Equations*, SIAM Journal On Numerical Analysis, Vol. 8, pp. 786–795, 1975.
8. TAVERINI, L., *Linear Multi-Step Methods for the Numerical Solution of Volterra Functional Differential Equations*, Applicable Analysis, Vol. 1, pp. 169–185, 1973.
9. BOWNDS, J. M., and WOOD, B., *On Numerically Solving Non-linear Volterra Equations with Fewer Computations*, SIAM Journal on Numerical Analysis, Vol. 13, pp. 705–719, 1976.
10. GAREY, L., *Solving Nonlinear Second Kind Volterra Equations by Modified Increment Methods*, SIAM Journal on Numerical Analysis, Vol. 12, pp. 501–508, 1975.
11. WEISS, R., *Product Integration for the Generalized Abel Equation*, Mathematics of Computation, Vol. 26, pp. 177–190, 1972.
12. WEISS, R., and ANDERSSEN, R. S., *A Product Integration Method for a Class of Singular First Kind Volterra Equations*, Numerische Matematik, Vol. 18, pp. 442–456, 1972.
13. ANDERSSEN, R. S., DEHOOG, F., and WEISS, R., *On the Numerical Solution of Brownian Motion Processes*, Journal of Applied Probability, Vol. 10, pp. 409–418, 1973.
14. DEHOOG, F., and WEISS, R., *High Order Methods for Volterra Integral Equations of the First Kind*, SIAM Journal on Numerical Analysis, Vol. 10, pp. 647–664, 1973.
15. MALINA, L., *A-Stable Methods of High Order for Volterra Integral Equations*, Aplikace Matematiky, Vol. 20, pp. 336–344, 1975.
16. EL-TOM, M. E. A., *Efficient Computing Algorithms for Volterra Integral Equations of the Second Kind*, Computing, Vol. 14, pp. 153–166, 1975.
17. GOURSAT, E., *Determination de la Resolvante d'une d'Equation*, Bulletin des Sciences et Matematiques, Vol. 57, pp. 144–150, 1933.

18. BOWNDS, J. M., and CUSHING, J. M., *Some Stability Criteria for Linear Systems of Volterra Integral Equations*, Funkcialaj Ekvacioj, Vol. 15, pp. 101–117, 1972.

19. BOWNDS, J. M., and CUSHING, J. M., *A Representation Formula for Linear Volterra Integral Equations*, Bulletin of the American Mathematical Society, Vol. 79, pp. 532–536, 1973.

20. CERHA, J., *A Note on Volterra Integral Equations with Degenerate Kernels*, Commentari Mathematica Universita Carolinae, Vol. 13, pp. 659–672, 1972.

21. CASTI, J., KALABA, R., and UENO, S., *A Cauchy System for a Class of Nonlinear Fredholm Integral Equations*, Applicable Analysis, Vol. 3, pp. 107–115, 1973.

22. KANTOROVICH, L., and KRYLOV, V., *Approximation Methods of Higher Analysis*, John Wiley and Sons (Interscience Publishers), New York, New York, 1965.

23. BOWNDS, J. M., *On Solving Weakly Singular Volterra Equations of the First Kind with Galerkin Approximations*, Mathematics of Computation, Vol. 30, pp. 747–757, 1976.

24. BOWNDS, J. M., and WOOD, B., *A Note on Solving Volterra Integral Equations with Convolution Kernels*, Applied Mathematics and Computation, Vol. 3, pp. 307–315, 1977.

25. TRICOMI, F. G., *Integral Equations*, John Wiley and Sons (Interscience Publishers), New York, New York, 1957.

26. GOLBERG, M. A., *On a Method of Bownds for Solving Volterra Integral Equations*, Chapter 10, this volume.

27. POGOROZELSKI, W., *Integral Equations and Their Applications*, Pergamon Press and Polish Scientific Publishers, Warsaw, Poland, 1966.

28. BOWNDS, J. M., and DEFRANCO, R. J., *On Converting Multiple Fredholm Equations to Systems of Partial Differential Equations* (to appear).

29. DEFRANCO, R. J., *Stability Results for Multiple Volterra Integral Equations*, University of Arizona, PhD Thesis, 1973.

30. BOWNDS, J. M., and DEFRANCO, R. J., Unpublished Results, 1977.

10

On a Method of Bownds for Solving Volterra Integral Equations[1]

M. A. GOLBERG[2]

Abstract. An initial-value method of Bownds for solving Volterra integral equations is reexamined using a variable-step integrator to solve the differential equations. It is shown that such equations may be easily solved to an accuracy of $O(10^{-8})$, the error depending essentially on that incurred in truncating expansions of the kernel to a degenerate one.

1. Introduction

In several recent articles, Bownds and Wood (Refs. 1–2) and Bownds (Ref. 3) have developed a new method of solving Volterra integral equations based on integrating a finite set of differential equations. The principal advantage of this technique is that it requires only $O(N)$ kernel evaluations (Ref. 1), in contrast to most other methods which require $O(N^2)$ kernel evaluations (Refs. 4–5), where N is the number of points at which the solution is desired. The method tends to be fast; however, the numerical results given in Refs. 1–2 appear to indicate that, unless very sophisticated kernel approximation methods are used, it is hard to achieve an accuracy of greater than $O(10^{-3})$. Although the theoretical error bounds given in Ref. 1 show that a tradeoff exists between the truncation error in approximating the kernel and the integration error, it appears to be extremely difficult to use this bound to see if the 10^{-3} figure is all that can be achieved.

The error estimates given in Refs. 1–2 were developed under the assumption that a fixed stepsize integrator was being used to solve the systems of differential equations, and the numerical results were obtained using a standard fourth-order Runge–Kutta integration method. It was the

---

[1] This work was sponsored by a University of Nevada at Las Vegas Research Grant.
[2] Professor of Mathematics, University of Nevada at Las Vegas, Las Vegas, Nevada.

stated intent of the authors to provide a fast algorithm, which was easily implementable. In this, they succeeded; however, in not choosing a more sophisticated integration routine, we believe that they prejudiced their technique, which appears to have great potential value.

The purpose of this paper is to present the results of resolving several of the equations given in Refs. 1–2 using a variable-step Runge–Kutta–Fehlberg algorithm (Ref. 6). As shown in Section 3, the improvement in accuracy is substantial. In all cases, we were able to solve the given integral equation to within the tolerance of the integrator. It was possible, then, to achieve accuracies of 10^{-8}, with solution times typically under 2 seconds on a CDC 6400 computer.

In our papers (Refs. 7–8), it has been shown how Bownds' method is related to the technique of invariant imbedding (Ref. 9). Since most of the numerical work for solving integral equations by this method has also been done using fixed-step algorithms for numerical integration (Refs. 8–10), it is our opinion that much of this area should be reexamined in the light of the more recent developments in the solution technology for initial-value problems (Refs. 6 and 11). Although we have not strived for maximal efficiency, the available *off the shelf* codes, to be found in Allen and Shampine's book (Ref. 6) or in Shampine and Gordon's book (Ref. 11), make highly sophisticated codes available at the cost of punching some cards. With this fact in mind, we think that our methods are in accordance with the philosophy of Bownds in producing an efficient, easily implemented algorithm for solving a variety of Volterra integral equations.

In the next section, we review several of the procedures given in Refs. 1–3 and present a modification of Bownds' method for convolution equations. This result, first given in Ref. 7, allows us to reduce the number of kernel evaluations to one in this case. In Section 3, our numerical results are given. The paper ends with a discussion of future work and of several outstanding problems in this area.

2. Initial-Value Methods

We consider the class of Volterra equations

$$u(t) = g(t) + \int_a^t K(t, s, u(s)) \, ds, \tag{1}$$

where $K(t, s, u)$ has the form

$$K(t, s, u) = \sum_{i=1}^N a_i(t) b_i(s, u). \tag{2}$$

We assume that

$$g(t), a_i(t), \qquad i = 1, 2, \ldots, N, \quad \text{and} \quad b_i(s, u), \qquad i = 1, 2, \ldots, N,$$

are continuous functions and that (1) has a unique solution for

$$a \le t \le b.$$

Define

$$y_i(t) = \int_a^t b_i(s, u(s)) \, ds, \qquad i = 1, 2, \ldots, N. \tag{3}$$

Then, as was shown in Ref. 1, $u(t)$ has the representation

$$u(t) = g(t) + \sum_{i=1}^{N} a_i(t)y_i(t), \tag{4}$$

where

$$y_i(t), \qquad i = 1, 2, \ldots, N,$$

solve the initial-value problem

$$y_i'(t) = b_i\left(t, g(t) + \sum_{i=1}^{N} a_i(t)y_i(t)\right), \tag{5}$$

$$y_i(0) = 0, \qquad i = 1, 2, \ldots, N. \tag{6}$$

To solve (1), Eqs. (5)–(6) are integrated numerically, and $u(t)$ is obtained from (4).

To use the above procedure to solve more general equations, where $K(t, s, u)$ does not have the special form (2), necessitates approximating $K(t, s, u)$ by such a finite sum. The method, then, is the analogue of the classical procedure of using degenerate kernel approximations for solving Fredholm equations (Ref. 12). Rather than dealing with the most general case, we concentrate on functions $K(t, s, u)$ of the particular form

$$K(t, s, u) = K(t, s)\phi(s, u), \tag{7}$$

where the kernel $K(t, s)$ has either an exact or approximate representation

$$K(t, s) = \sum_{i=1}^{N} a_i(t)b_i(s). \tag{8}$$

In this case, Eqs. (4)–(6) become

$$u(t) = g(t) + \sum_{i=1}^{N} a_i(t)y_i(t), \tag{9}$$

where

$$y_i(t) = \int_a^t b_i(s)\phi(s, u(s))\, ds, \tag{10}$$

and

$$y_i'(t) = b_i(t)\left[\phi\left(t, g(t) + \sum_{i=1}^{N} a_i(t)y_i(t) \right) \right], \tag{11}$$

$$y_i(0) = 0, \qquad i = 1, 2, \ldots, N. \tag{12}$$

If the representation (8) is not exact, then it was shown in Ref. 1 that the maximum error E_N incurred in using (9) to compute $u(t)$ has the form

$$E_N = \tau(N) + \iota(N), \tag{13}$$

where $\tau(N)$ is the truncation error due to the kernel approximation and $\iota(N)$ is the integration error. In general, $\tau(N)$ is a decreasing function, and $\iota(N)$ is an increasing function of N, thus indicating that there is an N for which E_N is minimized. As stated in the introduction, it is very difficult to determine $E_N^{\min}$ from (13), so experimentation is necessary. The estimate $\iota(N)$ in Ref. 13 was given for a fixed stepsize integrator, for a fixed, given stepsize. However, if a variable-step algorithm is used instead, then the error tolerance is supplied by the user (Refs. 6 and 11), and the code determines whether or not the integration can be carried out. If the integration cannot be performed without exceeding a certain number of arithmetic operations, then the problem is aborted, and the effective error is *infinite* (Ref. 6). In this case, a more appropriate error estimate for the above method is

$$E_N = \tau(N) + \text{Tol}, \tag{14}$$

where Tol is the prescribed integration error tolerance, or ∞ if the equation cannot be integrated to the given accuracy.

The results obtained in Refs. 1–2 and 13 tend to show that, for a fixed stepsize integrator,

$$E_N^{\min} = O(10^{-3}),$$

which occurs for N around 7 for a variety of kernel approximation methods. In contrast to this, our calculations, using a variable-step integrator, show that (14) is satisfied.

Before detailing our numerical results, we present a modification of Bownds' method for the case where

$$K(t, s) = \mathcal{K}(t - s),$$

that is, when (1) is a convolution equation.

Let

$$\mathcal{K}(t) = \sum_{i=1}^{N} \alpha_i k_i(t), \tag{15}$$

where

$$k_i(t), \qquad i = 1, 2, \ldots, N,$$

solve the system of ordinary differential equations

$$k'_i(t) = \sum_{j=1}^{N} a_{ij} k_j(t), \qquad i = 1, 2, \ldots, N. \tag{16}$$

Consider (7) with

$$K(t, s) = \mathcal{K}(t - s).$$

If the representation (14) is not exact, then we assume that $\mathcal{K}(t)$ can be approximated to within a given degree of accuracy by an expression of that form. The solution to (1) is given by

$$u(t) = g(t) + \int_{a}^{t} \mathcal{K}(t - s) \phi(s, u(s)) \, ds, \tag{17}$$

where

$$u(t) = g(t) + \sum_{i=1}^{N} \alpha_i y_i(t), \tag{18}$$

and

$$y_i(t) = \int_{a}^{t} k_i(t - s) \phi(s, u(s)) \, ds, \qquad i = 1, 2, \ldots, N, \tag{19}$$

are the solutions of the Cauchy problem

$$y'_i(t) = k_i(0) \phi\left(t, g(t) + \sum_{j=1}^{N} \alpha_j y_j(t)\right) + \sum_{j=1}^{N} a_{ij} y_j(t), \tag{20}$$

$$y_i(0) = 0, \qquad i = 1, 2, \ldots, N. \tag{21}$$

In particular, if

$$\phi(s, u) = u,$$

then

$$y_i(t), \qquad i = 1, 2, \ldots, N,$$

solve

$$y'_i(t) = \sum_{j=1}^{N} (a_{ij} + k_i(0)\alpha_j) y_j(t) + k_i(0) g(t), \tag{22}$$

$$y_i(0) = 0, \qquad i = 1, 2, \ldots, N. \tag{23}$$

Equations (20)–(21) and (22)–(23) are easily obtained by differentiation of (19); see Ref. 7.

Since a_{ij}, $i, j = 1, 2, \ldots, N$, are assumed known, then, in solving (20)–(21) or (22)–(23), the kernel $\mathcal{K}(t)$ needs to be evaluated only at $t = 0$. Since this can usually be done *by hand*, the above procedure essentially requires no kernel evaluations in the course of the computations. However, if (15) is not exact, then, in approximating $\mathcal{K}(t)$ by such a sum, the kernel will certainly need to be evaluated, usually at fewer than $O(N)$ points; see Bownds and Wood for discussion. The method thus achieves a substantial reduction in arithmetic operations when compared to the formulation in Ref. 2.

We now turn to the results of several sets of numerical computations. Our aim is to show that, if the above systems are integrated using an accurate and efficient integrator, then (1) can be solved rapidly by the above procedures to an order of accuracy equal to the error incurred by approximating the kernel by a degenerate one.

3. Numerical Examples[3]

This section presents the results of three sets of calculations on integral equations drawn from the paper of Bownds appearing in this volume (Ref. 13). To minimize the effect of the kernel approximations, we chose to solve convolution equations, using our version [Eqs. (22)–(23)] of Bownds' method. Since our main interest is to observe the variation in the solution error with increasing system size, this technique tends to remove the effects of any errors in the evaluation of the equations themselves.

Example 3.1. Our first equation is

$$u(t) = 1 + \int_0^t \exp(t - s)u(s)\,ds, \qquad 0 \le t \le 1. \tag{24}$$

The exact solution is

$$u(t) = [\exp(2t) + 1]/2.$$

Although this equation (along with the other examples) has a separable kernel, we treat it as if this were not the case, and so it is approximated by the first $N + 1$ terms of its Taylor expansion. That is, we consider the

---

[3] The author would like to acknowledge the help of Ms. Terri Goodman, who programmed and carried out the following calculations.

sequence of approximations $\{u_N(t)\}$ given by

$$u_N(t) = 1 + \int_0^t \left(\sum_{j=0}^N (t-s)^j/j! \right) u_N(s) \, ds. \tag{25}$$

It is straightforward to show that $u_N(t)$ converges uniformly to $u(t)$ on $[0, 1]$. To solve (25), we use (22)–(23). This gives

$$u_N(t) = 1 + \sum_{j=0}^N y_j^N(t), \tag{26}$$

where

$$dy_0(t)/dt = 1 + \sum_{j=0}^N y_j^N(t), \tag{27}$$

$$dy_j^N(t)/dt = y_{j-1}^N(t), \qquad j = 1, 2, \ldots, N, \tag{28}$$

$$y_j^N(0) = 0, \qquad j = 0, 1, \ldots, N. \tag{29}$$

Equations (27)–(29) were integrated using the Runge–Kutta–Fehlberg code given in Allen and Shampine's book (Ref. 6). This is a variable-step integrator which is a modification of the standard fourth-order Runge–Kutta scheme. The error tolerance Tol was taken as 10^{-6}, and an initial stepsize of 10^{-1} was chosen. The equations were integrated for $N = 1$ to $N = 12$. The estimated error was

$$\max\{t \in [0, 1], |u(t) - u_N(t)|\} \le 100/(N+1)!.$$

As an additional check, we observe that it follows from Gronwall's inequality (Ref. 1) that

$$u_N(t) \le u(t).$$

From this, since both $u(t)$ and $u_N(t)$ are increasing functions, one expects that the maximum error should occur at $t = 1$. This took place in all cases.

From the error estimate and the prescribed tolerance on the integrator, it was expected that the error in approximating $u(t)$ would decrease monotonely from $N = 1$ to about $N = 9$ or 10, and would not show any decrease for $N > 10$. The results given in Table 1 bear this out.

Example 3.2. Our second example is similar to the first. We solve

$$u(t) = 1 + \int_0^t \exp(-(t-s))u(s) \, ds. \tag{30}$$

The solution of (30) is

$$u(t) = 1 + t.$$

Table 1. Summary of numerical solution of Eq. (24).

N	Predicted error $100/N+2! \pm 10^{-6}$	Actual error	CPU time (sec)
1	$\sim 1.7 \times 10^1$	4×10^{-1}	1.65
2	$\sim 4 \times 10^0$	8×10^{-1}	1.70
3	$\sim 9 \times 10^{-1}$	2×10^{-2}	1.80
4	$\sim 1.4 \times 10^{-2}$	2×10^{-3}	1.86
5	$\sim 2.5 \times 10^{-3}$	3×10^{-4}	1.88
6	$\sim 2.5 \times 10^{-3}$	3×10^{-5}	1.91
7	$\sim 2.5 \times 10^{-4}$	2×10^{-6}	2.00
8	$\sim 2.5 \times 10^{-5}$	1×10^{-6}	2.00
9	$\sim 2.5 \times 10^{-6}$	2×10^{-6}	2.00
10	$\sim 1 \times 10^{-6}$	2×10^{-6}	2.00
11	$\sim 1 \times 10^{-6}$	2×10^{-6}	2.00
12	$\sim 1 \times 10^{-6}$	2×10^{-6}	2.00

Equation (30) was solved by making a Taylor series approximation to $\exp(-(t-s))$ and proceeding as for Example 3.1. The sequence of approximations to $u(t)$ are defined by

$$u_N(t) = 1 + \sum_{j=0}^{N} (-1)^j y_j^N(t)/j!, \qquad (31)$$

where

$$dy_0^N(t)/dt = 1 + \sum_{j=0}^{N} (-1)^j y_j^N(t)/j!, \qquad (32)$$

$$dy_j^N(t)/dt = y_{j-1}^N(t), \qquad j = 1, 2, \ldots, N, \qquad (33)$$

$$y_j^N(0) = 0, \qquad j = 0, 1, 2, \ldots, N. \qquad (34)$$

Equations (32)–(34) were integrated exactly as those equations in Example 3.1. The results are given in Table 2.

Table 2. Summary of numerical solution of Eq. (30), Tol = 10^{-6}.

N	Predicted error $1/(N+2)! \pm 10^{-6}$	Actual error	CPU time (sec)
3	$\sim 1 \times 10^{-2}$	5×10^{-2}	1.66
4	$\sim 4 \times 10^{-3}$	9.8×10^{-3}	1.70
5	$\sim 2 \times 10^{-3}$	2×10^{-3}	1.74
6	$\sim 2 \times 10^{-4}$	2×10^{-4}	1.80
7	$\sim 3 \times 10^{-5}$	2×10^{-3}	1.81
8	$\sim 3 \times 10^{-6}$	2×10^{-3}	1.84
9	$\sim 1 \times 10^{-6}$	2×10^{-3}	1.91
10	$\sim 1 \times 10^{-6}$	2×10^{-3}	1.91

Table 3. Summary of numerical solution of (30), Tol $= 10^{-8}$.

N	Maximum error	CPU time (sec)
5	1.6×10^{-3}	1.81
6	3.4×10^{-4}	1.83
7	2.8×10^{-5}	1.85
8	3×10^{-6}	1.95
9	1×10^{-8}	1.95
10	1×10^{-8}	1.97

Note that, in contrast to what one might expect from the results of Example 3.1, the error decreased to a minimum value of 2×10^{-4} for $N = 6$, and then increased back up to 2×10^{-3}. This was well above the value of Tol, and was about the same order of accuracy achieved by Bownds and Wood. Thus, we seemed to have run into a similar problem. However, when all the components of $y_N^i(t)$ were examined, the source of the difficulty became apparent. What was observed was that, for $N \geq 7$, all the components of $y_j^N(t)$ were equal to $y_j^5(y)$, $j = 0, 1, \ldots, 5$; the remaining components were zero. In effect, for $N \geq 7$, the solution $u_N(t)$ was being computed as $u_5(t)$, and thus the errors were the same. The fact that the higher-order components were being given as zero indicated that some or all were smaller than 10^{-6}, and so Tol was reduced to 10^{-8}. The results obtained were in accordance with the above observations and are given in Table 3.

Example 3.3. Our third example is

$$u(t) = \exp(-t) + \int_0^t \exp(-(t-s))[u(s) + \exp(-u(s))] \, ds, \qquad 0 \leq t \leq 1. \quad (35)$$

The solution of (35) is

$$u(t) = \log(t + e).$$

It was solved in the same manner as Examples 3.1 and 3.2 by making Taylor series approximations to the kernel. Tol was taken to be 10^{-8}, and the equation was solved to this accuracy for $N = 9$. Table 4 summarizes our results.

As in the previous examples, the maximum error occurred for $t = 1$. No difficulty was encountered with small components.

Table 4.   Summary of the numerical solution of (35), Tol $= 10^{-8}$.

N	Maximum error
2	2×10^{-1}
3	6×10^{-3}
4	1×10^{-3}
5	2×10^{-4}
6	9×10^{-5}
7	3×10^{-5}
8	1×10^{-6}
9	1×10^{-8}
10	1×10^{-8}

4. Conclusions

We have shown how a variable-step integrator can be used to substantially improve the accuracy of the initial-value method given by Bownds for the *fast* solution of Volterra integral equations of the second kind. By a suitable choice of the error tolerance, it was possible to solve such equations to within the error produced by approximating the kernel by a degenerate one. It has also been demonstrated that an improvement in efficiency can be made for equations with convolution kernels.

Many interesting questions remain concerning the general implementation of this procedure. First, is it possible to predict in advance when the difficulties occurring in Example 3.2 will take place? To do this, one should certainly have rough *a priori* estimates of the order of magnitude of each of the components of the solution. A reasonable thing to do, then, would be to scale the components so that they all have the same order of magnitude. However, this may not be necessary (or sufficient), since, in Example 3.3, we encountered components as small as 10^{-13} and no integration difficulties appeared. Possibly, there is a subtle relation between the equations and the method of stepsize control determined by Tol. This should be investigated further.

The method of kernel approximation to be used is also of great importance. Since this paper has been directed toward the effects of integration errors on the solution, we have deliberately not been concerned with this problem. In Bownds' work (Refs. 1–2), several techniques were employed, with expansions in Chebyshev polynomials generally proving to be the best. His concern, of course, was to keep the size of the system of differential equations to a minimum. It is interesting to note that, in all of the problems that we solved, the total running time consisted of about two-thirds compilation time; therefore, it was rather insensitive to the number of equations being integrated. There was only about a 20% variation in execution time for a range of 2 to 12 equations being integrated. In view of this, if the kernel is smooth, the efficiency of the method will probably be affected only slightly by the method used to approximate the kernel.

The remaining unsolved problem is to extend the method so that it is capable of solving equations with singular, but integrable, kernels. An interesting approach to obtaining degenerate kernel approximations for Fredholm equations has been given by Sloan (Ref. 14), and it would be useful, in this respect, to extend the method to Volterra equations as well.

Lastly, we observe that the above methods can be extended to solve integro-differential equations, Fredholm equations, and more general classes of functional equations. These topics are discussed in more detail in Refs. 7–8 and 15 and Chapter 13.

References

1. BOWNDS, J. M., and WOOD, B., *On Numerically Solving Nonlinear Volterra Integral Equations with Fewer Computations*, SIAM Journal of Numerical Analysis, Vol. 13, pp. 705–719, 1976.
2. BOWNDS, J. M., and WOOD, B., *A Note on Solving Volterra Integral Equations with Convolution Kernels*, Applied Mathematics and Computation, Vol. 3, pp. 307–315, 1977.
3. BOWNDS, J. M., *On Solving Weakly Singular Volterra Equations of the First Kind with Galerkin Approximations* (to appear).
4. NOBLE, B., *The Numerical Solution of Nonlinear Integral Equations and Related Topics*, Nonlinear Integral Equations, Edited by P. M. Anselone, University of Wisconsin Press, Madison, Wisconsin, 1964.
5. GAREY, L., *Solving Nonlinear Second Kind Volterra Equations by Modified Increment Methods*, SIAM Journal on Numerical Analysis, Vol. 12, No. 6, 1975.
6. ALLEN, R. C., and SHAMPINE, L. F., *Numerical Computing: An Introduction*, W. B. Saunders Company, Philadelphia, Pennsylvania, 1973.
7. GOLBERG, M. A., *The Conversion of Fredholm Integral Equations to Equivalent Cauchy Problems*, Applied Mathematics and Computation, Vol. 2, No. 1, 1976.
8. GOLBERG, M. A., *The Conversion of Fredholm Integral Equations to Equivalent Cauchy Problems, II: Computation of Resolvents*, Applied Mathematics and Computation, Vol. 1, No. 1, 1977.

9. KAGIWADA, H., and KALABA, R. E., *Integral Equations via Imbedding Methods*, Addison-Wesley Publishing Company, Reading, Massachusetts, 1974.
10. CASTI, J., and KALABA, R. E., *Imbedding Methods in Applied Mathematics*, Addison-Wesley Publishing Company, Reading, Massachusetts, 1973.
11. SHAMPINE, L. F., and GORDON, M. K., *The Computer Solution of Ordinary Differential Equations: The Initial-Value Problem*, W. H. Freeman, San Francisco, California, 1975.
12. ATKINSON, K., *A Survey of Numerical Methods for the Numerical Solution of Fredholm Integral Equations of the Second Kind*, Society for Industrial and Applied Mathematics, Philadelphia, Pennsylvania, 1976.
13. BOWNDS, J. M., *An Initial-Value Method for Quickly Solving Volterra Integral Equations*, Journal of Optimization Theory and Applications, Vol. 24, No. 1, 1978.
14. SLOAN, I. H., *Error Analysis for Integral Equation Methods* (to appear).
15. GOLBERG, M. A., *Boundary and Initial-Value Methods for Solving Fredholm Integral Equations with Semidegenerate Kernels*, Chapter 13, this volume.

11

Resolvent Kernels of Green's Function Kernels and Other Finite-Rank Modifications of Fredholm and Volterra Kernels[1]

L. B. RALL[2]

Abstract. Many important Fredholm integral equations have separable kernels which are finite-rank modifications of Volterra kernels. This class includes Green's functions for Sturm–Liouville and other two-point boundary-value problems for linear ordinary differential operators. It is shown how to construct the Fredholm determinant, resolvent kernel, and eigenfunctions of kernels of this class by solving related Volterra integral equations and finite, linear algebraic systems. Applications to boundary-value problems are discussed, and explicit formulas are given for a simple example. Analytic and numerical approximation procedures for more general problems are indicated.

1. A Classical Problem in the Theory of Integral Equations

The equation

$$y(x) - \lambda \int_0^1 K(x, t) y(t)\, dt = f(x), \qquad 0 \le x \le 1, \qquad (1)$$

for the unknown function $y(x)$ is called a *linear integral equation of second kind*. Equations of this form arise in the solution of initial- and boundary-value problems for ordinary differential equations and in other areas of applied analysis. In (1), the function $f(x)$, the *kernel* $K(x, t)$, and the

---

[1] This research was sponsored by the United States Army under Contract No. DAA29-75-C-0024.
[2] Professor, Mathematics Research Center, University of Wisconsin, Madison, Wisconsin.

parameter λ are assumed to be given. In case

$$K(x, t) = 0 \qquad \text{for } t > x,$$

Eq. (1) is said to be of *Volterra type*, and the interval of integration is actually $0 \le t \le x$; otherwise, (1) is called an integral equation of *Fredholm type*.

A central problem in the classical theory of linear integral equations of second kind is to determine the values of λ for which the solution $y(x)$ of (1) exists and is unique, and to express this solution in the form

$$y(x) = f(x) + \lambda \int_0^1 R(x, t; \lambda) f(t) \, dt, \qquad 0 \le x \le 1, \qquad (2)$$

where $R(x, t; \lambda)$ is called the *resolvent kernel* of $K(x, t)$ (see Ref. 1). The investigation of the unique solvability of Eq. (1) can thus be reduced to the problem of existence and construction of $R(x, t; \lambda)$.

Using operator notation, the kernel $K(x, t)$ may be taken to define the *linear integral operator* K on the space of functions considered. If I denotes the *identity operator*, then (1) can be written in the form

$$(I - \lambda K)y = f. \qquad (3)$$

The solution y of (3) exists and is unique if the operator $I - \lambda K$ is invertible for the given value of λ. To obtain the expression corresponding to (2), the inverse of $I - \lambda K$ is represented in the form

$$(I - \lambda K)^{-1} = I + \lambda R(\lambda), \qquad (4)$$

where the *resolvent operator* $R(\lambda)$ of K is the linear integral operator with kernel $R(x, t; \lambda)$. Equation (4) leads directly to the relationships

$$R(\lambda) = K + \lambda K R(\lambda),$$
$$R(\lambda) = K + \lambda R(\lambda) K, \qquad (5)$$

by the definition of the inverse operator. These are the so-called *resolvent equations*. In terms of the corresponding kernels, Eqs. (5) become

$$R(x, t; \lambda) = K(x, t) + \lambda \int_0^1 K(x, s) R(s, t; \lambda) \, ds,$$
$$R(x, t; \lambda) = K(x, t) + \lambda \int_0^1 R(x, s; \lambda) K(s, t) \, ds. \qquad (6)$$

If $K(x, t)$ is a Volterra kernel, then the intervals of integration in (6) reduce to $t \le s \le x$.

In the classical setting of the theory of integral equations, one is concerned with kernels which are bounded and at least square-integrable.

The representation (4) allows one to eliminate the identity operator and obtain relationships (5) between linear integral operators with kernels (6) of this type. This simplifies the analysis considerably, as the identity operator cannot be represented as a linear integral operator with a bounded kernel on the spaces of continuous or square-integrable functions (Refs. 2–4).

2. Volterra Resolvent Kernels

In case that $K(x, t)$ is a Volterra kernel, (1) takes the form

$$y(x) - \lambda \int_0^x K(x, t)y(t)\, dt = f(x), \qquad 0 \le x \le 1. \tag{7}$$

It is well-known (Refs. 1 and 3–5) that the resolvent kernel $R(x, t; \lambda)$ of $K(x, t)$ always exists, under the assumptions of the classical theory, and is given by the *Neumann series*

$$R(x, t; \lambda) = \sum_{n=1}^{\infty} \lambda^{n-1} K^{(n)}(x, t), \qquad 0 \le t \le x \le 1, \tag{8}$$

where

$$K^{(1)}(x, t) = K(x, t),$$

$$K^{(n+1)}(x, t) = \int_t^x K(x, s)K^{(n)}(s, t)\, ds = \int_t^x K^{(n)}(x, s)K(s, t)\, ds, \tag{9}$$

$n = 1, 2, \ldots .$

The convergence of (8) for all λ with finite modulus is easy to establish by mathematical induction for bounded kernels $K(x, t)$. If

$$|K(x, t)| \le M, \qquad 0 \le t \le x \le 1,$$

then

$$|K^{(n)}(x, t)| \le M \cdot [M^{n-1}/(n-1)!], \qquad n = 1, 2, \ldots, \tag{10}$$

from which the desired result follows. The kernels $K^{(2)}(x, t)$, $K^{(3)}(x, t)$, ... are sometimes called the *iterated kernels* of $K(x, t)$ (see Ref. 6).

It will be useful later to consider also the *transposed* Volterra integral equation corresponding to (7),

$$z(t) - \lambda \int_t^1 z(x)K(x, t)\, dx = g(t), \qquad 0 \le t \le 1. \tag{11}$$

The solution $z(t)$ of (11) is given in terms of $g(t)$, and the resolvent kernel $R(x, t; \lambda)$ of $K(x, t)$ as

$$z(t) = g(t) + \lambda \int_t^1 g(x) R(x, t; \lambda) \, dx, \qquad 0 \le t \le 1. \tag{12}$$

3. Fredholm Resolvent Kernels

For the more general case of a Fredholm kernel $K(x, t)$, an expression for the resolvent kernel is sought in the form

$$R(x, t; \lambda) = N(x, t; \lambda) / \Delta(\lambda), \tag{13}$$

with the numerator and denominator having series expansions

$$N(x, t; \lambda) = \sum_{n=1}^{\infty} \lambda^{n-1} K_n(x, t), \tag{14}$$

and

$$\Delta(\lambda) = 1 + \sum_{n=1}^{\infty} c_n \lambda^n, \tag{15}$$

respectively, which converge for all λ with finite modulus. The resolvent kernel $R(x, t; \lambda)$ will then exist for all values of λ for which the *Fredholm determinant* $\Delta(\lambda)$ of $K(x, t)$ does not vanish. This is analogous to Cramer's rule for the inversion of a finite-dimensional matrix.

Formulas for the so-called *associated kernels* $K_1(x, t), K_2(x, t), \ldots$ of $K(x, t)$ (Ref. 6) appearing in (14) may be obtained by substituting (13)–(15) into the resolvent equations (6). This gives

$$K_1(x, t) = K(x, t),$$

$$K_{n+1}(x, t) = c_n K(x, t) + \int_0^1 K(x, s) K_n(s, t) \, ds$$

$$= c_n K(x, t) + \int_0^1 K_n(x, s) K(s, t) \, ds, \qquad n = 1, 2, \ldots, \tag{16}$$

a result which satisfies (6) formally, independently of the values assigned to $c_1, c_2, \ldots$. For example, if

$$c_1 = c_2 = \cdots = 0,$$

then (14) becomes the Neumann series (8), which does not converge in general for $|\lambda|$ large. In order for (14) and (15) to be entire functions of λ, one chooses

$$c_n = -(1/n) \operatorname{tr} K_n = -(1/n) \int_0^1 K_n(x, x) \, dx, \qquad n = 1, 2, \ldots, \tag{17}$$

where the quantity $\operatorname{tr} K_n$ is called the *trace* of the kernel $K_n(x, t)$. The construction (16)–(17) of the resolvent kernel is in the form given by

Lalesco (Ref. 5). Fredholm (Ref. 7) originally obtained the formulas

$$K_n(x, t) = [(-1)^{n-1}/(n-1)!] \int_0^1 \cdots \int_0^1$$

$$\times \begin{vmatrix} K(s, t) & K(x, s_1) & \cdots & K(x, s_{n-1}) \\ K(s_1, t) & K(s_1, s_1) & \cdots & K(s_1, s_{n-1}) \\ \vdots & \vdots & & \vdots \\ K(s_{n-1}, t) & K(s_{n-1}, s_1) & \cdots & K(s_{n-1}, s_{n-1}) \end{vmatrix} ds_1 \cdots ds_{n-1}, \quad (18)$$

and the similar expressions corresponding to (17) for c_n, $n = 1, 2, \ldots$. The satisfaction of (16) is easily verified by mathematical induction.

Using Fredholm's formulas, the convergence of the series (14) and (15) can be established on the basis of Hadamard's inequality for determinants (see Ref. 4 for an elegant proof) and the ratio test. If

$$|K(x, t)| \leq M,$$

then Hadamard's inequality applied to (18) yields the estimates

$$|K_n(x, t)| \leq M \cdot [M^{n-1} n^{n/2}/(n-1)!], \qquad n = 1, 2, \ldots, \quad (19)$$

and similar bounds for $|c_1|, |c_2|, \ldots$. The rate of convergence which can be predicted for the series (14) and (15) on the basis of (19) is, of course, much slower than that given by the estimates (10) applied to the Neumann series expansion (8) of a Volterra resolvent kernel. For example, if given Volterra and Fredholm kernels are bounded in absolute value by M, then, for $n = 100$, the right-hand side of (19) exceeds that of (10) by a factor of 10^{100}. From a computational point of view, the relationships (16)–(17) would appear to be preferable to the equivalent expressions (18) involving determinants. It has been observed, however, that the formulas corresponding to (16)–(17) for finite-dimensional matrices are unstable numerically (Ref. 8).

Other important relationships which follow from the formulation (13) of the resolvent kernel and the resolvent equations (6) are

$$N(x, t; \lambda) = \Delta(\lambda)K(x, t) + \lambda \int_0^1 K(x, s)N(s, t; \lambda) \, ds,$$

$$N(x, t; \lambda) = \Delta(\lambda)K(x, t) + \lambda \int_0^1 N(x, s; \lambda)K(s, t) \, ds. \quad (20)$$

These are immediately evident for $\Delta(\lambda) \neq 0$. In the framework of the classical theory, they can also be extended to the case that $\lambda = \lambda^*$ is an eigenvalue of the kernel $K(x, t)$; that is,

$$\Delta(\lambda^*) = 0.$$

Assuming that $N(x, t; \lambda^*)$ does not vanish identically, a point (ξ, τ) in the square $0 \le x, t \le 1$ exists such that the functions

$$
\begin{aligned}
y^*(x) &= N(x, \tau; \lambda^*), & 0 \le x \le 1, \\
z^*(t) &= N(\xi, t; \lambda^*), & 0 \le t \le 1,
\end{aligned}
\tag{21}
$$

are also not identically zero. Furthermore, $y^*(x)$ satisfies the *homogeneous* integral equation

$$
y^*(x) = \lambda^* \int_0^1 K(x, t) y^*(t) \, dt, \qquad 0 \le x \le 1, \tag{22}
$$

and is said to be a *right eigenfunction* of $K(x, t)$ corresponding to the eigenvalue λ^*. Similarly, the function $z^*(t)$ satisfies the transposed homogeneous integral equation

$$
z^*(t) = \lambda^* \int_0^1 z^*(x) K(x, t) \, dx, \qquad 0 \le t \le 1, \tag{23}
$$

and is called a *left eigenfunction* of $K(x, t)$ corresponding to λ^*.

4. Symmetric Separable Kernels

Attention will now be devoted to the construction of resolvent kernels of a special class of Fredholm kernels. In the symmetric case, a kernel of the form

$$
G(x, t) = \begin{cases} u(t)v(x), & 0 \le t \le x \le 1, \\ u(x)v(t), & 0 \le x \le t \le 1, \end{cases} \tag{24}
$$

will be called a *simple separable kernel*. It is assumed that the functions $u(x)$ and $v(x)$ are linearly independent; otherwise, $G(x, t)$ would be a *degenerate* kernel of rank one (Ref. 3, pp. 37–40). In general, a *symmetric separable kernel* is a finite sum of linearly independent kernels (24).

Before dealing with the general case, the resolvent kernel of the symmetric simple separable kernel (24) will be constructed. To do this, the Fredholm integral equation

$$
y(x) - \lambda \int_0^1 G(x, t) y(t) \, dt = f(x), \qquad 0 \le x \le 1, \tag{25}
$$

will be solved. This is essentially the approach used by Drukarev (Ref. 9), Brysk (Ref. 10), and Aalto (Ref. 11). Using the definition (24) of $G(x, t)$,

Eq. (25) can be written as

$$y(x)-\lambda \int_0^x u(t)v(x)y(t)\,dt = f(x)+\lambda \int_x^1 u(x)v(t)y(t)\,dt. \qquad (26)$$

Adding the quantity

$$\lambda \int_0^x u(x)v(t)y(t)\,dt$$

to both sides of (26) gives

$$y(x)-\lambda \int_0^x K(x,t)y(t)\,dt = f(x)+\lambda cu(x), \qquad (27)$$

where

$$c = \int_0^1 v(t)y(t)\,dt \qquad (28)$$

is to be determined, and $K(x,t)$ is the Volterra kernel

$$K(x,t)=u(t)v(x)-u(x)v(t), \qquad 0\le t\le x\le 1. \qquad (29)$$

The system of equations (27)–(29) is easily seen to be equivalent to the original integral equation (25).

As was shown in Section 2, the kernel $K(x,t)$ has the Volterra resolvent kernel $R(x,t;\lambda)$, given by (8) for all λ. Define

$$F(x)=f(x)+\lambda \int_0^x R(x,t;\lambda)f(t)\,dt,$$
$$U(x)=u(x)+\lambda \int_0^x R(x,t;\lambda)u(t)\,dt, \qquad (30)$$

where the dependence of $F(x)$ and $U(x)$ on λ has been suppressed for clarity of notation. From (27),

$$y(x)=F(x)+\lambda cU(x); \qquad (31)$$

and, from (28),

$$c = \int_0^1 v(t)F(t)\,dt +\lambda c \int_0^1 v(t)U(t)\,dt. \qquad (32)$$

Thus, (32) has a unique solution for c if

$$\Delta(\lambda)\equiv 1-\lambda \int_0^1 v(t)U(t)\,dt \neq 0, \qquad (33)$$

in which case

$$c = [1/\Delta(\lambda)] \int_0^1 v(t)F(t)\,dt. \tag{34}$$

The expression (34) for c may be written in terms of $f(t)$ by introducing the function

$$V(t) = v(t) + \lambda \int_t^1 v(x)R(x, t; \lambda)\,dx. \tag{35}$$

It then follows from (30) and (34) that

$$c = [1/\Delta(\lambda)] \int_0^1 V(t)f(t)\,dt; \tag{36}$$

and, from (31),

$$y(x) = f(x) + \lambda \int_0^x R(x, t; \lambda)f(t)\,dt + [\lambda U(x)/\Delta(\lambda)] \int_0^1 V(t)f(t)\,dt. \tag{37}$$

By definition, the resolvent kernel $\Gamma(x, t; \lambda)$ of $G(x, t)$ can be obtained directly from (37) as

$$\Gamma(x, t; \lambda) = \begin{cases} R(x, t; \lambda) + [1/\Delta(\lambda)]U(x)V(t), & 0 \le t \le x \le 1, \\ [1/\Delta(\lambda)]U(x)V(t), & 0 \le x \le t \le 1, \end{cases} \tag{38}$$

provided, of course, that

$$\Delta(\lambda) \ne 0.$$

Another expression for $\Gamma(x, t; \lambda)$ can be obtained by making use of the fact that the resolvent kernel of a symmetric kernel $G(x, t)$ must also be symmetric in x and t; that is,

$$\Gamma(x, t; \lambda) = \Gamma(t, x; \lambda).$$

This gives

$$\Gamma(x, t; \lambda) = \begin{cases} [1/\Delta(\lambda)]U(t)V(x), & 0 \le t \le x \le 1, \\ R(t, x; \lambda) + [1/\Delta(\lambda)]U(t)V(x), & 0 \le x \le t \le 1, \end{cases} \tag{39}$$

which, when compared with (38), yields

$$\Delta(\lambda)R(x, t; \lambda) = U(t)V(x) - U(x)V(t), \qquad 0 \le t \le x \le 1, \tag{40}$$

and finally,

$$\Gamma(x, t; \lambda) = \begin{cases} [1/\Delta(\lambda)]U(t)V(x), & 0 \le t \le x \le 1, \\ [1/\Delta(\lambda)]U(x)V(t), & 0 \le x \le t \le 1. \end{cases} \tag{41}$$

These results give rise to the following observations.

Remark 4.1. Equation (41) shows that if the resolvent kernel of a symmetric simple separable kernel (24) exists, then it is also a symmetric simple separable kernel.

Remark 4.2. The expression (41) for the resolvent kernel $\Gamma(x, t; \lambda)$ of $G(x, t)$ can be obtained without finding the Volterra resolvent kernel $R(x, t; \lambda)$ of $K(x, t)$ explicitly; one need only solve the Volterra integral equation (7) with

$$f(x) = u(x)$$

for

$$y(x) = U(x),$$

and the transposed Volterra integral equation (11) with

$$g(t) = v(t)$$

for

$$z(t) = V(t).$$

Remark 4.3. An equivalent expression for the Fredholm determinant $\Delta(\lambda)$ of $G(x, t)$ is

$$\Delta(\lambda) = 1 - \lambda \int_0^1 V(x)u(x)\, dx, \tag{42}$$

which can be obtained directly from (33) by interchange of the order of integration and use of the definition (35) of $V(t)$. It is easy to show that the above expressions give expansions of $\Delta(\lambda)$ and $N(x, t; \lambda)$ in powers of λ with coefficients satisfying the relationships (16)–(17) (see Refs. 11–12). The advantage of the present approach is that the rate of convergence of the series for $\Delta(\lambda)$,

$$U(x) = U(x; \lambda),$$

and

$$V(t) = V(t; \lambda)$$

can be predicted on the basis of (10), rather than (19).

Remark 4.4. The definitions of the Volterra kernel $K(x, t)$ and its resolvent kernel $R(x, t; \lambda)$ extend to the entire square $0 \le x, t \le 1$, with

$$K(x, t) = 0, \qquad R(x, t; \lambda) = 0, \qquad 0 \le x \le t \le 1. \tag{43}$$

This will result in continuous kernels for $u(x)$, $v(t)$ continuous, as

$$K(x, x) = R(x, x; \lambda) = 0, \qquad 0 \le x \le 1.$$

In terms of these extended kernels, one may write

$$G(x, t) = K(x, t) + u(x)v(t),$$
$$\Gamma(x, t; \lambda) = R(x, t; \lambda) + [1/\Delta(\lambda)]U(x)V(t). \tag{44}$$

Thus, the kernel $G(x, t)$ given by (24) is the sum of the Volterra kernel $K(x, t)$ and the degenerate kernel $u(x)v(t)$ of rank one. If

$$\Delta(\lambda) \ne 0,$$

then its resolvent kernel $\Gamma(x, t; \lambda)$ exists and is also a rank-one modification of the Volterra resolvent kernel $R(x, t; \lambda)$ of $K(x, t)$. By symmetry, one also has

$$G(x, t) = K(t, x) + u(t)v(x),$$
$$\Gamma(x, t; \lambda) = R(t, x; \lambda) + [1/\Delta(\lambda)]U(t)V(x). \tag{45}$$

In some applications, the expressions (44) or (45) for $G(x, t)$ arise more naturally than (24).

The next case to be examined is that

$$\lambda = \lambda^*$$

is an eigenvalue of $G(x, t)$; that is,

$$\Delta(\lambda^*) = 0.$$

Noting that

$$U(x) = U(x; \lambda) \qquad \text{and} \quad V(t) = V(t; \lambda)$$

depend on λ, define

$$U^*(x) = U(x; \lambda^*), \qquad 0 \le x \le 1,$$
$$V^*(t) = V(t; \lambda^*), \qquad 0 \le t \le 1. \tag{46}$$

From (40), which can be extended easily to $\lambda = \lambda^*$, it follows that the functions $U^*(x)$ and $V^*(x)$ are linearly dependent. Hence, for

$$\lambda = \lambda^*,$$

the Fredholm numerator $N(x, t; \lambda)$ of $\Gamma(x, t; \lambda)$ becomes

$$N(x, t; \lambda^*) = \alpha U^*(x)V^*(t), \tag{47}$$

where $\alpha \ne 0$ is some constant. Thus, by (21)–(23), $U^*(x)$ will be a right eigenfunction of $G(x, t)$ corresponding to λ^*, and $V^*(t)$ is a corresponding left eigenfunction. As $G(x, t)$ is symmetric, the distinction between left and right eigenfunctions is inconsequential. However, it will be shown later that a similar approach gives these as distinct functions in the nonsymmetric case.

The implications of the above results for the solvability of the integral equation (25) may be summed up in the following familiar language.

Theorem 4.1. *Fredholm Alternative.* If the transposed homogeneous integral equation

$$z(t) - \lambda \int_0^1 z(x)G(x, t)\, dx = 0, \qquad 0 \le t \le 1, \tag{48}$$

has only the *trivial solution* $z(t) = 0$, $0 \le t \le 1$, then Eq. (25) has the unique solution

$$y(x) = F(x) + [\lambda U(x)/\Delta(\lambda)] \int_0^1 v(t)F(t)\, dt, \qquad 0 \le x \le 1. \tag{49}$$

On the other hand, if (48) has the nontrivial solution

$$z(t) = V^*(t),$$

then (25) has a solution only if $f(x)$ is *orthogonal* to $V^*(x)$, that is,

$$\int_0^1 V^*(t)f(t)\, dt = 0. \tag{50}$$

If (50) is satisfied, then (25) has the solutions

$$y(x) = F(x) + \alpha U^*(x), \tag{51}$$

where α is arbitrary, and

$$y(x) = U^*(x)$$

is a nontrivial solution of the homogeneous integral equation

$$y(x) - \lambda \int_0^1 G(x, t)y(t)\, dt = 0, \qquad 0 \le x \le 1. \tag{52}$$

As derived above, of course, Theorem 4.1 applies only to simple kernels (24). The same technique, however, applies to general symmetric separable kernels. Suppose that

$$G(x, t) = \sum_{j=1}^n G_j(x, t) = \begin{cases} \displaystyle\sum_{j=1}^n u_j(t)v_j(x), & 0 \le t \le x \le 1, \\[2ex] \displaystyle\sum_{j=1}^n u_j(x)v_j(t), & 0 \le x \le t \le 1. \end{cases} \tag{53}$$

Then, the integral equation (25) with the symmetric separable kernel (53) may be reduced to

$$y(x)-\lambda \int_0^x K(x, t)y(t)\, dt = f(x)+\lambda \sum_{j=1}^n c_j u_j(x), \tag{54}$$

where the numbers

$$c_j = \int_0^1 v_j(t)y(t)\, dt, \qquad j=1, 2, \ldots, n, \tag{55}$$

are to be determined, and $K(x, t)$ is the Volterra kernel

$$K(x, t) = \sum_{j=1}^n [u_j(t)v_j(x)-u_j(x)v_j(t)], \qquad 0 \le t \le x \le 1. \tag{56}$$

As before, let $R(x, t; \lambda)$ denote the Volterra resolvent kernel of $K(x, t)$, and define

$$F(x)=f(x)+\lambda \int_0^1 R(x, t; \lambda)f(t)\, dt,$$

$$U_i(x)=u_i(x)+\lambda \int_0^1 R(x, t; \lambda)u_i(t)\, dt, \qquad i=1, 2, \ldots, n. \tag{57}$$

Equation (54) is then equivalent to

$$y(x)=F(x)+\lambda \sum_{i=1}^n c_i U_i(x). \tag{58}$$

Multiplying (58) by $v_1(x), v_2(x), \ldots, v_n(x)$ in turn and integrating with respect to x from 0 to 1 gives the system of equations

$$c_i - \lambda \sum_{j=1}^n c_j \int_0^1 v_i(x)U_j(x)\, dx = \int_0^1 v_i(x)F(x)\, dx, \qquad i=1, 2, \ldots, n, \tag{59}$$

for the unknowns $c_1, c_2, \ldots, c_n$ defined by (55).

Remark 4.5. It follows from (55), (58), and (59) that the solution of the integral equation (25) with the symmetric separable kernel (53) is equivalent to solving the Volterra integral equation (7) with kernel (56) and right-hand sides $f(x)$, $u_1(x), \ldots, u_n(x)$ for $F(x)$, $U_1(x), \ldots, U_n(x)$, forming the n^2+n coefficients of the system (59), and then solving this system of n linear equations for the n unknowns $c_1, c_2, \ldots, c_n$.

It is also possible to obtain explicit expressions for the Fredholm determinant $\Delta(\lambda)$ of $G(x, t)$ and its resolvent kernel $\Gamma(x, t; \lambda)$ in terms of the determinant and the inverse of the coefficient matrix of the system (59).

It will be convenient to introduce the functions

$$V_i(t) = v_i(t) + \lambda \int_t^1 v_i(x) R(x, t; \lambda) \, dx, \qquad i = 1, 2, \ldots, n, \qquad (60)$$

which are the solutions of the transposed Volterra integral equations (11) with right-hand sides $v_1(t), v_2(t), \ldots, v_n(t)$. Let the coefficients

$$\alpha_{ij} = \int_0^1 v_i(x) U_j(x) \, dx = \int_0^1 V_i(t) u_j(t) \, dt, \qquad i, j = 1, 2, \ldots, n, \qquad (61)$$

define the $n \times n$ matrix

$$A = [\alpha_{ij}].$$

Then, the coefficient matrix of the system (59) has the form

$$I - \lambda A = (\delta_{ij} - \lambda \alpha_{ij}), \qquad (62)$$

where δ_{ij} is the *Kronecker delta*,

$$\delta_{ij} = 0 \qquad \text{if } i \neq j, \qquad \delta_{ii} = 1,$$

and

$$I = [\delta_{ij}]$$

is the $n \times n$ identity matrix. If the determinant

$$\Delta(\lambda) = \det(I - \lambda A) \qquad (63)$$

does not vanish, then the inverse of the matrix $I - \lambda A$ exists, and can be written as

$$(I - \lambda A)^{-1} = [1/\Delta(\lambda)] B(\lambda) = [\beta_{ij}(\lambda)/\Delta(\lambda)], \qquad (64)$$

by the use of Cramer's rule. In this case, the system (59) has the unique solutions

$$c_i = [1/\Delta(\lambda)] \sum_{j=1}^n \beta_{ij}(\lambda) \int_0^1 v_j(x) F(x) \, dx$$

$$= [1/\Delta(\lambda)] \sum_{j=1}^n \beta_{ij}(\lambda) \int_0^1 V_j(t) f(t) \, dt, \qquad (65)$$

for $i = 1, 2, \ldots, n$. Thus, the integral equation (25) will have the unique solution $y(x)$ given by (58), which may be written in the form

$$y(x) = f(x) + \lambda \int_0^x R(x, t; \lambda) f(t) \, dt$$

$$+ \lambda \sum_{i=1}^n \sum_{j=1}^n U_i(x) [\beta_{ij}(\lambda)/\Delta(\lambda)] \int_0^1 V_j(t) f(t) \, dt. \qquad (66)$$

From (66), it is possible to derive several expressions for the resolvent kernel $\Gamma(x, t; \lambda)$ of the symmetric separable kernel $G(x, t)$, provided that the Fredholm determinant $\Delta(\lambda)$ of $G(x, t)$ is nonzero. Extending the Volterra kernel $K(x, t)$ and its resolvent kernel $R(x, t; \lambda)$ to the entire square $0 \le x, t \le 1$, as before, one may write

$$G(x, t) = K(x, t) + \sum_{j=1}^{n} u_j(x)v_j(t),$$

$$\Gamma(x, t; \lambda) = R(x, t; \lambda) + [1/\Delta(\lambda)] \sum_{i=1}^{n} \sum_{j=1}^{n} U_i(x)\beta_{ij}(\lambda)V_j(t).$$

(67)

Remark 4.6. The symmetric separable kernel (53) is a rank n modification of the Volterra kernel (56). If

$$\Delta(\lambda) \ne 0,$$

the resolvent kernel $\Gamma(x, t; \lambda)$ is likewise a rank n modification of the Volterra resolvent kernel $R(x, t; \lambda)$ of $K(x, t)$.

Integral operators corresponding to degenerate kernels of rank n are sometimes called *n-term dyads* (Ref. 2). By symmetry,

$$G(x, t) = K(t, x) + \sum_{j=1}^{n} u_j(t)v_j(x),$$

$$\Gamma(x, t; \lambda) = R(t, x; \lambda) + [1/\Delta(\lambda)] \sum_{i=1}^{n} \sum_{j=1}^{n} U_i(t)\beta_{ij}(\lambda)V_j(x).$$

(68)

Comparison of (68) and (67) gives

$$\Delta(\lambda)R(x, t; \lambda) = \sum_{i=1}^{n} \sum_{j=1}^{n} \beta_{ij}(\lambda)[U_i(t)V_j(x) - U_i(x)V_j(t)],$$

(69)

and thus $\Gamma(x, t; \lambda)$ may also be written in the form

$$\Gamma(x, t; \lambda) = \begin{cases} \sum_{i=1}^{n} \sum_{j=1}^{n} [\beta_{ij}(\lambda)/\Delta(\lambda)]U_i(t)V_j(x), & 0 \le t \le x \le 1, \\ \sum_{i=1}^{n} \sum_{j=1}^{n} [\beta_{ij}(\lambda)/\Delta(\lambda)]U_i(x)V_j(t), & 0 \le x \le t \le 1. \end{cases}$$

(70)

Remark 4.7. If it exists, the Fredholm resolvent kernel $\Gamma(x, t; \lambda)$ of the symmetric separable kernel (53) is likewise the sum of n symmetric simple separable kernels.

For $n > 1$, there are a number of ways in which the kernel (70) may be written in the form (53). For example, defining the functions

$$\Psi_j(x) = \sum_{i=1}^{n} U_i(x)\beta_{ij}(\lambda), \qquad j = 1, 2, \ldots, n,$$

$$\Phi_i(t) = \sum_{j=1}^{n} \beta_{ij}(\lambda)V_j(t), \qquad i = 1, 2, \ldots, n,$$

(71)

one obtains the equivalent representations

$$\Gamma(x, t; \lambda) = \begin{cases} [1/\Delta(\lambda)] \sum_{j=1}^{n} \Psi_j(t)V_j(x), & 0 \le t \le x \le 1, \\[2mm] [1/\Delta(\lambda)] \sum_{j=1}^{n} \Psi_j(x)V_j(t), & 0 \le x \le t \le 1, \end{cases}$$

(72)

and

$$\Gamma(x, t; \lambda) = \begin{cases} [1/\Delta(\lambda)] \sum_{i=1}^{n} U_i(t)\Phi_i(x), & 0 \le t \le x \le 1, \\[2mm] [1/\Delta(\lambda)] \sum_{i=1}^{n} U_i(x)\Phi_i(t), & 0 \le x \le t \le 1, \end{cases}$$

(73)

for the resolvent kernel of $G(x, t)$.

Suppose now that

$$\lambda = \lambda^*$$

is an eigenvalue of the kernel (53), that is,

$$\Delta(\lambda^*) = 0.$$

By the same method as used in the nonhomogeneous case, the homogeneous equation (52) can be shown to be equivalent to the system

$$y(x) = \lambda^* \sum_{i=1}^{n} c_i U_i(x),$$

$$c_i - \lambda^* \sum_{j=1}^{n} \alpha_{ij}c_j = 0, \qquad i = 1, 2, \ldots, n,$$

(74)

corresponding to

$$F(x) = 0$$

in (58) and (59). As

$$\Delta(\lambda^*) = 0,$$

the homogeneous system

$$(I - \lambda^* A)c = 0 \tag{75}$$

has $m \leq n$ linearly independent solutions

$$c^{(k)} = (c_1^{(k)}, c_2^{(k)}, \ldots, c_n^{(k)})^T, \qquad k = 1, 2, \ldots, m. \tag{76}$$

Corresponding to these solutions, which are right eigenvectors of the matrix

$$A = [\alpha_{ij}]$$

corresponding to the reciprocal eigenvalue λ^*, one obtains m linearly independent eigenfunctions

$$y_k^*(x) = \lambda^* \sum_{i=1}^{n} c_i^{(k)} U_i(x), \qquad k = 1, 2, \ldots, n, \tag{77}$$

of the kernel $G(x, t)$ corresponding to the eigenvalue λ^*. Similarly, the transposed homogeneous equation (48) for $\lambda = \lambda^*$ has m linearly independent solutions

$$z_k^*(t) = \lambda^* \sum_{i=1}^{n} d_j^{(k)} V_i(t), \qquad k = 1, 2, \ldots, m, \tag{78}$$

corresponding to the m linearly independent solutions

$$d^{(k)} = (d_1^{(k)}, d_2^{(k)}, \ldots, d_n^{(k)}), \qquad k = 1, 2, \ldots, m, \tag{79}$$

of the transposed homogeneous system

$$d(I - \lambda^* A) = 0, \tag{80}$$

that is,

$$d_j - \lambda^* \sum_{i=1}^{n} d_i \alpha_{ij} = 0, \qquad j = 1, 2, \ldots, n. \tag{81}$$

It follows from (59) that the nonhomogeneous integral equation (25) will have no solutions unless the orthogonality conditions

$$\sum_{j=1}^{n} d_j^{(k)} \int_0^1 v_j(x) F(x) \, dx = \int_0^1 \sum_{j=1}^{n} d_j^{(k)} V_j(t) f(t) \, dt = \int_0^1 z_k^*(t) f(t) \, dt = 0 \tag{82}$$

hold for $k = 1, 2, \ldots, m$; that is, the right-hand side of (25) must be orthogonal to all solutions of the transposed homogeneous equation (48) with

$$\lambda = \lambda^*.$$

If (82) holds, then (25) is satisfied by the family of solutions

$$y(x) = f(x) + \lambda^* \int_0^x R(x, t; \lambda^*) f(t) \, dt + \sum_{k=1}^{m} \gamma_k y_k^*(x), \qquad (83)$$

with $\gamma_1, \gamma_2, \ldots, \gamma_m$ arbitrary.

The above results can be stated as the corresponding generalization of the Fredholm alternative theorem to kernels (53) with $n > 1$. Another method for computing resolvent kernels of separable kernels (symmetric or not) will be indicated in a later section.

5. Applications to Boundary-Value Problems

Symmetric separable kernels appear frequently as *Green's functions* for two-point boundary-value problems for ordinary differential operators (Ref. 13). More precisely, suppose that $L[\,\cdot\,]$ is a linear ordinary differential operator, and a solution $y(x)$ of the differential equation

$$L[y(x)] = h(x) \qquad (84)$$

is sought which, together with its derivatives of lower order than the order of $L[\,\cdot\,]$, satisfies given conditions at

$$x = 0 \quad \text{and} \quad x = 1.$$

If this boundary-value problem has a unique solution $y(x)$ which can be represented as

$$y(x) = \int_0^1 G(x, t) h(t) \, dt \qquad (85)$$

for all functions $h(x)$ from some class such as continuous functions, then $G(x, t)$ is said to be the *Green's function* for $L[\,\cdot\,]$ corresponding to the given boundary conditions. If

$$h(x) = \lambda y(x) + g(x), \qquad (86)$$

then the boundary-value problem for the differential equation (84) is equivalent to the Fredholm integral equation

$$y(x) - \lambda \int_0^1 G(x, t) y(t) \, dt = f(x), \qquad 0 \le x \le 1, \qquad (87)$$

where

$$f(x) = \int_0^1 G(x, t) g(t) \, dt, \qquad 0 \le x \le 1. \qquad (88)$$

Thus, the techniques of Section 4 apply to the solution of (87) if $G(x, t)$ is a symmetric separable kernel; they also apply to finding eigenvalues and eigenfunctions in the homogeneous case

$$g(x) \equiv 0.$$

In many cases, it is possible to proceed directly from the boundary-value problem to representations for the Green's function, its Fredholm determinant, resolvent kernel, and eigenfunctions. For example, consider the *Sturm–Liouville* operator (Ref. 13)

$$L[y(x)] = -(p(x)y'(x))', \tag{89}$$

with $p(x) > 0$, subject to the boundary conditions

$$\begin{aligned} ap(0)y'(0) + by(0) &= 0, \\ cp(1)y'(1) + dy(1) &= 0. \end{aligned} \tag{90}$$

A simple way to find the Green's function for this problem is to start directly from the differential equation

$$(p(x)y'(x))' = -h(x). \tag{91}$$

One integration gives

$$p(x)y'(x) = p(0)y'(0) - \int_0^x h(t)\, dt. \tag{92}$$

Dividing (92) by $p(x)$ and integrating again yields

$$y(x) = y(0) + p(0)y'(0) \int_0^x [1/p(t)]\, dt - \int_0^x \int_0^s [h(t)/p(s)]\, dt\, ds. \tag{93}$$

By defining

$$F(x) = \int_0^x [1/p(t)]\, dt, \tag{94}$$

and noting that change of order of integration results in

$$\int_0^x \int_0^s [h(t)/p(s)]\, dt\, ds = \int_0^x \left\{ \int_t^x [1/p(s)]\, ds \right\} h(t)\, dt = \int_0^x [F(x) - F(t)]h(t)\, dt, \tag{95}$$

one obtains

$$y(x) = y(0) + p(0)y'(0)F(x) - \int_0^x [F(x) - F(t)]h(t)\, dt \tag{96}$$

from (93). The boundary conditions (90) will now be used to express $y(0)$ and $p(0)y'(0)$ in terms of integral transforms of $h(x)$. From (92) and (96),

$$p(1)y'(1)=p(0)y'(0)-\int_0^1 h(t)\,dt,$$

$$y(1)=y(0)+p(0)y'(0)F(1)-\int_0^1 [F(1)-F(t)]h(t)\,dt. \tag{97}$$

Multiplying the first equation of (97) by c, the second by d, and adding the results gives

$$0=dy(0)+\gamma p(0)y'(0)-\int_0^1 [\gamma-dF(t)]h(t)\,dt, \tag{98}$$

by the second of the boundary conditions (90), where

$$\gamma=c+dF(1). \tag{99}$$

Now, multiplying (98) by a and using the first boundary condition of (90) results in

$$0=(ad-b\gamma)y(0)-\int_0^1 a[\gamma-dF(t)]h(t)\,dt; \tag{100}$$

or, if

$$\delta=ad-b\gamma\neq 0, \tag{101}$$

then

$$y(0)=\int_0^1 (a/\delta)[\gamma-dF(t)]h(t)\,dt. \tag{102}$$

Similarly, multiplication of (98) by b and use of the first boundary condition gives

$$p(0)y'(0)=-\int_0^1 (b/\delta)[\gamma-dF(t)]h(t)\,dt. \tag{103}$$

Equations (102) and (103) may be substituted into (96) to obtain

$$y(x)=-\int_0^x [F(x)-F(t)]h(t)\,dt+\int_0^1 (1/\delta)[a-bF(x)][\gamma-dF(t)]h(t)\,dt. \tag{104}$$

Comparison of (104) with (85) gives the following results.

Remark 5.1. If $\delta \neq 0$, then the Green's function $G(x, t)$ of the Sturm–Liouville problem (89)–(90) is the rank-one modification

$$G(x, t) = K(x, t) + u(x)v(t) \tag{105}$$

of the Volterra kernel

$$K(x, t) = -[F(x) - F(t)], \qquad 0 \leq t \leq x \leq 1, \tag{106}$$

with

$$u(x) = (1/\delta)[a - bF(x)],$$
$$v(t) = \gamma - dF(t). \tag{107}$$

Remark 5.2. If $\delta \neq 0$, then the Green's function (105) is the symmetric simple separable kernel

$$G(x, t) = \begin{cases} (1/\delta)[a - bF(t)][\gamma - dF(x)], & 0 \leq t \leq x \leq 1, \\ (1/\delta)[a - bF(x)][\gamma - dF(t)], & 0 \leq x \leq t \leq 1. \end{cases} \tag{108}$$

This follows directly from (105)–(106) and the simple calculation

$$[a - bF(x)][\gamma - dF(t)] - \delta F(x) + \delta F(t) = [a - bF(t)][\gamma - dF(x)]. \tag{109}$$

Thus, the evaluation of the Fredholm determinant $\Delta(\lambda)$ of $G(x, t)$ and the calculation of its resolvent kernel and eigenfunctions depends only on being able to calculate the functions $U(x)$, $V(t)$ either in terms of the Volterra resolvent kernel $R(x, t; \lambda)$ of the kernel (106), or directly by solving the Volterra integral equations cited in Remark 4.2.

For certain simple examples, it is possible to give explicit formulas for these results. Taking

$$p(x) \equiv 1$$

gives

$$L[y(x)] = -y''(x), \tag{110}$$

and hence

$$F(x) = x, \qquad K(x, t) = -(x - t), \tag{111}$$

from which

$$R(x, t; \lambda) = -\sin[\sqrt{\lambda}(x - t)]/\sqrt{\lambda}. \tag{112}$$

The boundary conditions

$$y(0) = 0, \qquad y'(1) = 0 \tag{113}$$

correspond to (90) with

$$a = 0, \qquad b = 1, \qquad c = 1, \qquad d = 0,$$

and hence

$$\gamma = 1, \qquad \delta = -1,$$

and thus

$$u(x) = x, \qquad v(t) = 1. \tag{114}$$

It follows that the Green's function $G(x, t)$ for the differential operator (110) with boundary conditions (113) is

$$G(x, t) = -(x - t)_+ + x, \tag{115}$$

where

$$(x - t)_+ = 0 \qquad \text{for } 0 \le x \le t \le 1,$$

or

$$G(x, t) = \begin{cases} t, & 0 \le t \le x \le 1, \\ x, & 0 \le x \le t \le 1. \end{cases} \tag{116}$$

Using (112) and (114), one obtains

$$U(x) = \sin(\sqrt{\lambda}x)/\sqrt{\lambda}, \qquad V(t) = \cos[\sqrt{\lambda}(1-t)], \tag{117}$$

from which the Fredholm determinant $\Delta(\lambda)$ of $G(x, t)$ is seen to be

$$\Delta(\lambda) = \cos \sqrt{\lambda}. \tag{118}$$

If

$$\Delta(\lambda) \ne 0,$$

then the resolvent kernel $\Gamma(x, t; \lambda)$ of $G(x, t)$ may be written as

$$\Gamma(x, t; \lambda) = -\sin[\sqrt{\lambda}(x-t)_+]/\sqrt{\lambda} + \sin(\sqrt{\lambda}x)\cos[\sqrt{\lambda}(1-t)]/\sqrt{\lambda}\cos\sqrt{\lambda}, \tag{119}$$

corresponding to (115), or as the symmetric simple separable kernel

$$\Gamma(x, t; \lambda) = \begin{cases} \sin(\sqrt{\lambda}t)\cos[\sqrt{\lambda}(1-x)]/\sqrt{\lambda}\cos\sqrt{\lambda}, & 0 \le t \le x \le 1, \\ \sin(\sqrt{\lambda}x)\cos[\sqrt{\lambda}(1-t)]/\sqrt{\lambda}\cos\sqrt{\lambda}, & 0 \le x \le t \le 1, \end{cases} \tag{120}$$

of the same form as (116). It also follows immediately from (118) that the eigenvalues of $G(x, t)$ are

$$\lambda_n^* = \{[(2n-1)/2]\pi\}^2, \qquad n = 1, 2, 3, \ldots, \tag{121}$$

and the corresponding eigenfunctions are proportional to

$$\sqrt{\lambda_n^*}U_n^*(x) = \sin[(2n-1)\pi x/2], \qquad n = 1, 2, 3, \dots. \qquad (122)$$

It is also possible to write down explicit formulas for the more general second-order boundary-value problem

$$L[y(x)] = -y''(x),$$
$$ay'(0) + by(0) = 0, \qquad (123)$$
$$cy'(1) + dy(1) = 0,$$

by the use of (107), (111), and (112). The Green's function $G(x, t)$ is obtained from

$$u(x) = (1/\delta)[a - bx],$$
$$v(t) = c + d(1-t), \qquad (124)$$

provided

$$\delta = ad - b(c + d) \neq 0,$$

the resolvent kernel $\Gamma(x, t; \lambda)$ of $G(x, t)$ from

$$U(x) = (1/\delta)[a \cos(\sqrt{\lambda}x) - (b/\sqrt{\lambda}) \sin(\sqrt{\lambda}x)],$$
$$V(t) = c \cos[\sqrt{\lambda}(1-t)] + (d/\sqrt{\lambda}) \sin[\sqrt{\lambda}(1-t)], \qquad (125)$$

and the Fredholm determinant is

$$\Delta(\lambda) = (1/\delta)[(ad - bc) \cos\sqrt{\lambda} - (ac\sqrt{\lambda} + bd/\sqrt{\lambda}) \sin\sqrt{\lambda}]. \qquad (126)$$

The eigenvalues of $G(x, t)$ may thus be found by solving the simple transcendental equation

$$\tan\sqrt{\lambda}/\sqrt{\lambda} = (ad - bc)/(\lambda ac + bd), \qquad (127)$$

for $\lambda_1, \lambda_2, \dots$, and the results substituted into (125) to obtain the corresponding eigenfunctions. To simplify these expressions, one may introduce the angles

$$\phi = \phi(\lambda) = \tan^{-1}(b/a\sqrt{\lambda}), \qquad \theta = \theta(\lambda) = \tan^{-1}(c\sqrt{\lambda}/d), \qquad (128)$$

which gives

$$U(x) = (1/\delta)\sqrt{(a^2 + b^2/\lambda)} \cos(\sqrt{\lambda}x + \phi),$$
$$V(t) = \sqrt{(c^2 + d^2/\lambda)} \sin[\sqrt{\lambda}(1-t) + \theta], \qquad (129)$$

and

$$\Delta(\lambda) = (1/\delta)\sqrt{[(a^2\lambda + b^2)(c^2\lambda + d^2)/\lambda]} \cos(\sqrt{\lambda} + \phi + \theta). \qquad (130)$$

Thus,

$$\Gamma(x, t; \lambda) = \begin{cases} \cos(\sqrt{\lambda}t + \phi) \sin[\sqrt{\lambda}(1 - x) + \theta]/\sqrt{\lambda} \cos(\sqrt{\lambda} + \phi + \theta), \\ \qquad\qquad 0 \le t \le x \le 1, \\ \cos(\sqrt{\lambda}x + \phi) \sin[\sqrt{\lambda}(1 - t) + \theta]/\sqrt{\lambda} \cos(\sqrt{\lambda} + \phi + \theta), \\ \qquad\qquad 0 \le x \le t \le 1, \end{cases} \tag{131}$$

provided, of course, that

$$\Delta(\lambda) \ne 0.$$

From (130), the eigenvalues $\lambda_1, \lambda_2, \ldots$ must satisfy

$$\sqrt{\lambda_n} + \phi(\lambda_n) + \theta(\lambda_n) = [(2n - 1)/2]\pi, \qquad n = 1, 2, \ldots, \tag{132}$$

or

$$\sqrt{\lambda_n} + \cot^{-1}[(ad - bc)\sqrt{\lambda_n}/(\lambda_n ac + bd)] = [(2n - 1)/2]\pi, \qquad n = 1, 2, \ldots, \tag{133}$$

which is equivalent to (127).

The method given above extends readily to two-point boundary-value problems of arbitrary order. In general, the Volterra kernel $K(x, t)$ is obtained by integrating (84) as an initial-value problem with zero initial conditions; for example, for

$$L[y(x)] = -y^{iv}(x), \tag{134}$$

one obtains

$$K(x, t) = -(x - t)^3/3!, \qquad 0 \le t \le x \le 1. \tag{135}$$

The functions $u_1(x), \ldots, u_n(x)$ and $v_1(t), \ldots, v_n(t)$ which give the Green's function $G(x, t)$ as a finite-rank modification of $K(x, t)$ are then found by solving for the initial conditions, in terms of integral transforms of $h(x)$, to satisfy the given boundary conditions, it being assumed that the resulting system of equations has a unique solution. Of course, only self-adjoint boundary-value problems give rise to symmetric Green's functions. In the next section, nonsymmetric separable kernels will be discussed.

For *nonlinear boundary-value problems*,

$$h(x) = f(x, y(x)), \tag{136}$$

the use of the Green's function $G(x, t)$ leads to a *Hammerstein integral equation*

$$y(x) - \int_0^1 G(x, t)f(t, y(t))\, dt = 0. \tag{137}$$

If $G(x, t)$ is of the form (24), then (137) is equivalent to the nonlinear system

$$y(x) - \int_0^x K(x, t) f(t, y(t))\, dt = \alpha u(x),$$

$$\alpha = \int_0^1 v(t) f(t, y(t))\, dt. \tag{138}$$

The first equation of (138) is a nonlinear Volterra integral equation. If this can be solved for

$$y(x) = y(x; \alpha), \tag{139}$$

then substitution into the second equation gives the single nonlinear scalar equation

$$\alpha = \int_0^1 v(t) f(t, y(t; \alpha))\, dt = \phi(\alpha) \tag{140}$$

for α. In general, this procedure for solution cannot be carried out explicitly, as in the linear case, so various approximation methods, usually based on iterations, have been studied for this problem. In particular, it may happen that

$$\alpha = y'(0),$$

in which case the iterative determination of α, and hence $y(x)$, is sometimes referred to as a *shooting method* (Ref. 14). If $G(x, t)$ is of the form (53), then a similar construction leads to a single nonlinear Volterra integral equation for

$$y(x) = y(x; \alpha_1, \ldots, \alpha_n)$$

and the nonlinear scalar system

$$\alpha_i = \int_0^1 v_i(t) f(t, y(t; \alpha_1, \ldots, \alpha_n))\, dt, \qquad i = 1, 2, \ldots, n, \tag{141}$$

for $\alpha_1, \alpha_2, \ldots, \alpha_n$. Further discussion of nonlinear boundary-value problems is outside the scope of this paper.

6. Nonsymmetric Case

In certain problems, one may have to deal with separable kernels which are nonsymmetric, such as finite linear combinations of kernels of the form $r(x) G(x, t) w(t)$, where $G(x, t)$ is a symmetric, simple separable kernel (24). As a prototype of separable kernels in the general case, consider the simple kernel

$$G(x, t) = \begin{cases} u(t)v(x), & 0 \le t \le x \le 1, \\ p(x)q(t), & 0 \le x \le t \le 1, \end{cases} \tag{142}$$

subject to the condition

$$u(x)v(x) = p(x)q(x), \qquad 0 \le x \le 1, \tag{143}$$

which ensures that the traces of $G(x, t)$ and its associated kernels are uniquely defined. Formulas will now be developed for the Fredholm determinant $\Delta(\lambda)$ and the resolvent kernel $\Gamma(x, t; \lambda)$ of the simple separable kernel (142). The general separable kernel, which is a finite linear combination of kernels of the form (142), can then be handled by the technique given in Section 4, or by the method to be discussed in Section 7.

Writing the Fredholm integral equation (25) as the system

$$y(x) - \lambda \int_0^x K(x, t)y(t)\, dt = f(x) + \lambda c p(x),$$

$$c = \int_0^1 q(t)y(t)\, dt, \tag{144}$$

where

$$K(x, t) = u(t)v(x) - p(x)q(t), \tag{145}$$

one obtains, as before,

$$\Delta(\lambda) = 1 - \lambda \int_0^1 q(x)P(x)\, dx = 1 - \lambda \int_0^1 Q(t)p(t)\, dt, \tag{146}$$

with the functions $P(x)$, $Q(t)$ given by

$$P(x) = p(x) + \lambda \int_0^x R(x, t; \lambda)p(t)\, dt,$$

$$Q(t) = q(t) + \lambda \int_t^1 q(x)R(x, t; \lambda)\, dx, \tag{147}$$

in terms of the Volterra resolvent kernel $R(x, t; \lambda)$ of $K(x, t)$. If

$$\Delta(\lambda) \ne 0,$$

then Eq. (25) has the unique solution

$$y(x) = f(x) + \lambda \int_0^x R(x, t; \lambda)f(t)\, dt + [\lambda/\Delta(\lambda)] \int_0^1 P(x)Q(t)f(t)\, dt, \tag{148}$$

from which it follows that the Fredholm resolvent kernel of the kernel (142) is

$$\Gamma(x, t; \lambda) = \begin{cases} R(x, t; \lambda) + P(x)Q(t)/\Delta(\lambda), & 0 \le t \le x \le 1, \\ P(x)Q(t)/\Delta(\lambda), & 0 \le x \le t \le 1. \end{cases} \tag{149}$$

Alternatively, one may write

$$y(x) + \lambda \int_x^1 K(x, t)y(t)\, dt = f(x) + \lambda\, dv(x),$$

$$d = \int_0^1 u(t)y(t)\, dt. \tag{150}$$

This leads to the expressions

$$\Delta(\lambda) = 1 - \lambda \int_0^1 u(x)V(x)\, dx = 1 - \lambda \int_0^1 U(t)v(t)\, dt \tag{151}$$

for the Fredholm determinant of $G(x, t)$ in terms of the functions

$$V(x) = v(x) - \lambda \int_x^1 R(x, t; \lambda)v(t)\, dt,$$

$$U(t) = u(t) - \lambda \int_0^t u(x)R(x, t; \lambda)\, dx; \tag{152}$$

and, if

$$\Delta(\lambda) \ne 0,$$

this leads to the unique solution

$$y(x) = f(x) - \lambda \int_x^1 R(x, t; \lambda)f(t)\, dt + [\lambda/\Delta(\lambda)] \int_0^1 V(x)U(t)f(t)\, dt \tag{153}$$

of (25). Thus,

$$\Gamma(x, t; \lambda) = \begin{cases} U(t)V(x)/\Delta(\lambda), & 0 \le t \le x \le 1, \\ -R(x, t; \lambda) + U(t)V(x)/\Delta(\lambda), & 0 \le x \le t \le 1, \end{cases} \tag{154}$$

provided that the Fredholm determinant does not vanish. In (152)–(154), use is made of the extension of the function $R(x, t; \lambda)$, defined by (8), to the triangle $0 \le x \le t \le 1$.

Comparison of (154) and (149) yields

$$\Delta(\lambda)R(x, t; \lambda) = U(t)V(x) - P(x)Q(t), \qquad 0 \le x, t \le 1, \tag{155}$$

and finally,

$$\Gamma(x, t; \lambda) = \begin{cases} U(t)V(x)/\Delta(\lambda), & 0 \le t \le x \le 1, \\ P(x)Q(t)/\Delta(\lambda), & 0 \le x \le t \le 1. \end{cases} \tag{156}$$

Theorem 6.1. If $\Delta(\lambda) \ne 0$, then the simple separable kernel (156) is the Fredholm resolvent kernel of the simple separable kernel (142).
If

$$\Delta(\lambda^*) = 0,$$

then it follows from (155) that

$$U^*(t)V^*(x) = P^*(x)Q^*(t), \tag{157}$$

where

$$\begin{aligned} U^*(t) &= U(t; \lambda^*), & Q^*(t) &= Q(t; \lambda^*), \\ V^*(x) &= V(x; \lambda^*), & P^*(x) &= P(x; \lambda^*) \end{aligned} \tag{158}$$

are obtained from (147) and (152) with

$$\lambda = \lambda^*.$$

It follows from (157) that $V^*(x)$ and $P^*(x)$ are linearly dependent, and either may be taken as a right eigenfunction of $G(x, t)$ corresponding to the eigenvalue λ^*; similarly, $U^*(t)$ and $Q^*(t)$ are proportional and furnish a corresponding left eigenfunction of the kernel (142).

Of course, the functions $U(t)$, $V(x)$, $P(x)$, $Q(t)$ may be found by solving the appropriate Volterra integral equations if it is desired to avoid the explicit computation of $R(x, t; \lambda)$.

7. Alternative Computational Method

Suppose that $G(x, t)$ is a Fredholm or Volterra kernel with known resolvent kernel $\Gamma(x, t; \lambda)$; and suppose that it is desired to construct the resolvent kernel $\Gamma_n(x, t; \lambda)$ of the modified kernel

$$G_n(x, t) = G(x, t) + \sum_{j=1}^{n} u_j(x)v_j(t). \tag{159}$$

In the special case that

$$G(x, t) = K(x, t)$$

is a Volterra kernel, one form of the solution of this problem is given by (67), which requires the inversion of an $n \times n$ matrix. It is also possible to obtain the resolvent kernel by a step-by-step process similar to an elimination method for matrix inversion. First of all, consider the rank-one modification

$$G_1(x, t) = G(x, t) + u_1(x)v_1(t). \qquad (160)$$

By applying the method of Section 4 [solving Eq. (25) with the kernel (160)], one obtains

$$\Delta_1(\lambda) = 1 - \lambda \int_0^1 v_1(x)U_1(x)\,dx = 1 - \lambda \int_0^1 V_1(t)u_1(t)\,dt \qquad (161)$$

for the Fredholm determinant of $G_1(x, t)$, where

$$U_1(x) = u_1(x) + \lambda \int_0^1 \Gamma(x, t; \lambda)u_1(t)\,dt,$$
$$\qquad\qquad (162)$$
$$V_1(t) = v_1(x) + \lambda \int_0^1 v_1(x)\Gamma(x, t; \lambda)\,dx.$$

Theorem 7.1. If $\Delta_1(\lambda) \neq 0$, then the resolvent kernel $\Gamma_1(x, t; \lambda)$ of the rank-one modification $G_1(x, t)$ of $G(x, t)$ is

$$\Gamma_1(x, t; \lambda) = \Gamma(x, t; \lambda) + U_1(x)V_1(t)/\Delta_1(\lambda), \qquad (163)$$

which is a rank-one modification of the resolvent kernel $\Gamma(x, t; \lambda)$.

This result has been exploited a number of times previously; see, for example, (38), (39), (44), (55), (119), (149), and (154). It is analogous to the Sherman–Morrison–Woodbury formula for finite matrices (Ref. 15, pp. 123–124). Now, instead of (159), one may consider the sequence of kernels

$$G_0(x, t) = G(x, t),$$
$$\qquad\qquad (164)$$
$$G_k(x, t) = G_{k-1}(x, t) + u_k(x)v_k(t), \qquad k = 1, 2, \ldots, n,$$

each of which is a rank-one modification of the previous kernel. Taking

$$\Gamma_0(x, t; \lambda) = \Gamma(x, t; \lambda), \qquad (165)$$

one may construct the corresponding sequence of resolvent kernels

$$\Gamma_k(x, t; \lambda) = \Gamma_{k-1}(x, t; \lambda) + U_k(x)V_k(t)/\Delta_k(\lambda), \qquad k = 1, 2, \ldots, n, \qquad (166)$$

where

$$U_k(x) = u_k(x) + \lambda \int_0^1 \Gamma_{k-1}(x, t; \lambda)u_k(t)\,dt,$$
$$\qquad\qquad (167)$$
$$V_k(t) = v_k(t) + \lambda \int_0^1 v_k(x)\Gamma_{k-1}(x, t; \lambda)\,dx,$$

and

$$\Delta_k(\lambda) = 1 - \lambda \int_0^1 v_k(x) U_k(x)\, dx = 1 - \lambda \int_0^1 V_k(t) u_k(t)\, dt, \qquad (168)$$

and obtain $\Gamma_n(x, t; \lambda)$, provided, of course, that

$$\Delta_k(\lambda) \neq 0, \qquad k = 1, 2, \dots, n. \qquad (169)$$

8. Numerical Implications

For a Fredholm integral equation with a separable kernel $G(x, t)$, it has been shown that methods appropriate to Volterra integral equations can be used to obtain the Fredholm determinant, resolvent kernel, and eigenfunctions of $G(x, t)$. As better estimates are available for convergence of the resulting expansions than in the general Fredholm case, effective analytic or approximate computations can be carried out. Although explicit formulas can be obtained only for very simple problems, it may be that the Volterra kernel $K(x, t)$ is a polynomial or other simple function of x and t, in which case a computer can be programmed to find the coefficients in the expansion (8) of the resolvent kernel $R(x, t; \lambda)$ to obtain any desired degree of accuracy. Eigenvalues of $G(x, t)$ can also be obtained by computing zeros of the entire function $\Delta(\lambda)$. This is a more difficult problem, but can once again be done with any desired accuracy.

If the analytic or semi-analytic approach appears fruitless or uneconomical, then strictly numerical methods based on numerical integration may be used. For the Volterra integral equations considered, these lead to lower triangular systems of equations which can be solved quickly and accurately, even if large (Refs. 16–17). As numerical integration is usually much more accurate than numerical differentiation, this technique offers an alternative to finite-difference methods for two-point boundary-value problems for ordinary differential operators.

References

1. LOVITT, W. V., *Linear Integral Equations*, Dover Publications, New York, New York, 1950.
2. FRIEDMAN, B., *Principles and Techniques of Applied Mathematics*, John Wiley and Sons, New York, New York, 1956.
3. HOCHSTADT, H., *Integral Equations*, John Wiley and Sons, New York, New York, 1973.
4. WIDOM, H., *Lectures on Integral Equations*, Van Nostrand–Reinhold Company, New York, New York, 1969.

5. LALESCO, T., *Introduction a la Theorie des Equations Integrales*, A. Herman et Fils, Paris, France, 1912.
6. HELLINGER, E., and TOEPLITZ, O., *Integralgleichungen und Gleichungen mit unendlichvielen Unbekannten*, Chelsea Publishing Company, New York, New York, 1953.
7. FREDHOLM, I., *Sur une Classe d'Equations Fonctionnelles*, Acta Mathematica, Vol. 27, pp. 152–156, 1955.
8. FORSYTH, G. E., and STRAUS, L. W., *The Souriau–Frame Characteristic Equation Algorithm on a Digital Computer*, Journal of Mathematics and Physics, Vol. 34, pp. 152–156, 1955.
9. DRUKAREV, G. F., *The Theory of Collisions of Electrons with Atoms*, Soviet Physics JEPT, Vol. 4, pp. 309–320, 1957.
10. BRYSK, H., *Determinantal Solution of the Fredholm Equation with Green's Function Kernel*, Journal of Mathematical Physics, Vol. 4, pp. 1536–1538, 1963.
11. AALTO, S. K., *Reduction of Fredholm Integral Equations with Green's Function Kernels to Volterra Equations*, Oregon State University, MS Thesis, 1966.
12. MANNING, I., *Theorem on the Determinantal Solution of the Fredholm Equation*, Journal of Mathematical Physics, Vol. 5, pp. 1223–1225, 1964.
13. YOSIDA, K., *Lectures on Differential and Integral Equations*, John Wiley and Sons (Interscience Publishers), New York, New York, 1960.
14. KELLER, H. B., *Numerical Methods for Two-Point Boundary Value Problems*, Blaisdell Publishing Company, Waltham, Massachusetts, 1968.
15. HOUSEHOLDER, A. S., *The Theory of Matrices in Numerical Analysis*, Blaisdell Division of Ginn and Company, New York, New York, 1964.
16. RALL, L. B., *Numerical Inversion of Green's Matrices*, University of Wisconsin, Madison, Wisconsin, Mathematics Research Center, Technical Summary Report No. 1149, 1971.
17. RALL, L. B., *Numerical Inversion of a Class of Large Matrices*, Proceedings of the 1975 Army Numerical Analysis and Computers Conference, US Army Research Office, Research Triangle Park, North Carolina, 1975.

12

On the Algebraic Classification of Fredholm Integral Operators

J. CASTI[1]

Abstract. Using the well-known and specific connections between Fredholm integral equations, two-point boundary-value problems, and linear dynamics–quadratic cost control processes, we present a complete, independent set of algebraic invariants suitable for classifying a wide range of Fredholm integral operators with respect to a certain group of transformations. The group, termed the *Riccati group*, is naturally suggested by the control theoretic setting, but seems nonintuitive from a purely integral-equations point of view. Computational considerations resulting from this classification are discussed.

1. Introduction

One of the mainstreams of twentieth-century mathematics has been the *algebraicization* of classical notions developed in quite different contexts; witness the proliferation of work in algebraic geometry, algebraic topology, algebraic-xxx, . . . as evidence of this trend. Recently, the algebraic spirit of the times has been taken up by theoretical engineers and mathematical system theorists as they come to realize that the particular type of economy of thought provided by the constructions of abstract algebra shed considerable light on previously shadowy areas of system and control theory. A good account of some of these developments is given in Refs. 1–2.

One of the central themes of modern *algebraic* system theory has been the *translation* and *interpretation* of classical (i.e., 19th-century) work on the theory of invariants. System theorists have found that much of the

[1] Professor, Department of Computer Applications and Information Systems, and Department of Quantitative Analysis, New York University, New York, New York.

pioneering work by Hilbert, Sylvester, Gordan, and others has deep system-theoretic significance when viewed within the context of 20th-century linear system theory. For example, the celebrated Hilbert basis theorem may be interpreted in system-theoretic terms as stating that "the ring of invariants of a dynamical system Σ with respect to linear coordinate transformations of the state space is equivalent to the external (i.e., input–output) behavior of Σ." Further details along these lines are given in Refs. 3–4. Far deeper results connecting various structural aspects of linear systems with modern concepts of moduli, projective imbeddings, and so forth are treated in Refs. 5–6.

All of the algebraic studies cited above have one basic feature in common: they attempt, either explicitly or implicitly, to categorize or classify a set of objects (read: linear systems) into mutually disjoint and exhaustive classes by means of a fixed set of operations (read: transformations) on the objects. The motivation underlying this strategy is clear, of course: if a group of objects are abstractly *equivalent*, then investigations should be confined to only one member of the group, preferably the *simplest*. Thus, the algebraic machinery has been set up to decide the questions: (i) when two objects are equivalent and, partially, (ii) to interpret, mathematically speaking, what constitutes a *simple* object. These are the aims of invariant theory. More precise definitions, details, and examples are found in Section 3.

The principal goal of this chapter is to investigate the use of algebraic invariant theory in the study of Fredholm integral operators. Since there are well-known and quite specific connections between Fredholm integral equations, two-point boundary-value problems, and linear dynamics–quadratic cost (LQG) control processes, we shall exploit these connections to provide some algebraic rationale to classify Fredholm integral operators. More specifically, using results developed in Ref. 4 for LQG control processes, we shall present a complete, independent set of algebraic invariants suitable for classifying a wide range of Fredholm integral operators with respect to a certain group of transformations. The group under consideration, first termed the *Riccati group* in Ref. 4, is naturally suggested by the control-theoretic setting, but seems to be fairly nonintuitive from a purely integral-equations standpoint. The class of kernels under consideration were first studied in Ref. 7 and were shown to form a dense set in the class of all continuous kernels.

Following the presentation of the main classification theorem in Section 5, we present some illustrative examples showing the intimate connection between the algebraic invariants, certain frequently occurring integral operators, and certain equally important system-theoretic canonical forms.

2. Fundamental Kernels, Matrix Riccati Equations, and LQG Problems

We consider the class of Fredholm integral equations

$$u(t) = f(t) + \int_0^a K(t, s)u(s)\, ds, \tag{1}$$

where u and f are n-dimensional vector functions of t, assumed to be continuous on the interval $[0, a]$. The kernel $K(t, s)$ is an $n \times n$ matrix function of t and s possessing the structure

$$K(t, s) = Hk(t, s)J, \tag{2}$$

where H and J are $n \times r$ and $r \times n$ constant matrices, respectively, while $k(t, s)$ satisfies the equations

$$\partial k(t, s)/\partial t = Bk(t, s), \qquad t > s, \tag{3}$$

$$\partial k(t, s)/\partial s = k(t, s)C, \qquad t < s, \tag{4}$$

$$(d/dt)k(t, t) = A + Bk(t, t) + k(t, t)C, \tag{5}$$

$$k(a, a) = N, \tag{6}$$

with A, B, C, N constant matrices of size $r \times r$. Such a matrix kernel K is called *fundamental*.

In Ref. 7, the equivalence between the Fredholm resolvent associated with (1) and the matrix Riccati equation

$$\dot{R}(x) = A + BR + RC + RDR, \tag{7}$$

$$R(a) = N, \tag{8}$$

where

$$D = JH,$$

is demonstrated. Thus, for all practical purposes the study of the Fredholm integral equation (1) is equivalent to the investigation of the properties of the matrix Riccati equation (7). We shall exploit systematically this equivalence in this paper.

The introduction of the matrix Riccati equation (7) also suggests the study of optimal control processes with linear dynamics and quadratic costs. Specifically, the problem of minimizing (over u) the quadratic form

$$\int_t^T [(x, Qx) + (u, Ru)]\, ds + (P_0 x(T), x(T)),$$

with x and u connected by the differential equations

$$\dot{x} = Fx + Gu, \qquad x(t) = c, \qquad \text{and } Q \geq 0, R > 0,$$

leads to the well-known feedback solution

$$u_{\min}(t) = -R^{-1}G'P(t)x(t) = -K(t)x(t),$$

where $P(t)$ is the solution of the matrix Riccati equation

$$-\dot{P}(t) = Q + PF + F'P - PGR^{-1}G'P, \tag{9}$$

$$P(T) = P_0. \tag{10}$$

The identifications

$$A \leftrightarrow -Q, \qquad B \leftrightarrow -F', \qquad C \leftrightarrow -F,$$
$$D \leftrightarrow GR^{-1}G', \qquad N \leftrightarrow P_0, \tag{11}$$

generate, via Ref. 7, the equivalence between the Fredholm resolvent for (1) and the LQG control problem specified by the matrices (F, G, Q, R, P_0). The algebraic classification theorem, which is the main result of this paper, will be a consequence of the above identifications; i.e., we shall develop first the result for LQG control problems, then use the identifications (11) to translate to the matrix Riccati equation (7). The results previously cited, relating the solution of (7) to the Fredholm resolvent for (1), will then complete the chain.

Note that the symmetry conditions of the LQG problem now limit our study to fundamental kernels with

$$A = A', \qquad B = C', \qquad D = D', \qquad N = N'.$$

3. Invariant Theory and Riccati Group

Before proceeding to the central topic of this chapter, we pause briefly to sketch a skeletal outline of the theory of algebraic invariants in order to motivate the later sections of the paper. In addition, this section also serves to introduce terminology facilitating presentation of our classification theorem.

In the simplest possible terms, invariant theory deals with the action of a group of transformations $\mathcal{G}$ upon a collection of mathematical objects $\mathcal{Q}$; i.e., we have the action

$$\mathcal{G} \times \mathcal{Q} \to \mathcal{Q}.$$

We say that two objects $q, q' \in \mathcal{Q}$ are *equivalent mod* $\mathcal{G}$ if there exists a $g \in \mathcal{G}$ such that

$$gq = q'.$$

It now becomes of interest to ask how the equivalence classes of $\mathscr{D}$ may be characterized in a finite, even algorithmic, way. The algebraic theory of invariants provides one procedure to answer this question.

In essence, the algebraic theory of invariants is used to find a set of polynomials (or rational functions) in the *entries* of q whose values remain *invariant* under application of transformations from $\mathscr{G}$. If $\{P_\alpha\}$, $\alpha = 1, 2, \ldots, N^*$, is such a set (with N^* finite, positive integer), then in the classical terminology the set $\{P_\alpha\}$ is called

(i) *independent* if, for each N-tuple k in the range of $\{P_\alpha\}$, there exists *some* $q \in \mathscr{D}$ such that

$$(P_1(q), P_2(q), \ldots, P_{N^*}(q)) = k;$$

(ii) *complete* if

$$P_\alpha(q) = P_\alpha(q')$$

for all α implies

$$q = gq'$$

for some $g \in \mathscr{G}$.

Assuming that a set of complete, independent invariants $\{P_\alpha\}$ can be found, they then provide a scheme for categorizing the equivalence classes of $\mathscr{D}$; in fact, the set $\{P_\alpha\}$ then enables us to study geometrically, as well as algebraically, the equivalence classes by means of the tools of algebraic geometry. Thus, our goal will be to find just such a set of rational invariants for the LQG problem discussed in Section 2, then transfer the result via the matrix Riccati equation to the integral equation (1).

In order to make headway toward the stated goal, the abstract entities $\mathscr{D}$ and the abstract group $\mathscr{G}$ will have to be made more specific. Since the LQG problem of Section 2 is completely specified by the five matrices F, G, Q, R, P_0, it is natural to choose the set $\mathscr{D}$ as

$$\mathscr{D} = \{(F, G, Q, R, P_0): Q = Q', R > 0, P_0 = P'_0, \text{ and } (F, G) \text{ reachable}\}.$$

Appropriate choice of the group $\mathscr{G}$ is not quite so straightforward.

In order to focus upon a meaningful group of transformations $\mathscr{G}$, we take our cue from the fact that the optimal feedback control law $u_{\min}(t)$ is totally characterized by the solution of the matrix Riccati equation (9). Thus, we seek a group of transformations which leave the Riccati equation invariant in form. It is not difficult to verify that the following set of transformations, termed the *Riccati group* in Ref. 8, satisfy the invariance requirement:

(I) linear coordinate transformations of the state variable x,

$$x \to Tx, \det T \neq 0;$$

(II) linear coordinate transformations of the control variable u,

$$u \to Vu, \det V \neq 0;$$

(III) application of an arbitrary, but fixed, feedback law L,

$$K(t) \to K(t) - L;$$

(IV) addition of a constant symmetric matrix M to P,

$$P \to P + M.$$

Since the set of transformations (I)–(IV) leave the form of (9) invariant, it is easy to calculate the effect of applying an element of the Riccati group to a representative element from $\mathscr{Q}$. Specifically, we have

$$(F, G, Q, R, P_0) \xrightarrow{\text{(I)}} (TFT^{-1}, TG, T'^{-1}QT^{-1}, R, T'^{-1}P_0T^{-1}),$$

$$(F, G, Q, R, P_0) \xrightarrow{\text{(II)}} (F, GV^{-1}, Q, V'^{-1}RV^{-1}, P_0),$$

$$(F, G, Q, R, P_0) \xrightarrow{\text{(III)}} (F - GL, G, Q + L'RL, R, P_0),$$

$$(F, G, Q, R, P_0) \xrightarrow{\text{(IV)}} (F, G, Q - F'M - MF, R, P_0 + M).$$

In what follows, we shall work with the set $\mathscr{Q}$ and the Riccati group of transformations (I)–(IV) as the basic elements in our invariant-theoretic classification of LQG problems and Fredholm integral operators.

4. Linear Matrix Inequality

In other works (Refs. 8–9), it has been shown that the solution to the LQG problem posed in Section 2 may be obtained by minimizing the expression

$$\text{trace}\left\{ [I - K] \begin{bmatrix} F'P + PF + Q + \dot{P} & PG \\ G'P & R \end{bmatrix} \begin{bmatrix} I \\ -K \end{bmatrix} W \right\},$$

over all matrix functions $K(t)$ and $P(t)$, with $W \geq 0$ being arbitrary. This result follows by substituting the original system dynamics into the cost function and performing an integration by parts.

What is of interest in this approach from our point of view is the matrix

$$\mathcal{M}(P) = \begin{bmatrix} Q + PF + F'P & PG \\ G'P & R \end{bmatrix},$$

which plays an essential role in the selection of the particular elements from the Riccati group needed to obtain a convenient canonical form for LQG problems.

It can be shown that the condition

$$\mathcal{M}(P) \geq 0 \tag{12}$$

is necessary and sufficient for a matrix P to be the unique, nonnegative-definite solution of the *algebraic* Riccati equation

$$Q + PF + F'P - PGR^{-1}G'P = 0,$$

showing the connection between the solution of the *infinite-interval* LQG problem and the matrix $\mathcal{M}(P)$. Relation (12) is generally termed the linear matrix inequality (Ref. 9). Since our development is purely algebraic, we shall have no need of the result (12), reserving our sole interest for the structure of $\mathcal{M}(P)$.

5. Main Classification Theorem

The principal task in obtaining canonical forms for LQG problems is to select judiciously a set of transformations from the Riccati group which, when applied to the defining data (F, G, Q, R, P_0) from $\mathcal{Q}$, in some sense *simplify* the problem. At this point, we invoke a result from Ref. 9, which states that the optimal feedback law $K(t)$ may be obtained by inspection from the triangular factors of $\mathcal{M}(P)$. Thus, we shall choose the transformations to make this factorization as easy as possible, i.e., we wish to choose the element of the Riccati group to (block) diagonalize $\mathcal{M}(P)$, if possible.

Fortunately, in Ref. 8 it has been shown that the following choice of transformations will achieve the desired goal:

(a) Choose the Type (I) transformations T such that

$$T'^{-1}QT^{-1} = \text{diag}_p(\pm 1),$$

where the subscript p indicates that the first p positions equal ± 1, the remainder being zero,

$$p = \text{rank } Q.$$

Such a choice of T is always possible, since by assumption

$$Q = Q'.$$

(b)   Select the Type (II) transformation V so that

$$V'^{-1}RV^{-1} = I.$$

As with T, such a choice is possible, since

$$R > 0.$$

(c)   The feedback law L is now selected in order to *zero out* the off-diagonal blocks of $\mathcal{M}$. This involves selecting L such that

$$R_{\text{I-II}}L = V'^{-1}(G'P_0)_{\text{I-II}}T^{-1},$$

where the subscript denotes the form resulting after application of the indicated transformation. Since

$$R_{\text{I-II}} = J$$

we have

$$L = V'^{-1}(G'P_0)_{\text{I-II}}T^{-1}.$$

(d)   For any choice of M, we have

$$\mathcal{M}_{\text{I-IV}}(P_0) = \left[\begin{array}{c|c} \begin{array}{c} \text{diag}_p(\pm 1) + T'^{-1}(P_0F + F'P_0 - L'L)T^{-1} \\ -MF_{\text{I-III}} - F'_{\text{I-III}}M \\ -G'_{\text{I-III}}M \end{array} & \begin{array}{c} -MG_{\text{I-III}} \\ \\ I \end{array} \end{array}\right].$$

Thus, we choose M such that

(i) $MG_{\text{I-III}} = 0$,

(ii) $-MF_{\text{I-III}} - F'_{\text{I-III}}M + T'^{-1}(P_0F + F'P_0 - L'L)T^{-1}$

$$= -\left[\begin{array}{c|c} \text{diag}_{n-m}(\pm 1) & X_1 \\ \hline X'_1 & X_2 \end{array}\right],$$

where the entries in X_1 and X_2 are, in general, nonzero elements determined by M. Note that (i) implies that

$$\text{rank } M \le n - m.$$

Thus, the final canonical form for $\mathcal{M}$ is

$$\mathcal{M}_{\text{I-IV}}(P_0) = \left[\begin{array}{c|c} \text{diag}_p(\pm 1) - \left[\begin{array}{c|c} \text{diag}_{n-m}(\pm 1) & X_1 \\ \hline X'_1 & X_2 \end{array}\right] & 0 \\ \hline 0 & I \end{array}\right],$$

with the appropriate canonical structure for F, G, Q, R, P_0 being induced by the above choices of T, V, L, M, and the relations of Section 3.

The algebraic invariants of $\mathscr{Q}$, under the action of the Riccati group, can now be seen to be (A) the entries of the matrices X_1 and X_2, (B) the entries of the matrix $(P_0)_{\text{I-IV}}$, i.e., the numbers

$$T'^{-1}P_0T^{-1}+M.$$

It is verified in Ref. 8 that these elements do indeed form a complete, independent set of invariants for the set $\mathscr{Q}$, relative to the Riccati group.

In summary, we have established the following theorem.

Theorem 5.1. *Algebraic Invariant Theorem for the LQG Problem.* Given the LQG problem data

$$\mathscr{Q} = (F, G, Q, R, P_0),$$

a complete independent set of algebraic invariants for $\mathscr{Q}$, with respect to the Riccati group, is given by the elements of the three matrices X_1, X_2, and $(P_0)_{\text{I-IV}}$.

An interesting corollary of this theorem, which is an immediate consequence of the proof in Ref. 8, is the following.

Corollary 5.1. If P_0 is a solution of the algebraic Riccati equation

$$Q + PF + F'P - PGR^{-1}G'P = 0,$$

the invariants X_1 and X_2 vanish identically, and the only nontrivial invariants are the elements of $(P_0)_{\text{I-IV}}$.

6. Invariants and Fredholm Integral Operators

The foregoing classification theorem, coupled with the discussion of Section 2, now makes it evident how to algebraically classify Fredholm integral operators with fundamental kernels: first, form the relevant matrix Riccati equation from the matrices defining the kernel; translate the Riccati equation into an equivalent LQG problem; then, apply the algebraic classification theorem for LQG problems. We cite an example.

Example 6.1. We assume that the kernel of the integral operator has the *displacement form*

$$K(t, s) = \sum_{i=1}^{n} \exp[\lambda_i|t - s|], \qquad \lambda_i \neq \lambda_j, \qquad \lambda_i \neq 0.$$

The relevant matrices for this fundamental kernel are

$$B = C = \operatorname{diag}(\lambda_1, \ldots, \lambda_n), \qquad H = J' = (1, 1, \ldots, 1),$$

$$A = -2B, \qquad N = I.$$

The corresponding LQG problem has the matrices

$$F = -\operatorname{diag}(\lambda_1, \ldots, \lambda_n), \qquad G = (1, 1, \ldots, 1)', \qquad Q = 2\operatorname{diag}(\lambda_1, \ldots, \lambda_n),$$

$$R = I, \qquad P_0 = I.$$

Except for an inessential change of sign in F, this system is in the so-called Lurie–Lefschetz–Letov controller canonical form (Ref. 10). Note also that the seemingly more general kernel

$$K(t, s) = \sum_{i=1}^{n} \alpha_i \exp[\lambda_i |t - s|]$$

is identical to the above, with the exception that

$$R = \operatorname{diag}(1/\alpha_1, \ldots, 1/\alpha_n).$$

However, such an R is equivalent [via a Type (II) transformation] to the original problem, so there is no loss of generality in considering the first case.

Following the steps of Section 5, we compute the invariants. The choice of

$$T = \operatorname{diag}(\sqrt{(2\lambda_1)}, \ldots, \sqrt{(2\lambda_n)}) = T'$$

gives

$$T'^{-1} Q T^{-1} = I.$$

Since

$$R = I,$$

V may be chosen to be any orthogonal matrix. For simplicity, we take

$$V = I.$$

The above choices of T and V lead to the selection of L as

$$L = G' T^{-1} = (1/\sqrt{(2\lambda_1)}, \ldots, 1/(\sqrt{2\lambda_n})).$$

The determination of M is somewhat more complicated. The prescription of Section 5 states that M is chosen such that

$$-MF_{\text{I–III}} - F'_{\text{I–III}} M + T'^{-1}(P_0 F + F' P_0 - L'L) T^{-1} = -\left[\begin{array}{c|c} \operatorname{diag}_{n-1}(\pm 1) & X_1 \\ \hline X'_1 & X_2 \end{array} \right],$$

$$MG_{\text{I–III}} = 0.$$

An easy calculation shows that the matrices $F_{\text{I-III}}$ and $G_{\text{I-III}}$ are

$$F_{\text{I-III}} = (f_{ij})_{\text{I-III}},$$

$$(f_{ij})_{\text{I-III}} = \begin{cases} \lambda_i + 1, & i = j, \\ \sqrt{(\lambda_i/\lambda_j)}, & i \neq j, \end{cases}$$

$$G_{\text{I-III}} = \sqrt{2} \begin{bmatrix} \sqrt{\lambda_1} \\ \sqrt{\lambda_2} \\ \vdots \\ \sqrt{\lambda_n} \end{bmatrix}.$$

After some algebra, we find that

$$MF_{\text{I-III}} + F'_{\text{I-III}}M =$$

$$\begin{bmatrix} \frac{1}{4}[1/\lambda_i\lambda_j] & \frac{1}{4}\begin{pmatrix} 1/\lambda_1\lambda_n \\ \vdots \\ 1/\lambda_{n-1}\lambda_n \end{pmatrix} - X_1 \\ \hline \frac{1}{4}(1/\lambda_1\lambda_n, \dots, 1/\lambda_{n-1}\lambda_n) - X'_1 & 1 + 1/4\lambda_n - X_2 \end{bmatrix}. \qquad (13)$$

Upon partitioning $F_{\text{I-III}}$, $G_{\text{I-III}}$, and M as

$$F_{\text{I-III}} = \begin{bmatrix} A_{11} & A_{12} \\ A_{21} & A_{22} \end{bmatrix}, \qquad G_{\text{I-III}} = \begin{bmatrix} G_1 \\ G_2 \end{bmatrix}, \qquad M = \begin{bmatrix} M_{11} & M_{12} \\ M_{12} & M_{22} \end{bmatrix},$$

we obtain the matrix M by solving the equations

$$M_{11}A_{11} + M_{12}A_{21} + A'_{11}M_{11} + A'_{21}M'_{12} = [1/4\lambda_i\lambda_j], \quad i, j = 1, 2, \dots, n-1,$$

$$M_{11}G_1 + M_{12}G_2 = 0, \qquad M'_{12}G_1 + M_{22}G_2 = 0.$$

Having obtained M, the invariants X_1 and X_2 are determined from (13).

Unfortunately, the algebra required for the general situation of distinct λ_i proves to be somewhat onerous. So, we consider the special case when

$$n = 3 \qquad \text{and} \qquad \lambda_1 = \lambda_2 = \lambda_3 = \lambda.$$

In this case, Eq. (4) becomes

$$\begin{bmatrix} 2(\lambda+1) & 2 & 2 & 0 & 0 & 0 \\ 1 & 2(\lambda+1) & 1 & 1 & 1 & 0 \\ 0 & 2 & 0 & 2(\lambda+1) & 2 & 0 \\ \sqrt{(2\lambda)} & \sqrt{(2\lambda)} & \sqrt{(2\lambda)} & 0 & 0 & 0 \\ 0 & \sqrt{(2\lambda)} & 0 & \sqrt{(2\lambda)} & \sqrt{(2\lambda)} & 0 \\ 0 & 0 & \sqrt{(2\lambda)} & 0 & \sqrt{(2\lambda)} & \sqrt{(2\lambda)} \end{bmatrix} \begin{bmatrix} M_{11} \\ M_{12} \\ M_{13} \\ M_{22} \\ M_{23} \\ M_{33} \end{bmatrix} = \begin{bmatrix} 1/4\lambda^2 \\ 1/4\lambda^2 \\ 1/4\lambda^2 \\ 0 \\ 0 \\ 0 \end{bmatrix}.$$

The unique solution to this system is

$$M_{11} = 1/8\lambda^3, \qquad M_{12} = -1/8\lambda^3, \qquad M_{13} = 0,$$
$$M_{22} = 1/8\lambda^3, \qquad M_{23} = 0, \qquad M_{33} = 0.$$

Thus, the invariants for the problem are

$$X_1 = \frac{1}{4}\begin{bmatrix} 1/\lambda^2 \\ 1/\lambda^2 \end{bmatrix}, \qquad X_2 = 1/4\lambda^2 + 1,$$

$$(P_0)_{\text{I-IV}} = \begin{bmatrix} (4\lambda^2+1)/8 & -1/8\lambda^3 & 0 \\ -1/8\lambda^3 & (4\lambda^2+1)/8\lambda^3 & 0 \\ 0 & 0 & 1/2\lambda \end{bmatrix}.$$

7. Generalized X and Y Functions

In earlier reports (Ref. 9), it has been noted that the triangular factorization

$$\mathcal{M}(P) = \begin{bmatrix} \phi & \Gamma \\ 0 & \psi \end{bmatrix}\begin{bmatrix} \phi' & 0 \\ \Gamma' & \psi' \end{bmatrix}$$

leads to the identifications of ϕ and Γ with the generalized X and Y functions discussed in Refs. 10–11. More specifically, ϕ is a generalized Y function, while Γ is a generalized X function (modulo an inessential orthogonal transformation θ).

Since the preceding discussion has shown that the canonical matrix is

$$\mathcal{M}_{\text{can}}(P) = \begin{bmatrix} \text{diag}_p(\pm 1) - \begin{bmatrix} \text{diag}_{n-m}(\pm 1) & | & X_1 \\ \hline X_1' & | & X_2 \end{bmatrix} & | & 0 \\ \hline 0 & | & I \end{bmatrix},$$

we see the connection between the invariants X_1 and X_2 and the induced canonical structure for X and Y, i.e.,

$$Y_{\text{can}} = \left\{ \text{diag}_p(\pm 1) - \left[\frac{\begin{array}{c|c} \text{diag}_{n-m}(\pm 1) & X_1 \\ \hline X_1' & X_2 \end{array}}{} \right] \right\}^{1/2} \mod \theta,$$

$$X_{\text{can}} = 0.$$

For instance, for the displacement kernel example given earlier, we have

$$Y_{\text{can}} = (i/2\lambda)(1, 1, 1),$$

$$X_{\text{can}} = 0.$$

The canonical X and Y functions may be of particular importance for various Wiener–Hopf type equations on the semi-infinite interval, since, in these cases, the X and Y functions are constant.

8. More General Kernels

The class of kernels termed *fundamental* in Section 2 does not correspond exactly to those of the same name in Ref. 7, the difference being our additional requirement that the defining matrices A, B, C, H, J be constant. Unfortunately, this requirement makes it difficult to represent some interesting classes of kernels, such as degenerate kernels. In the treatment of Ref. 7, such a restriction is unnecessary to develop the connection between the Fredholm resolvent and a matrix Riccati equation. The constraint has been imposed here in order to generate a time-invariant classification; however, at the expense of introducing time-varying matrices A, B, C, and/or D, a theorem similar to the main classification result can be stated, the principal difference being that the rational invariants X_1, X_2, and $(P_0)_{\text{I–IV}}$ may now turn out to be time-varying.

Probably, the most interesting and useful relaxation of the constancy condition would be to allow

$$D = HJ$$

to be time-varying, keeping the remaining matrices constant. In principle, such a situation would then allow us to consider degenerate kernels by the choices

$$A = B = C = 0, \qquad M = I,$$

$$H = (h_1(t), \dots, h_n(t)), \qquad J = H,$$

corresponding to the *scalar* symmetric kernel

$$K(t, s) = \sum_{i=1}^{n} h_i(t)h_i(s).$$

Unfortunately, such a kernel generates a system Σ having (F, G) unreachable, a violation of our original assumptions on $\mathcal{D}$. This violation shows up in the attempt to determine M. The one situation which causes no trouble is for an $n \times n$ *matrix* degenerate kernel, in which each entry of $K(t, s)$ has the above form, i.e.,

$$K_{ij}(t, s) = \sum_{k=1}^{n} g_{ik}(t)g_{kj}(s),$$

with the matrix

$$G = [g_{ij}(\cdot)], \qquad i, j = 1, \ldots, n,$$

having full rank for all t. In this event, we again have reachability, and can employ the earlier analysis.

References

1. BROCKETT, R., and MAYNE, D., *Proceedings of the NATO Advanced Study Institute on Geometric and Algebraic Methods for Nonlinear Systems*, D. Reidel Publishing Company, New York, New York, 1973.
2. IEEE Proceedings, Special Issue on Recent Trends in System Theory, Vol. 64, No. 1, 1976.
3. KALMAN, R., *System Theoretic Aspects of Invariant Theory*, University of Florida, Center for Mathematical Systems Theory, Gainesville, Florida, Private Communication, 1974.
4. KALMAN, R., and HAZEWINKEL, M., *On Invariants, Canonical Forms, and Moduli for Linear, Constant, Finite-Dimensional Dynamical Systems*, Proceedings of the CISM Conference on Algebraic System Theory, Edited by A. Marzollo and G. Marchessini, New York, New York, 1976.
5. HAZEWINKEL, M., *Representations of Quivers and Moduli of Linear Dynamical Systems*, University of Rotterdam, Rotterdam, Holland, Economic Institute of Erasmus, 1976.
6. BYRNES, C., and HURT, N., *On the Moduli of Linear Dynamical Systems* (to appear).
7. SHUMITZKY, A., *On the Equivalence Between Matrix Riccati Equations and Fredholm Resolvents*, Journal of Computer and System Sciences, Vol. 2, pp. 76–87, 1969.
8. CASTI, J., *Invariant Theory, the Riccati Group and Linear Control Problems*, IEEE Transactions on Automatic Control (to appear).

9. CASTI, J., *Generalized X–Y Functions, the Linear Matrix Inequality and Triangular Factorization for Linear Control Problems*, IIASA, Laxenburg, Austria, Research Memorandum No. RM-76-10, 1976.
10. CASTI, J., *Dynamical Systems and Their Applications: Linear Theory*, Academic Press, New York, New York, 1977.
11. KAILATH, T., *Some New Algorithms for Recursive Estimation in Constant Linear Systems*, IEEE Transactions on Information Theory, Vol. 19, pp. 750–760, 1973.

13

Boundary and Initial-Value Methods for Solving Fredholm Equations with Semidegenerate Kernels

M. A. GOLBERG[1]

Abstract. We develop a new approach to the theory and numerical solution of a class of linear and nonlinear Fredholm equations. These equations, which have semidegenerate kernels, are shown to be equivalent to two-point boundary-value problems for a system of ordinary differential equations. Application of numerical methods for this class of problems allows us to develop a new class of numerical algorithms for the original integral equation. The scope of the paper is primarily theoretical; developing the necessary Fredholm theory and giving comparisons with related methods. For convolution equations, the theory is related to that of boundary-value problems in an appropriate Hilbert space. We believe that the results here have independent interest. In the last section, our methods are extended to certain classes of integrodifferential equations.

1. Introduction

Since its inception, the theory of linear integral equations has been intimately related to the theory of linear algebraic equations. In particular, for equations of the second kind with *nice* kernels, the concept of a compact operator allows one to visualize both situations as special cases of a more general theory. In fact, compactness is related to the ability to approximate an operator by ones of finite rank. In the case of integral equations of the second kind, this has the consequence of having degenerate (separable) kernels play a vital role, with many theorems for equations with general kernels reduced to those involving degenerate ones (Ref. 1).

---

[1] Professor of Mathematics, University of Nevada at Las Vegas, Las Vegas, Nevada.

Similarly, in terms of solving equations numerically, one most often thinks of algorithms which solve the original equation by a sequence of algebraic ones. This approach has led to very general and efficient methods for solving a variety of equations (Refs. 1–2).

It is our purpose in this chapter to develop an alternate approach to the theory and numerical solution of a large class of integral equations of the second kind. This is done by relating them to initial- and boundary-value problems for ordinary differential equations. The class of kernels that we discuss has been termed semidegenerate (Refs. 3–4) and are kernels which have the same general properties as Green's functions for ordinary differential operators (Refs. 5–6). Equations with such kernels have been the object of intensive study over the past 30–35 years (Ref. 3); yet, to the author's knowledge, they do not appear to be singled out as a separate class in any of the standard treatises on integral equations. Much of the work on problems in this area seem to have been developed in the context of invariant imbedding (Refs. 3–4), particularly in relation to problems of radiation and neutron transport (Refs. 3 and 7–8). The only other papers that we are aware of that treat the Fredholm theory are Aalto's (Ref. 3) and the chapter of Rall in this volume. Their motivation was in part similar to ours. However, their development of the theory uses a reduction involving the solution of a Volterra equation; thus, our methods are somewhat different from theirs.

There are two problems which can, in some sense, be regarded as paradigms in numerical analysis: solving a set of linear algebraic equations and integrating a system of ordinary differential equations with given initial conditions. Traditional numerical analysis centers almost exclusively on the conversion of problems to ones in the first category. We believe, and will show, that looking at the conversion to ones in the second category should not be overlooked.

The chapter is divided into six sections. In Section 2, the basic definitions of the class of equations we discuss are given. Section 3 is devoted to the development of the Fredholm theory using a reduction of the given integral equation to an equivalent two-point boundary-value problem. This provides the necessary theoretical background for a class of numerical algorithms discussed here and elsewhere (Refs. 4, 9). In addition, we present a brief comparison of our method to that of Aalto and Rall.

In Section 4, the results of the previous sections are used to give several new expressions for the Fredholm resolvent, analogous to those presented by Rall in Ref. 6. These representations can be regarded as generalizations of those given by Goursat in Ref. 10 and Bownds and Cushing in Ref. 11 for Volterra equations.

In Section 5, the emphasis is changed to the consideration of equations with convolution kernels. Our purpose here is to show how the reduction of an integral equation to a two-point boundary-value problem is useful in developing important structural properties of this class of equations. In particular, new representations are given for the solutions of these equations which should lead to very efficient numerical algorithms.

In Section 6, a brief discussion is given of the possible extensions of our methods to nonlinear equations, Weiner–Hopf equations, and integro-differential equations.

2. Semidegenerate Kernels

We consider scalar integral equations of the form

$$u(t) = g(t) + \lambda \int_a^b K(t, s) u(s) \ ds, \tag{1}$$

where $K(t, s)$ is of the form

$$K(t, s) = \begin{cases} \sum\limits_{i=1}^{N} a_i(t) b_i(s), & a \leq s \leq t, \\ \sum\limits_{i=1}^{M} c_i(t) d_i(s), & t < s \leq b. \end{cases} \tag{2}$$

A kernel of the form in (2) is called semidegenerate. Note that it is not required that

$$\sum_{i=1}^{N} a_i(t) b_i(t) = \sum_{i=1}^{M} c_i(t) d_i(t),$$

so that we allow for discontinuities across the line $t = s$.

Let SD denote the class of semidegenerate kernels. These kernels are often said to be of *Green's function* type (Refs. 5–6) and contain the class of degenerate and Volterra degenerate kernels (Refs. 3–4). As a consequence, the SD class is extensive, and equations involving these kernels should be expected to have a rich theory.

A closely related class of kernels are the generalized semidegenerate kernels, that is, kernels of the form

$$K(t, s) = \begin{cases} \int_A a(t, \mu) b(s, \mu) \ dF_1(\mu), & a \leq s < t, \\ \int_A c(t, \mu) b(s, \mu) \ dF_2(\mu), & t < s \leq b, \end{cases} \tag{3}$$

where $F_1(\mu)$ and $F_2(\mu)$ are functions of bounded variation on $A \subset R$. These kernels have been discussed in detail in the literature on radiative transfer, and filtering theory for stochastic processes (Refs. 3, 8, 12). Since much of this work has been numerical in nature, it has been common practice to replace the integrals in (3) by finite sums, thus giving rise to kernels in SD. In Section 5, we return to a discussion of equations in this class. For now, we concentrate on (1) and (2).

3. Fredholm Theory

We now turn our attention to the existence theory for (1)–(2). Our basic assumptions are as follows:

(i) $a_i(t)$, $i = 1, 2, \ldots, N$, $b_i(t)$, $i = 1, 2, \ldots, N$, $c_i(t)$, $i = 1, 2, \ldots, M$, $d_i(t)$, $i = 1, 2, \ldots, M$, and $g(t)$ are continuous, real-valued functions on $[a, b]$;

(ii) $[a, b]$ is compact;
(iii) λ is complex.

With these assumptions, it is straightforward to show that solving (1)–(2) is equivalent to solving a two-point boundary-value problem for a system of $N + M$ ordinary differential equations. Of course, under the assumptions (i)–(iii), $u(t)$ is not in general differentiable, but certain integrals of it are. The following theorem, first established in Ref. 4, gives this basic equivalence result on which all our further developments are based.

Theorem 3.1. Assume that (1)–(2) has a continuous solution $u(t), t \in [a, b]$. Then, $u(t)$ has the representation

$$u(t) = g(t) + \lambda \left[\sum_{i=1}^{N} a_i(t)\alpha_i(t) + \sum_{i=1}^{M} c_i(t)\beta_i(t) \right], \tag{4}$$

where

$$\alpha_i(t) = \int_a^t b_i(s)u(s)\, ds, \qquad i = 1, 2, \ldots, N, \tag{5}$$

$$\beta_i(t) = \int_t^b c_i(s)u(s)\, ds, \qquad i = 1, 2, \ldots, M, \tag{6}$$

solve the two-point boundary-value problem

$$\alpha_i'(t) = \lambda \left[\sum_{j=1}^{N} b_i(t)a_j(t)\alpha_j(t) + \sum_{j=1}^{M} b_i(t)c_j(t)\beta_j(t) \right]$$
$$+ b_i(t)g(t), \qquad i = 1, 2, \ldots, N, \tag{7}$$

$$-\beta_i'(t) = \lambda \left[\sum_{j=1}^{N} d_i(t)a_j(t)\alpha_j(t) + \sum_{j=1}^{M} d_i(t)c_j(t)\beta_j(t) \right]$$
$$+ d_i(t)g(t), \qquad i = 1, 2, \ldots, M, \tag{8}$$

$$\alpha_i(a) = 0, \qquad i = 1, 2, \ldots, N, \qquad \beta_i(b) = 0, \qquad i = 1, 2, \ldots, M. \tag{9}$$

Conversely, if $\alpha_i(t)$, $i = 1, 2, \ldots, N$, and $\beta_i(t)$, $i = 1, 2, \ldots, M$, solve (7)–(9), then $u(t)$ given by (4) is a solution of (1).

Proof. The representation in (4) follows immediately from the definition of the α's and β's. To get (7)–(8), differentiate (5) and (6), giving

$$\alpha_i'(t) = b_i(t)u(t), \qquad i = 1, 2, \ldots, N,$$
$$\beta_i'(t) = -c_i(t)u(t), \qquad i = 1, 2, \ldots, M. \tag{10}$$

Substitution of (4) into (10) gives (7)–(8). Equations (9) are obvious. To obtain the converse, define $\psi(t)$ by

$$\psi(t) = g(t) + \lambda \left[\sum_{i=1}^{N} a_i(t)\alpha_i(t) + \sum_{i=1}^{M} b_i(t)\beta_i(t) \right]. \tag{11}$$

From (7) and (8), it is seen that

$$\alpha_i'(t) = b_i(t)\psi(t), \qquad i = 1, 2, \ldots, N,$$
$$\beta_i'(t) = -c_i(t)\psi(t), \qquad i = 1, 2, \ldots, M. \tag{12}$$

Integration of (12) using (9) gives

$$\alpha_i(t) = \int_a^t b_i(s)\psi(s)\, ds, \qquad \beta_i(t) = \int_t^b c_i(s)\psi(s)\, ds. \tag{13}$$

Substitution of (13) into the right-hand side of (11) shows that $\psi(t)$ solves (1). □

Theorem 3.1 shows that solving (1)–(2) is equivalent to finding solutions of the two-point boundary-value problem (7)–(9). In Refs. 4 and 9, this representation, and related ones, were discussed in detail in terms of algorithms for the numerical solution of (1)–(2). Of particular interest is the case where

$$c_i(t)d_i(t) = 0, \qquad i = 1, 2, \ldots, M,$$

so that (1) becomes a Volterra equation and the system (7)–(9) reduces to an initial-value problem. Extensive calculations reported in Refs. 13–15 for this case show the method to be a fast and accurate procedure for solving integral equations of this class. Since our interest in this chapter is in filling some of the theoretical gaps left in previous work, we now proceed to use Theorem 3.1 to characterize the Fredholm theory for (1)–(2).

For ease of exposition, (7)–(9) are rewritten in vector matrix form. Define matrices $A(t)$, $B(t)$, $C(t)$, $D(t)$ and vectors $\alpha(t)$, $\beta(t)$, $e(t)$, $f(t)$ by

$$A(t) = \{b_i(t)a_j(t)\}, \quad i = 1, 2, \ldots, N, \quad j = 1, 2, \ldots, N,$$

$$B(t) = \{b_i(t)c_j(t)\}, \quad i = 1, 2, \ldots, N, \quad j = 1, 2, \ldots, M,$$

$$C(t) = \{d_i(t)a_j(t)\}, \quad i = 1, 2, \ldots, M, \quad j = 1, 2, \ldots, N,$$

$$D(t) = \{d_i(t)c_j(t)\}, \quad i = 1, 2, \ldots, M, \quad j = 1, 2, \ldots, M,$$

$$e(t) = (g(t)b_i(t)), \quad i = 1, 2, \ldots, N,$$

$$f(t) = (g(t)d_i(t)), \quad i = 1, 2, \ldots, M,$$

$$\alpha(t) = (\alpha_i(t)), \quad i = 1, 2, \ldots, N,$$

$$\beta(t) = (\beta_i(t)), \quad i = 1, 2, \ldots, M.$$

In this notation, (7)–(9) become

$$\alpha'(t) = \lambda(A(t)\alpha(t) + B(t)\beta(t)) + e(t), \quad \alpha(a) = 0, \tag{14}$$

$$-\beta'(t) = \lambda(C(t)\alpha(t) + D(t)\beta(t)) + f(t), \quad \beta(b) = 0. \tag{15}$$

Now, define

$$z(t) = (\alpha(t), \beta(t)), \quad h(t) = (e(t), f(t)), \tag{16}$$

$$\mathscr{A}(t) = \begin{bmatrix} A(t) & B(t) \\ -C(t) & -D(t) \end{bmatrix}. \tag{17}$$

Then, (14)–(15) can be written in the canonical form for two-point boundary-value problems as (Refs. 16–17)

$$z'(t) = \lambda\mathscr{A}(t)z(t) + h(t), \quad Pz(a) + Qz(b) = 0, \tag{18}$$

where

$$P = \begin{bmatrix} I_N & 0 \\ 0 & 0 \end{bmatrix}, \quad Q = \begin{bmatrix} 0 & 0 \\ 0 & I_M \end{bmatrix}, \tag{19}$$

and where I_N and I_M are the $N \times N$ and $M \times M$ identity matrices, respectively.

We now present a summary of some of the theorems about two-point boundary-value problems necessary for the discussion of (18). Let $U(t, \lambda)$ be the fundamental solution for the equation

$$x'(t) = \lambda A(t) x(t).\tag{20}$$

That is, $U(t, \lambda)$ is the unique solution of the matrix equation

$$U'(t, \lambda) = \lambda \mathscr{A}(t) U(t, \lambda), \qquad U(a) = I_{N+M}.\tag{21}$$

Define the matrix $\mathscr{H}(t, \lambda)$ by

$$\mathscr{H}(t, \lambda) = P + Q U(t, \lambda).\tag{22}$$

Then, (18) has a unique solution iff $\mathscr{H}(b, \lambda)$ is nonsingular (Ref. 17). If $\mathscr{H}(b, \lambda)$ is singular, then there exists a nonzero vector

$$\zeta \in \mathbb{C}^{N+M}$$

such that

$$\mathscr{H}(b, \lambda)\zeta = 0.$$

Let

$$\eta(t) = U(t, \lambda)\zeta;$$

then, $\eta(t)$ satisfies

$$\eta'(t) = \lambda \mathscr{A}(t)\eta(t), \qquad P\eta(a) + Q\eta(b) = 0.\tag{23}$$

Using this we arrive at the Fredholm alternative for (1)–(2).

Theorem 3.2. Either the homogeneous version of (1),

$$g(t) = 0,$$

has a nonzero solution or the solution of (1)–(2) is unique. A unique solution exists iff $\mathscr{H}(b, \lambda)$ is nonsingular. If $\mathscr{H}(b, \lambda)$ is singular, the number of linearly independent solutions of

$$u(t) = \lambda \int_a^b K(t, s) u(s)\, ds\tag{24}$$

is equal to the dimension of the null space of $\mathscr{H}(b, \lambda)$. In particular, the solutions of (24) have the representation

$$u_i(t) = \sum_{j=1}^N a_j(t)\eta_{1j}(t) + \sum_{j=1}^M c_j(t)\eta_{2j}(t), \qquad i = 1, \ldots, \dim N(\mathscr{H}(b, \lambda)),\tag{25}$$

where

$$\eta(t) = U(t, \lambda)\zeta \quad \text{and} \quad \zeta \in N(\mathscr{H}(b, \lambda)).$$

Proof. By Theorem 3.1, the solution of (1)–(2) is unique iff (18) has a unique solution. From the preceding discussion, this is true iff $\mathscr{H}(b, \lambda)$ is nonsingular. Assume now that (24) has only the zero solution and that $\mathscr{H}(b, \lambda)$ is singular. It then follows that there exists a nonzero solution $\eta(t)$ of (23). Let

$$\psi(t) = \sum_{i=1}^{N} a_i(t)\eta_{1i}(t) + \sum_{i=1}^{M} c_i(t)\eta_{2i}(t). \tag{26}$$

Integration of (23) shows that

$$\eta_{1i}(t) = \int_a^t b_i(s)\psi(s)\,ds, \qquad i = 1, 2, \ldots, N, \tag{27}$$

$$\eta_{2i}(t) = \int_t^b c_i(s)\psi(s)\,ds, \qquad i = 1, 2, \ldots, M. \tag{28}$$

Thus, $\psi(t)$ is not identically zero; if it were, it would follow from (27) and (28) that $\eta(t) = 0$ for all $t \in [a, b]$, and this contradicts our assumptions.

A similar computation shows that the functions $u_i(t)$ in (25) are linearly independent iff

$$\eta_i(t), \qquad i = 1, 2, \ldots, \dim N(\mathscr{H}(b, \lambda)),$$

are linearly independent, and the last assertion of the theorem follows. $\square$

As usual, it is important to characterize solutions of (1) when the homogeneous equation (24) has nonzero solutions. This is done in terms of the adjoint problem to (18).

Definition 3.1. The two-point boundary-value problem

$$x'(t) = -\bar{\lambda}\mathscr{A}^t(t)x(t), \qquad Qx(a) + Px(b) = 0, \tag{29}$$

where $\bar{\lambda}$ is the complex conjugate of λ, is called the adjoint problem to (23). The superscript t denotes matrix transpose.

Lemma 3.1. Let $x(t)$ solve (29) and $z(t)$ solve (18). Then,

$$\int_a^b \langle h(t), x(t) \rangle \, dt = 0, \tag{30}$$

where $\langle \, , \, \rangle$ is the standard inner product on $\mathbb{C}^{N+M}$.

Proof. We have

$$\int_a^b [\langle x'(t), z(t)\rangle + \langle x(t), z'(t)\rangle]\, dt = \langle x(t), z(t)\rangle|_b - \langle x(t), z(t)\rangle|_a = 0, \qquad (31)$$

by the boundary conditions on $x(t)$ and $z(t)$. But

$$z'(t) = \lambda \mathscr{A}(t)z(t) + h(t) \quad \text{and} \quad x'(t) = -\bar{\lambda}\mathscr{A}^t(t)x(t),$$

so that the left-hand side of (31) becomes

$$\int_a^b [\langle -\bar{\lambda}A^t(t)x(t), z(t)\rangle + \langle x(t), \lambda A(t)z(t) + h(t)\rangle]\, dt = \int_a^b \langle x(t), h(t)\rangle\, dt = 0.$$

$$\square$$

From Lemma 3.1, it follows that a necessary condition for (18) to have a solution is that $h(t)$ be orthogonal to the solutions of the adjoint problem. If the solution of (18) is unique, then it follows from Theorem 3.2 that the only solution of (24) is zero. From Lemma 3.2 below, we then find that the only solution of the adjoint problem is zero, so that in this case the orthogonality condition is satisfied.

Our aim now is to show that the orthogonality condition (30) is not only necessary, but sufficient for the existence of a solution to (1). In keeping with our desire to have an independent existence theory for (1)–(2), a proof of this fact is given using only results from linear algebra.

Lemma 3.2. Let
$$\mathscr{H}^T(b, \lambda) = Q + PU^T(b, \lambda),$$

where $U^T(b, \lambda)$ is the fundamental solution for the adjoint equation (29). Then,

$$\dim [\text{nullspace } \mathscr{H}^T(b, \lambda)] = \dim [\text{nullspace } \mathscr{H}(b, \lambda)].$$

Proof. Note that, for a matrix A, $r(A)$ denotes its rank, $\nu(A)$ its nullity, and $N(A)$ its null space. Observe, first of all, that

$$U^T(t, \lambda) = (U^{-1}(t, \lambda))^*,$$

where $*$ denotes Hermitian conjugation. This gives

$$\mathscr{H}^T(b, \lambda) = Q + PU^T(b, \lambda) = (QU^*(b, \lambda) + P)U^T(b, \lambda).$$

Therefore,

$$r(\mathscr{H}^T(b, \lambda)) = r(QU^*(b, \lambda) + P) = r(U(b, \lambda)Q + P),$$

since

$$r(A^*) = r(A) \qquad \text{for all } A$$

and

$$P^* = P \quad \text{and} \quad Q^* = Q.$$

This gives

$$\nu(\mathcal{H}^T(b, \lambda)) = \nu(U(b, \lambda)Q + P).$$

By straightforward computation, we find that

$$QU(b, \lambda) + P = \begin{bmatrix} I_N & 0 \\ G & H \end{bmatrix}, \qquad U(b, \lambda)Q + P = \begin{bmatrix} I_N & F \\ 0 & H \end{bmatrix},$$

where $U(t, \lambda)$ is partitioned as

$$U(t, \lambda) = \begin{bmatrix} E & F \\ G & H \end{bmatrix},$$

and the dimensions of E, F, G, H conform to those in the partition of $\mathcal{A}(t)$. From this, it is easily seen that

$$\nu(QU(b, \lambda) + P) = \nu(U(b, \lambda)Q + P) = \nu(H);$$

and it follows that

$$r(\mathcal{H}^T(b, \lambda)) = r(QU(b, \lambda) + P) = r(\mathcal{H}(b, \lambda)),$$

so that

$$\nu(\mathcal{H}^T(b, \lambda)) = \nu(\mathcal{H}(b, \lambda)).$$

This establishes the lemma. $\qquad\qquad\qquad\qquad\qquad\qquad\qquad$ □

Theorem 3.3. A necessary and sufficient condition for (1)–(2) to have a solution is that

$$\int_a^b \langle h(t), x(t) \rangle \, dt = 0,$$

for all $x(t)$ solving the adjoint problem.

Proof. The necessity is just the statement of Lemma 3.1. To prove the sufficiency, we observe that, from the variation-of-constants formula (Ref. 18), that (18) has a solution iff

$$QU(b, \lambda) \int_a^b U^{-1}(s, \lambda) h(s) \, ds$$

is orthogonal to

$$N(\mathcal{H}^*(b, \lambda)).$$

Let

$$\eta \in N(\mathcal{H}^*(b, \lambda)).$$

Then,

$$\eta = (\eta_1, \eta_2),$$

where

$$H^*\eta_2 = 0 \quad \text{and} \quad \eta_1 = -F^*\eta_2.$$

Here, H, F are as in Lemma 3.2. From the orthogonality condition above, we get that

$$\left\langle QU(b, \lambda) \int_a^b U^{-1}(s, \lambda)h(s)\, ds, \eta \right\rangle$$

$$= \int_a^b \langle U^{-1}(s, \lambda)h(s), U^*(b, \lambda)Q\eta \rangle\, ds$$

$$= \int_a^b \langle h(s), (U^{-1}(s, \lambda))^* U^*(b, \lambda)Q\eta \rangle = 0.$$

Now,

$$(U^{-1}(s, \lambda))^* = U^T(s, \lambda),$$

so that

$$\eta(t) = U^T(t, \lambda)U^*(b, \lambda)Q\eta$$

solves the adjoint differential equation. To check the boundary condition, observe that

$$Q\eta(a) + P\eta(b) = QU^T(a, \lambda)U^*(b, \lambda)Q\eta + PQ\eta = QU^*(b, \lambda)Q\eta.$$

But

$$U^*(b, \lambda)Q\eta = (F^*\eta_2, H^*\eta_2) = (F^*\eta_2, 0) = (-\eta_1, 0),$$

and

$$Q(-\eta_1, 0) = (0, 0).$$

So, it follows that $\eta(t)$ satisfies (29), and thus

$$\int_a^b \langle h(s), \eta(s) \rangle\, ds = 0. \qquad \square$$

Theorem 3.4. A necessary and sufficient condition that (1)–(2) have a solution is that

$$\int_a^b g(t)u^*(t)\,dt = 0, \tag{32}$$

where $u^*(t)$ is a solution of the homogeneous adjoint equation

$$u^*(t) = \bar{\lambda} \int_a^b K^*(t,s)u^*(s)\,ds, \tag{33}$$

and

$$K^*(t,s) = \begin{cases} \displaystyle\sum_{i=1}^{M} c_i(s)d_i(t), & a \leq s < t, \\[2ex] \displaystyle\sum_{i=1}^{N} a_i(s)b_i(t), & t \leq s < b. \end{cases} \tag{34}$$

Proof. From Theorem 3.3, it is seen that (1)–(2) has a solution iff

$$\int_a^b \langle h(s),\eta(s)\rangle\,ds = 0, \tag{35}$$

for all $\eta(t)$ solving (29). Now,

$$h(t) = g(t)(b(t),\,-d(t)),$$

where

$$b(t) = (b_i(t)), \qquad i = 1, 2, \ldots, N,$$
$$d(t) = (d_i(t)), \qquad i = 1, 2, \ldots, M.$$

Using this in (34) gives

$$\int_a^b g(s)[\langle b(s),\eta_1(s)\rangle - \langle d(s),\eta_2(s)\rangle] = 0.$$

Let

$$\phi(t) = \langle b(t),\eta_1(t)\rangle - \langle d(t),\eta_2(t)\rangle.$$

To prove the theorem, it suffices to show that $\phi(t)$ solves (33). Since $\eta(t)$ solves (29), we find by integration that

$$\eta_1(t) = \bar{\lambda} \int_t^b [A^*(s)\eta_1(s) - C^*(s)\eta_2(s)]\,ds,$$

$$\eta_2(t) = \bar{\lambda} \int_a^b [-B^*(s)\eta_1(s) + D^*(s)\eta_2(s)]\,ds.$$

Therefore,

$$\langle b(t), \eta_1(t)\rangle = \bar{\lambda} \int_t^b \langle b(t), A^*(s)\eta_1(s) - C^*(s)\eta_2(s)\rangle \, ds$$

$$= \bar{\lambda} \int_t^b [\langle A(s)b(t), \eta_1(s)\rangle - \langle C(s)b(t), \eta_2(s)\rangle] \, ds,$$

$$\langle d(t), \eta_2(t)\rangle = \bar{\lambda} \int_a^t [\langle -B(s)d(t), \eta_1(s)\rangle + \langle D(s)d(t), \eta_2(s)\rangle] \, ds.$$

This gives

$$\phi(t) = \bar{\lambda} \int_a^t [\langle B(s)d(t), \eta_1(s)\rangle - \langle D(s)d(t), \eta_2(s)\rangle] \, ds$$

$$+ \bar{\lambda} \int_t^b [\langle A(s)b(t), \eta_1(s)\rangle - \langle c(s)b(t), \eta_2(s)\rangle] \, ds$$

$$= \bar{\lambda} \int_a^t \left[\sum_{i=1}^N \sum_{j=1}^M b_i(s)c_j(s)d_j(t)\eta_{1i}(s) - \sum_{i=1}^N \sum_{j=1}^M d_j(t)c_j(s)d_i(s)\eta_{2i}(s) \right] ds$$

$$+ \bar{\lambda} \int_t^b \left[\sum_{i=1}^M \sum_{j=1}^N b_i(s)a_j(s)b_i(t)\eta_{1i}(s) - \sum_{i=1}^M \sum_{j=1}^N d_i(s)a_j(s)b_j(t)\eta_{2i}(s) \right] ds$$

$$= \bar{\lambda} \int_a^t \left[\sum_{i=1}^M c_i(s)d_i(t) \right] \phi(s) \, ds + \bar{\lambda} \int_t^b \left[\sum_{j=1}^N a_j(s)b_j(t) \right] \phi(s) \, ds$$

$$= \bar{\lambda} \int_a^b K^*(t,s)\phi(s) \, ds. \qquad \square$$

From the above theorems, it is seen that the existence and uniqueness theory for (1)–(2) can be derived in an elementary way, independently of the use of Fredholm determinants (Ref. 19) or functional analytic techniques. In addition, as was shown in Ref. 4 and Ref. 9, the representation of the solution of (1)–(2) in terms of the solution to (18) leads to efficient procedures for numerically solving (1)–(2). The emphasis there was primarily on algorithms of the imbedding type; however, it is obvious that any efficient, stable, numerical method for solving two-point boundary-value problems may be used to solve (18), and thus (1). We will return to this again in Section 5.

Before leaving our discussion of the Fredholm theory, it is interesting to examine the relationship between the adjoint boundary-value problem and the usual adjoint integral equation. Since $K^*(t, s)$ is a semidegenerate kernel, one can find an equivalent two-point boundary-value problem for the adjoint equation as in Theorem 3.1. One expects that this boundary-value problem should be (29). But it is not. However, a modification of the representation given in Theorem 3.1 gives the appropriate relationship. This is summarized below.

Theorem 3.5.   Let $u^*(t)$ solve

$$u^*(t) = \bar{\lambda} \int_a^b K^*(t, s) u^*(s)\, ds + g(t). \tag{36}$$

Define $\alpha_i^*(t)$, $i = 1, 2, \ldots, N$, and $\beta_i^*(t)$, $i = 1, 2, \ldots, M$, by

$$\alpha_i^*(t) = -\int_t^b a_i(s) u^*(s)\, ds, \qquad \beta_i^*(t) = \int_a^t c_i(s) u^*(s)\, ds.$$

Then, $u^*(t)$ has the representation

$$u^*(t) = g(t) - \bar{\lambda} \left[\sum_{i=1}^N \alpha_i^*(t) b_i(t) - \sum_{i=1}^M \beta_i^*(t) c_i(t) \right], \tag{37}$$

where

$$\alpha^*(t) = (\alpha_i^*(t)), \qquad i = 1, 2, \ldots, N,$$
$$\beta^*(t) = (\beta_i^*(t)), \qquad i = 1, 2, \ldots, M,$$

satisfy the adjoint two-point boundary-value problem

$$z^{*\prime}(t) = -\bar{\lambda}\mathscr{A}^*(t) z^*(t) + h(t),$$
$$Q z^*(a) + P z^*(b) = 0,$$
$$z^*(t) = (\alpha^*(t), \beta^*(t)).$$

The remaining quantities in (37) are defined as before.

Proof.   The proof follows exactly along the lines as Theorem 3.1. We leave the details to the reader. $\square$

Using Theorem 3.5, it is seen that the two two-point boundary-value problems related to the adjoint equation (36) are related in the following way:

$$z_*(t) = \mathscr{S} z^*(t), \tag{38}$$

where $z_*(t)$ is the solution of the boundary-value problem associated with (36) coming from Theorem 3.1, $z^*(t)$ is given in Theorem 3.5, and

$$\mathscr{S} = \begin{vmatrix} 0 & I_M \\ -I_N & 0 \end{vmatrix}. \tag{39}$$

To further illustrate the utility of our approach, we indicate briefly how our theory may be used to obtain generalized solutions to (1) when the homogeneous equation has nonzero solutions. In addition, it is assumed that the orthogonality condition of Theorem 3.4 does not necessarily hold. Such *ill-posed problems* have recently been discussed in detail by Rall and Nashed in Ref. 20 using generalized inverses of operators in a Banach space. For the class of equations given by (1)–(2), the construction of generalized solutions can be reduced to computing the generalized inverse of a matrix.

Let the above assumptions concerning (1)–(2) hold. Then, it follows from Theorems 3.1 to 3.4 that the homogeneous boundary-value problem

$$z'(t) = \lambda \mathcal{A}(t) z(t), \qquad Pz(a) + Qz(b) = 0 \tag{40}$$

in general will have nonzero solutions, and the orthogonality condition (30) will not necessarily be satisfied. To find generalized solutions of (18), we proceed in the following way.

Let $v(t)$ satisfy

$$v'(t) = \lambda \mathcal{A}(t) v(t) + h(t), \qquad v(a) = 0.$$

Let $\mathcal{H}^+(b, \lambda)$ be the Moore–Penrose generalized inverse of $\mathcal{H}(b, \lambda)$ (Refs. 20–21) (other generalized inverses may be used; however, the Moore–Penrose one is commonly used in practice), and define

$$z^+(t) = -U(t, \lambda) \mathcal{H}^+(b, \lambda) Qv(b) + v(t). \tag{41}$$

Lemma 3.3. $z^+(t)$ solves the boundary-value problem

$$dz^+(t)/dt = \lambda \mathcal{A}(t) z^+(t) + h(t), \tag{42}$$

$$Pz^+(a) + Qz^+(b) = (I_{N+M} - P^R) Qv(b), \tag{43}$$

where P^R is the matrix of orthogonal projection onto the range of $\mathcal{H}(b, \lambda)$.

Proof. That $z^+(t)$ is a solution of (42) is an immediate consequence of the representation theory for linear differential equations. To verify the boundary condition, we compute

$$
\begin{aligned}
Pz^+(a) + Qz^+(b) &= -P\mathcal{H}^+(b, \lambda) Qv(b) \\
&\quad + Q[-U(b, \lambda) \mathcal{H}^+(b, \lambda) Qv(b) + v(b)] \\
&= -(P + QU(b, \lambda)) \mathcal{H}^+(b, \lambda) Qv(b) + Qv(b) \\
&= -\mathcal{H}(b, \lambda) \mathcal{H}^+(b, \lambda) Qv(b) + Qv(b) \\
&= (I_{N+M} - P^R) Qv(b),
\end{aligned}
$$

where we have used the fact that

$$\mathcal{H}^+(b, \lambda)\mathcal{H}(b, \lambda) = P^R$$

(see Ref. 20).   □

Note that, if the orthogonality condition (30) is satisfied, then

$$Qv(b) \in \text{Range}(\mathcal{H}(b, \lambda)),$$

so that

$$(I_{N+M} - P^R)Qv(b) = 0,$$

and $z^+(t)$ is then a true solution of (18). In general, $z^+(t)$ satisfies the boundary condition only in a *least-square sense*.

Now, let

$$u^+(t) = g(t) + \sum_{i=1}^{N} a_i(t)z_{1i}^+(t) + \sum_{i=1}^{M} c_i(t)z_{2i}^+(t), \tag{44}$$

where

$$z^+(t) = (z_1^+(t), z_2^+(t)).$$

$u^+(t)$ is then a generalized solution of (1)–(2).

We close this section by offering a brief comparison of our approach to that of Aalto (Ref. 5) and Rall (Ref. 6). As was pointed out in the introduction, they have also developed an independent Fredholm theory for (1)–(2), but in a somewhat different manner.

In Refs. 5–6, (1)–(2) are written in the form

$$u(t) = \lambda \int_a^t \left[\sum_{i=1}^{N} a_i(t)b_i(s) - \sum_{i=1}^{M} c_i(t)d_i(s) \right] u(s)\, ds$$

$$+ \lambda \int_a^b \left[\sum_{i=1}^{M} c_i(t)d_i(s) \right] u(s)\, ds + g(t)$$

$$= \lambda \int_a^t \hat{K}(t, s)u(s)\, ds + \gamma(t), \tag{45}$$

where

$$\hat{K}(t, s) = \sum_{i=1}^{N} a_i(t)b_i(s) - \sum_{i=1}^{M} c_i(t)d_i(s),$$

$$\gamma(t) = \lambda \int_a^b \left(\sum_{i=1}^{M} c_i(t)d_i(s) \right) u(s)\, ds + g(t).$$

Regarding (45) as a Volterra equation with kernel $\hat{K}(t, s)$ it can be solved by the formula (Ref. 22)

$$u(t) = \gamma(t) + \lambda \int_a^t R_{\hat{K}}(t, s, \lambda)\gamma(s)\, ds, \tag{46}$$

where $R_{\hat{K}}(t, s, \lambda)$ is the resolvent kernel of $\hat{K}(t, s)$. But

$$\int_a^t R_{\hat{K}}(t, s, \lambda)\gamma(s)\, ds$$

$$= \int_a^t R_{\hat{K}}(t, s, \lambda)\left\{ \int_a^b \left(\left[\sum_{i=1}^M c_i(s)d_i(\tau)\right]u(\tau)\, d\tau + g(s)\right)\right\}\, ds$$

$$= \sum_{i=1}^M \int_a^b \left[\left[\int_a^t R_{\hat{K}}(t, s, \lambda)c_i(s)\, ds\right]d_i(\tau)u(\tau)\, d\tau + \int_a^t R_{\hat{K}}(t, s, \lambda)g(s)\, ds. \tag{47}$$

Using (47) in (46), it is easily seen that (46) is a Fredholm equation with a separable kernel; thus, it can be reduced in the usual way (Ref. 1) to solving a system of M linear equations in M unknowns. Using this representation, Aalto and Rall have established the basic solvability conditions for (1)–(2). Note that the apparent discrepancy in the dimensions of $\mathcal{H}(b, \lambda)$ and Aalto's matrix disappears when we observe from Lemma 3.2 that

$$\text{rank}(\mathcal{H}(b, \lambda)) = \text{rank}(H)$$

and that H is an $M \times M$ matrix. It is not immediately clear if H is the same matrix that Aalto and Rall derive. A more detailed comparison of these two methods will be given in future work (Ref. 23).

4. Fredholm Representations

In this section, the results of Section 3 are applied to give determinantal representations of the resolvent kernel of $K(t, s)$, analogous to the ones obtained in the general Fredholm theory. These are then utilized to discuss the eigenvalue problem for $K(t, s)$. Here again, our results parallel those of Aalto and Rall. The representation that we obtain for the resolvent kernel is a generalization of one first given by Goursat (Ref. 10) for Volterra separable kernels and further discussed by Bownds and Cushing in Ref. 11.

It follows from Theorem 3.2 that, if λ is not an eigenvalue of $K(t, s)$, then (1) has a unique solution which can be represented in the form

$$u(t, \lambda) = g(t) + \lambda \int_a^b R(t, s, \lambda)g(s)\, ds, \tag{48}$$

where $R(t, s, \lambda)$ is the resolvent kernel of $K(t, s)$. Our purpose is to obtain an expression for $R(t, s, \lambda)$ in terms of the Green's matrix for the boundary-value problem (18). From this formula, one easily establishes that $R(t, s, \lambda)$ is a meromorphic function of λ which can be expressed in terms of the ratio of appropriately chosen determinants.

Theorem 4.1. There are at most countably many eigenvalues λ_n, $n = 1, 2, \ldots$, of $K(t, s)$. If $\lambda \neq \lambda_n$, then the resolvent kernel $R(t, s, \lambda)$ can be expressed by the formula

$$R(t, s, \lambda) = \langle a(t), P\mathcal{G}(t, s, \lambda)\hat{h}(s)\rangle_N + \langle c(t), Q\mathcal{G}(t, s, \lambda)\hat{h}(s)\rangle_M, \quad (49)$$

where $\mathcal{G}(t, s, \lambda)$ is the Green's matrix for (18) and is given by

$$\mathcal{G}(t, s, \lambda) = \begin{cases} U(t, \lambda)\mathcal{H}^{-1}(b, \lambda)PU^{-1}(s, \lambda), & a \leq s < t, \\ U(t, \lambda)\mathcal{H}^{-1}(b, \lambda)QU(b, \lambda)U^{-1}(s, \lambda), & t \leq s \leq b. \end{cases} \quad (50)$$

Here, $\langle,\rangle_N$ and $\langle,\rangle_M$ represent the standard inner products on R^N and R^M, respectively, and

$$\hat{h}(t) = (b(t), -d(t)).$$

Proof. It follows from Theorem 3.2 that the eigenvalues of $K(t, s)$ are given by the zeros of $\det(\mathcal{H}(b, \lambda))$. Since $U(t, \lambda)$ is an entire function of λ (Ref. 18), $\det(\mathcal{H}(b, \lambda))$ is also entire. From this, it is seen that $\det(\mathcal{H}(b, \lambda))$ has at most countably many zeros, and thus $K(t, s)$ has at most countably many eigenvalues.

If λ is not an eigenvalue of $K(t, s)$, then standard results for two-point boundary-value problems show that the solution of (18) is given by

$$z(t) = \int_a^b \mathcal{G}(t, s, \lambda)h(s)\, ds. \quad (51)$$

From Theorem 3.1., it follows that

$$u(t, \lambda) = g(t) + \langle a(t), Pz(t)\rangle_N + \langle c(t), Qz(t)\rangle_M. \quad (52)$$

Substitution of (51) into (52) gives

$$u(t, \lambda) = g(t) + \int_a^b [\langle a(t), P\mathcal{G}(t, s, \lambda)\hat{h}(s)\rangle_N$$

$$+ \langle c(t), Q\mathcal{G}(t, s, \lambda)\hat{h}(s)\rangle_M]g(s)\, ds. \quad (53)$$

Since the kernel in (53) is continuous, except possibly for $t = s$, and since (53) holds for all continuous $g(t)$, the kernel must equal $R(t, s, \lambda)$, except possibly for $t = s$. Redefining $R(t, s, \lambda)$ on $t = s$, if necessary a set of measure zero, gives (49).   □

Note that, if $K(t, s)$ is a Volterra degenerate kernel, then

$$\mathcal{G}(t, s, \lambda) = \begin{cases} U(t, \lambda)U^{-1}(s, \lambda), & a \le s < t, \\ 0, & t \le s < b, \end{cases}$$

and

$$R(t, s, \lambda) = \begin{cases} \langle a(t), PU(t, \lambda)U^{-1}(s, \lambda)\hat{h}(s)\rangle_N \\ \quad + \langle c(t)QU(t, \lambda)U^{-1}(s, \lambda)\hat{h}(s)\rangle_M, & a \le s < t, \\ 0, & t \le s \le b, \end{cases}$$

and thus agrees with the representation given by Bownds and Cushing in Refs. 11 and 24. It is also interesting to observe that it follows from (49) and (50) that $R(t, s, \lambda)$ is also a semidegenerate kernel. Rall has established the same result in Ref. 6, although his representation is different from ours.

Theorem 4.2.   The resolvent kernel can be factored as

$$R(t, s, \lambda) = D(t, s, \lambda)/\det(\mathcal{H}(b, \lambda)), \tag{54}$$

where $D(t, s, \lambda)$ and $\det(\mathcal{H}(b, \lambda))$ are entire functions of λ. In particular, every pole of $R(t, s, \lambda)$ is an eigenvalue of $K(t, s)$.

Proof.   From Cramer's rule, it follows that

$$\mathcal{H}^{-1}(b, \lambda) = \mathrm{Adj}(\mathcal{H}(b, \lambda))/\det(\mathcal{H}(b, \lambda)), \tag{55}$$

where both the numerator and denominator in (55) are entire in λ. Substitution of (55) into (50), the expression for $\mathcal{G}(t, s, \lambda)$, shows that it can be factored as

$$\mathcal{G}(t, s, \lambda) = \mathcal{J}(t, s, \lambda)/\det(\mathcal{H}(b, \lambda)), \tag{56}$$

where $\mathcal{J}(t, s, \lambda)$ is an entire matrix-valued function of λ. Using (56) in (49) establishes the theorem.   □

The representation of $R(t, s, \lambda)$ in (54) is our analogue of the Fredholm determinant representation given by the general theory (Ref. 19). It is not known at this writing if $\det(\mathcal{H}(b, \lambda))$ is in fact the Fredholm determinant of $K(t, s)$, although it can differ from it by at most a factor, entire in λ, having no zeros.

5. Convolution Equations

Although the theory of Sections 3 and 4 can be used to develop general computing algorithms for equations with semidegenerate kernels, we have found that, if the kernel has some additional structure, then it is possible to derive alternate boundary-value problems to calculate the solutions (Refs. 4, 9, 15). The most interesting case occurs when the kernel in (1) is a function of $t-s$. Then, as has been shown in Refs. 4, 9, 15, boundary-value problems with constant coefficients may be obtained for particular kernels, e.g., those of the form

$$K(t-s) = \sum_{i=1}^{N} \sigma_i \exp[-\lambda_i|t-s|]. \tag{57}$$

The theory is intimately related to the method of invariant imbedding introduced by Chandrasekhar (Ref. 7) and Sobolev (Ref. 3) for the equations of radiative transfer. In this work, and the more recent extensions by Bellman (Ref. 25), Kalaba and Kagiwada (Ref. 3), Casti (Ref. 26), and others (Ref. 27), the emphasis has generally been placed on the derivation of initial-value problems for integrodifferential equations equivalent to integral equations with kernels of the form

$$K(|t-s|) = \int_A \sigma(z) \exp[-z|t-s|] \, dz. \tag{58}$$

The integrals in these equations are discretized, giving rise to a finite set of ordinary differential equations which are then integrated to obtain the solution of the original integral equation (Ref. 3).

In this section, we consider equations with kernels of the type in (58) and show how one can find equivalent two-point boundary-value problems for their solutions. Our main purpose is to compare this procedure to the more standard imbedding approach. Several results are established. First, it is shown that the application of the generalized Riccati transformation (Refs. 28–29) gives an initial-value problem which structurally has many of the properties of the invariant imbedding equations. In particular, generalizing a result of Casti (Ref. 30) and Kailath (Ref. 31), we obtain analogues of the Chandrasekhar X and Y functions (Ref. 3). As in Ref. 3, the equations that are derived are integrodifferential ones. To obtain a feasible numerical algorithm, the integrals are replaced by quadratures, the result being a Cauchy problem for a system of ordinary differential equations. The solution of this Cauchy problem is then identical to that obtained by first approximating (58) by the same quadrature rule and then reducing the resulting integral equation to an initial-value problem via the usual Riccati transformation as in Refs. 4 and 9. This type of commutativity result is analogous to ones proved by us for other types of imbedding algorithms (Ref. 32). From a numerical standpoint, then, the direct reduction of (1) with the kernel given by (58) gives nothing new, although we feel that it does throw considerable light on the rich structure of these equations. A further consequence of our approach is that the initial-value problem that we obtain contains only functions of a single variable, in contrast to the functions of two or three variables that occur in the more traditional approaches (Ref. 3). The resulting reduction in the number of equations to be integrated appears to counteract some of the criticism recently made about such methods.

Now, consider (1) with the kernel (58). Throughout this section, we set $\lambda = 1$, for convenience. Assume that

$$\sigma(z) \in L_1[A].$$

Define

$$\alpha(t, z) = \int_a^t \exp[-z(t-s)]u(s) \, ds, \tag{59}$$

$$\beta(t, z) = \int_t^b \exp[-z(s-t)]u(s) \, ds. \tag{60}$$

Then, $u(t)$ has the representation

$$u(t) = g(t) + \int_A [\alpha(t, z) + \beta(t, z)]\sigma(z) \, dz. \tag{61}$$

Differentiation of (59) and (60) with respect to t gives

$$\alpha_t(t, z) = u(t) - z \int_a^t \exp[-z(t-s)]u(s) \, ds$$

$$= g(t) + \int_A [\alpha(t, z') + \beta(t, z')]\sigma(z') \, dz'$$

$$- z\alpha(t, z), \qquad \alpha(a, z) = 0, \tag{62}$$

$$\beta_t(t, z) = -g(t) - \int_A [\alpha(t, z') + \beta(t, z')]\sigma(z') \, dz'$$

$$- z\beta(t, z), \qquad \beta(b, z) = 0. \tag{63}$$

Assume now that, for each $t \in [a, b]$, that $\alpha(t, z)$ and $\beta(t, z)$ are elements of an appropriate Banach space of functions on A. Since our interest in this paper is primarily focused on the use of semidegenerate kernels, the following calculations are mostly formal; rigorous justification will be given elsewhere (Ref. 33). With these assumptions, (62)–(63) can be formulated as a two-point boundary-value problem in B. Define vector-valued functions

$$\alpha(t): [a, b] \rightarrow B \quad \text{and} \quad \beta(t): [a, b] \rightarrow B$$

by

$$\alpha(t) = \alpha(t, z), \qquad \beta(t) = \beta(t, z), \qquad z \in A.$$

In addition, every function $\psi(t)$ of t alone will be considered as a constant B-valued function. Let operators

$$\Sigma: B \to B \quad \text{and} \quad \mathscr{Z}: B \to B$$

be defined by

$$(\Sigma u)(z) = \int_A \sigma(z')u(z')\,dz', \qquad (64)$$

$$(\mathscr{Z} u)(z) = zu(z), \qquad z \in A. \qquad (65)$$

Then, taking

$$V = B \otimes B$$

($\otimes$ denotes direct product), (62)–(63) take the form

$$\alpha'(t) = (\Sigma - \mathscr{Z})\alpha(t) + \Sigma\beta(t) + g(t), \qquad \alpha(a) = 0, \qquad (66)$$

$$-\beta'(t) = (\Sigma - \mathscr{Z})\beta(t) + \Sigma\alpha(t) + g(t), \qquad \beta(b) = 0. \qquad (67)$$

To solve (66) and (67), we consider converting it to an equivalent initial-value problem by the use of the generalized Riccati transformation (Refs. 28–29). That is, let

$$\alpha(t) = R(t)\beta(t) + \rho(t), \qquad (68)$$

where $R(t)$ is assumed to be a bounded operator on B for each $t \in [a, b]$. Using the results in Ref. 28, it is seen that $R(t)$ and $\rho(t)$ solve

$$R'(t) = \Sigma + (\Sigma - \mathscr{Z})R(t) + R(t)(\Sigma - \mathscr{Z}) + R(t)\Sigma R(t), \qquad (69)$$

$$R(a) = 0, \qquad (70)$$

$$\rho'(t) = (\Sigma - \mathscr{Z} + R(t)\Sigma)\rho(t) + R(t)g(t) - g(t), \qquad (71)$$

$$\rho(a) = 0. \qquad (72)$$

In principle then, the boundary-value problem (66)–(67) can be solved by integrating (69)–(72) from a to b, obtaining $\rho(t)$, substituting (68) into (67), and then integrating the equation for $\beta(t)$ backward from b to a. $\alpha(t)$ is then obtained from (68) and $u(t)$ from (61). This approach is the generalization of the well-known double-sweep method (Ref. 34) for solving finite-dimensional two-point boundary-value problems.

Of course, for the above procedure to be numerically practical, a representation is needed for $R(t)$. If it is assumed that

$$(R(t)u)(z) = \int_A r(t, z, z')u(z')\, dz', \tag{73}$$

then it can be shown that (69)–(72) are equivalent to a pair of integrodifferential equations for $r(t, z, z')$ and $\rho(t, z)$ (Ref. 29). When the integrals are replaced by quadratures, one gets a finite set of ordinary differential equations which can be integrated to obtain $u(t)$.

In the standard imbedding approach, the most interesting structural feature is the existence of the Chandrasekhar X and Y functions. Numerically, their availability leads to a smaller system of equations than using the above scheme directly. Although the above algorithm can be shown to possess similar features, a more useful and direct comparison with the usual imbedding methods can be given if a different transformation is used. To this end, we make the additional assumption that

$$\sigma(z) \geq 0, \qquad z \in A,$$

a condition frequently encountered in practice (Ref. 3). Then, introduce

$$\gamma(t, z) = \sqrt{[\sigma(z)]} \int_a^t \exp[-z(t-s)]u(s)\, ds, \tag{74-1}$$

$$\delta(t, z) = \sqrt{[\sigma(z)]} \int_t^b \exp[-z(s-t)]u(s)\, ds. \tag{74-2}$$

Proceeding as before, we find that

$$u(t) = g(t) + \int_A \sqrt{[\sigma(z')]}[\gamma(t, z') + \delta(t, z')]\, dz', \tag{75}$$

where $\gamma(t, z)$ and $\delta(t, z)$ solve the boundary-value problem

$$\gamma_t(t, z) = \int_A \sqrt{[\sigma(z)\sigma(z')]}\gamma(t, z')\, dz' - z\gamma(t, z) + \sqrt{[\sigma(z)]}g(t)$$

$$+ \int_A \sqrt{[\sigma(z)\sigma(z')]}\delta(t, z')\, dz', \qquad \gamma(a, z) = 0, \tag{76}$$

$$-\delta_t(t, z) = \int_A \sqrt{[\sigma(z)\sigma(z')]}\gamma(t, z')\, dz' + \int_A \sqrt{[\sigma(z)\sigma(z')]}\delta(t, z')\, dz'$$

$$- z\delta(t, z) + \sqrt{[\sigma(z)]}g(t), \qquad \delta(b, z) = 0. \tag{77}$$

Equations (76)–(77) are now put into the form of an abstract two-point boundary-value problem. To do this, B above is taken to be $L_2[A]$; and, as

326

M. A. Golberg

before, operators Σ and $\mathscr{Z}$ are defined by

$$(\Sigma u)(z) = \int_0^1 \sqrt{[\sigma(z)\sigma(z')]} u(z') \, dz', \qquad (\mathscr{Z}u)(z) = zu(z). \qquad (78)$$

Since

$$\sigma(z) \in L_1[A],$$

it is easily shown that both Σ and $\mathscr{Z}$ are bounded operators on $L_2[A]$. In addition, Σ and $\mathscr{Z}$ are self-adjoint and Σ has rank one. With these observations, (76)–(77) take the form

$$\gamma_t(t) = (\Sigma - \mathscr{Z})\gamma(t) + \Sigma\delta(t) + G(t), \qquad \gamma(a) = 0, \qquad (79)$$

$$-\delta_t(t) = \Sigma\gamma(t) + (\Sigma - \mathscr{Z})\delta(t) + G(t), \qquad \delta(b) = 0, \qquad (80)$$

where

$$G(t) = \sqrt{[\sigma(z)]}g(t).$$

As before, the Riccati transformation is used to solve (79) and (80). Therefore, let

$$\gamma(t) = R(t)\delta(t) + \rho(t), \qquad (81)$$

where $R(t)$ and $\rho(t)$ solve the initial-value problems

$$R'(t) = \Sigma + (\Sigma - \mathscr{Z})R(t) + R(t)(\Sigma - \mathscr{Z}) + R(t)\Sigma R(t), \qquad R(a) = 0, \qquad (82)$$

$$\rho'(t) = ((\Sigma - \mathscr{Z}) + R(t)\Sigma)\rho(t) + R(t)G(t) + G(t), \qquad \rho(a) = 0. \qquad (83)$$

As we pointed out above, the most important aspect of the invariant imbedding approach to solving (79)–(80) is the existence of the X and Y functions of Chandrasekhar (Refs. 3, 7). In effect, they allow one to bypass solving an operator-Riccati equation analogous to (82) by solving a pair of vector equations instead. It is shown that a similar reduction exists for (82) as well.

From standard results for differential equations in a Banach space, it follows that the initial-value problem for $R(t)$ has a unique solution for $a \le t < \bar{a}$ (Ref. 18). Assume that $b < \bar{a}$.

To proceed, it is necessary to introduce several algebraic constructions.

Definition 5.1. Let $\mathscr{H}$ be a Hilbert space with inner product $\langle \, , \, \rangle$. The tensor product $u \otimes v$ of $u \in \mathscr{H}$ and $v \in \mathscr{H}$ is the linear operator defined by

$$(u \otimes v)(w) = \langle v, w \rangle u, \qquad w \in \mathscr{H}. \qquad (84)$$

It follows easily from the Cauchy–Schwarz inequality that $u \otimes v$ is a bounded operator on $\mathcal{H}$. Denote the function $\sqrt{[\sigma(z)]}$ by A, and observe that it follows from (84) that

$$\Sigma = A \otimes A. \tag{85}$$

Before establishing our main factorization theorem for $R(t)$, several properties of it and the tensor product are established. First, observe that $R(t)$ is self-adjoint on the interval $[a, \bar{a})$. This follows from the self-adjointness of Σ and $\mathcal{L}$ and the fact that $R^*(t)$ satisfies

$$(R^*(t))' = \Sigma + R^*(t)(\Sigma - \mathcal{L}) + (\Sigma - \mathcal{L})R^*(t) + R^*(t)\Sigma R^*(t), \qquad R^*(a) = 0,$$

which is the same initial-value problem as that for $R(t)$. By uniqueness (Ref. 18),

$$R(t) = R^*(t).$$

Second, if C is a bounded operator on $\mathcal{H}$, then

$$(Cu) \otimes v = C(u \otimes v), \qquad (u \otimes C^*v) = (u \otimes v)C. \tag{86}$$

Here, * denotes the adjoint of a bounded operator on $\mathcal{H}$.

Theorem 5.1. Let $R(t)$ solve the initial-value problem (82), and let $\mathcal{P}(t)$ be determined by

$$\mathcal{P}'(t) = ((\Sigma - \mathcal{L}) + R(t)\Sigma)\mathcal{P}(t), \qquad \mathcal{P}(a) = I. \tag{87}$$

Then, the vectors

$$\mathcal{H}(t) = R(t)A \qquad \text{and} \qquad \mathcal{L}(t) = \mathcal{P}(t)A$$

solve the initial-value problems.

$$\mathcal{L}'(t) = ((\Sigma - \mathcal{L}) + \mathcal{H}(t) \otimes A))\mathcal{L}(t), \qquad \mathcal{L}(a) = A, \tag{88}$$

$$\mathcal{H}'(t) = (\mathcal{L}(t) \otimes \mathcal{L}(t))A, \qquad \mathcal{H}(a) = 0. \tag{89}$$

Here, I denotes the identity operator on $L_2[A]$.

Proof. Since $R(t)$ is a C^∞ function (Ref. 18), differentate (82) to give

$$R''(t) = (\Sigma - \mathcal{L})R'(t) + R'(t)(\Sigma - \mathcal{L}) + R'(t)\Sigma R(t) + R(t)\Sigma R'(t). \tag{90}$$

Let

$$S(t) = R'(t).$$

Then, $S(t)$ solves

$$S'(t) = CS(t) + SD(t), \qquad S(a) = \Sigma, \tag{91}$$

where

$$C = \Sigma - \mathscr{L} + R(t)\Sigma \quad \text{and} \quad D = \Sigma - \mathscr{L} + \Sigma R(t). \tag{92}$$

Since $S(t)$ solves a linear initial-value problem, a unique solution to (91) exists for $t \in [a, \bar{a})$ (Ref. 18). Since

$$\Sigma = A \otimes A,$$

It can be shown by differentiation and the use of (86) that:

$$S(t) = (\mathscr{P}(t)A) \otimes (Q^*(t)A), \tag{93}$$

where $Q(t)$ solves

$$Q'(t) = Q(t)D, \qquad Q(a) = I. \tag{94}$$

Now,

$$\mathscr{L}(t) = \mathscr{P}(t)A,$$

so that differentiation gives

$$\begin{aligned}
\mathscr{L}'(t) &= \mathscr{P}'(t)A = C\mathscr{P}(t)A = C\mathscr{L}(t) \\
&= (\Sigma - \mathscr{L} - R(t)(A \otimes A))\mathscr{L}(t).
\end{aligned} \tag{95}$$

But

$$R(t)A \otimes A = (R(t)A) \otimes A,$$

so that

$$\mathscr{L}'(t) = (\Sigma - \mathscr{L} - \mathscr{K}(t) \otimes A)\mathscr{L}(t).$$

Also,

$$\mathscr{K}'(t) = R'(t)A = S(t)A = (\mathscr{L}(t) \otimes (Q^*(t)A))A. \tag{96}$$

To complete the proof, it suffices to show that

$$Q^*(t)A = \mathscr{L}(t).$$

This is established by showing that $Q^*(t)A$ satisfies the same initial-value problem as $\mathscr{L}(t)$. Let

$$M(t) = Q^*(t)A,$$

and differentiate, giving

$$M'(t) = (Q^*(t))'A = D^*Q^*(t)A = D^*M(t).$$

But

$$D^* = \Sigma - \mathscr{L} + R^*(t)\Sigma = \Sigma - \mathscr{L} + R(t)\Sigma = C,$$

so that

$$M'(t) = CM(t) \quad \text{and} \quad M(a) = I.$$

By uniqueness (Ref. 18),

$$M(t) = \mathscr{L}(t),$$

and the theorem is proved. $\qquad\qquad\qquad\qquad\qquad\qquad\qquad\qquad\qquad\quad$ □

The functions $\mathscr{L}(t)$ and $\mathscr{K}(t)$ are the analogues of the Chandrasekhar X and Y functions and generalize those given by us for the finite-dimensional Riccati equations in Ref. 4. The importance of $\mathscr{L}(t)$ and $\mathscr{K}(t)$ is that their existence allows one to completely bypass the computation of $R(t)$, leading (as we shall see) to numerical algorithms with a minimal number of equations to be integrated.

For this, observe that $\rho(t)$ solves the equation

$$\rho'(t) = [\Sigma - \mathscr{L} + R(t)(A \otimes A)]\rho(t) + \mathscr{K}(t) \circ g(t) + A \circ g(t)$$

$$= [\Sigma - \mathscr{L} + \mathscr{K}(t) \otimes A]\rho(t) + \mathscr{K}(t) \circ g(t) + A \circ g(t), \qquad \rho(a) = 0, \qquad (97)$$

where $\circ$ denotes pointwise multiplication of functions. Having obtained $\mathscr{L}(t)$, $\mathscr{K}(t)$, $\rho(t)$, then $\delta(t)$ may be determined by solving the initial-value problem

$$-\delta'(t) = \Sigma(R(t)\delta(t) + \rho(t)) + (\Sigma - \mathscr{L})\delta(t) + G(t), \qquad (98)$$

$$\delta(b) = 0. \qquad (99)$$

Since $R(t)$ is self-adjoint,

$$\Sigma R(t)\delta(t) = (A \otimes A)R(t)\delta(t) = (A \otimes \mathscr{K}(t))\delta(t).$$

Thus, $\delta(t)$ solves

$$-\delta'(t) = (A \otimes K(t) + \Sigma - \mathscr{L})\delta(t) + \Sigma\rho(t) + G(t), \qquad (100)$$

and can be obtained without explicitly knowing $R(t)$. From (76), it is seen that $u(t)$ is determined by

$$u(t) = g(t) + \langle A, \gamma(t) + \delta(t) \rangle$$

$$= g(t) + \langle A, R(t)\delta(t) + \rho(t) + \delta(t) \rangle$$

$$= g(t) + \langle \mathscr{K}(t), \delta(t) \rangle + \langle A, \rho(t) + \delta(t) \rangle, \qquad (101)$$

where the self-adjointness of $R(t)$ has been used.

Summarizing, then, $u(t)$ can be found by integrating the four vector equations for $\mathscr{L}(t)$, $\mathscr{K}(t)$, $\rho(t)$, $\delta(t)$, and then evaluating it from (101). In contrast to previous initial-value methods based on interval-length imbedding, only functions of a single variable appear in our procedure, whereas the usual methods typically involve functions of t and b (Refs. 3, 8, 35).

To compare the results of this paper to those based on kernel approximations of the form (57), we first write out the equations for $\mathcal{L}(t)$, $\mathcal{K}(t)$, $\rho(t)$, $\delta(t)$, $u(t)$ explicitly as functions of t and z.

Theorem 5.2. The solution to (1) with the kernel (58), where $\sigma(z) \ge 0$, can be obtained by solving the following Cauchy problems:

$$L_t(t, z) = \int_A \sqrt{[\sigma(z')]} L(t, z')\, dz' [\sqrt{[\sigma(z)]} + K(t, z)]$$

$$- zL(t, z), \qquad L(a, z) = \sqrt{[\sigma(z)]}, \tag{102}$$

$$K_t(t, z) = \left[\int_A \sqrt{[\sigma(z')]} L(t, z')\, dz' \right] L(t, z), \qquad K(a, z) = 0, \tag{103}$$

$$\rho_t(t, z) = \int_A \sqrt{[\sigma(z')]} \rho(t, z')\, dz' [\sqrt{[\sigma(z)]} - K(t, z)] - z\rho(t\, z)$$

$$+ g(t)[K(t, z) + \sqrt{[\sigma(z)]}], \qquad \rho(a, z) = 0, \tag{104}$$

$$- \delta_t(t, z) = \left[\int_A K(t, z')\delta(t, z')\, dz' \right] \sqrt{[\sigma(z)]} - z\delta(t, z)$$

$$+ \sqrt{[\sigma(z)]} \int_A \sqrt{[\sigma(z')]} [\delta(t, z') + \rho(t, z')]\, dz'$$

$$+ \sqrt{[\sigma(z)]} g(t), \qquad \delta(b, z) = 0, \tag{105}$$

$$u(t) = g(t) + \int_A K(t, z')\delta(t, z')\, dz'$$

$$+ \int_A \sqrt{[\sigma(z')]} [\delta(t, z') + \rho(t, z')]\, dz'. \tag{106}$$

Proof. Equations (102)–(106) follow directly from (88), (89), (97), (98), and (99). (106) comes from (101). □

To obtain a feasible numerical algorithm from Theorem 5.2, the integrals in Eqs. (102)–(106) are discretized using a quadrature rule of the form

$$\int_A f(z)\, dz = \sum_{i=1}^{N} w_i f(z_i) + E_N, \tag{107}$$

where E_N is the integration error. If (107) is used to evaluate the right-hand sides of (102)–(105) and the error term neglected, we arrive at a finite system of ordinary differential equations with given initial conditions. These can then be integrated, and the results substituted into (106), to obtain a numerical approximation $u_N(t)$ to $u(t)$. Written out in full these equations (in an obvious notation) are

$$L_t(t, z_i) = [\sqrt{[\sigma(z_i)]} - K(t, z_i)]$$

$$\times \left[\sum_{j=1}^{N} w_j \sqrt{[\sigma(z_j)]} L(t, z_j) \right] - z_i L(t, z_i), \qquad (108)$$

$$L(a, z_i) = \sqrt{[\sigma(z_i)]}, \qquad i = 1, 2, \ldots, N, \qquad (109)$$

$$K_t(t, z_i) = \left[\sum_{j=1}^{N} w_j \sqrt{[\sigma(z_j)]} L(t, z_j) \right] L(t, z_i), \qquad (110\text{-}1)$$

$$K(a, z_i) = 0, \qquad i = 1, 2, \ldots, N, \qquad (110\text{-}2)$$

$$\rho_t(t, z_i) = [\sqrt{[\sigma(z_i)]} + K(t, z_i)] \left[\sum_{j=1}^{N} w_j \sqrt{[\sigma(z_j)]} \rho(t, z_j) \right]$$

$$- z_i \rho(t, z_i) + g(t)[K(t, z_i) + \sqrt{[\sigma(z_i)]}], \qquad (111)$$

$$\rho(a, z_i) = 0, \qquad i = 1, 2, \ldots, N, \qquad (112)$$

$$-\delta_t(t, z_i) = \sqrt{[\sigma(z_i)]} \left[\sum_{j=1}^{N} w_j K(t, z_j) \delta(t, z_j) \delta(t, z_j) \right.$$

$$\left. + \sqrt{[\sigma(z_j)]} \rho(t, z_j) \right] - z_i \delta(t, z_i) + \sqrt{[\sigma(z_i)]} g(t), \qquad (113)$$

$$\delta(b, z_i) = 0, \qquad i = 1, 2, \ldots, N, \qquad (114)$$

$$u_N(t) = g(t) + \sum_{j=1}^{N} w_j K(t, z_j) \delta(t, z_j)$$

$$+ \sqrt{[\sigma(z_j)]} [\delta(t, z_j) + \rho(t, z_j)]. \qquad (115)$$

In Refs. 4 and 9, (1) was treated by first replacing $K(t-s)$ by an approximate kernel of the form (57) and then utilizing the usual finite-dimensional Riccati transformation (Ref. 4) to obtain an equivalent initial-value problem. Using the results of Theorem 5.2 and Eqs. (108)–(115), we now show that both procedures lead to the same numerical approximation for $u(t)$. This type of *commutativity* result appears to be typical of imbedding methods and is analogous to those proved by us in Ref. 32. for parameter-imbedding techniques.

Theorem 5.3. Let $u_N(t)$ be given by Eq. (115) and $\hat{u}_N(t)$ by the solution (assumed unique) to

$$\hat{u}_N(t) = g(t) + \int_a^b K_N(t-s) \hat{u}_N(s) \, ds, \qquad (116)$$

where

$$\hat{K}_N(t-s) = \sum_{i=1}^{N} w_i \sigma(z_i) \exp[-z_i|t-s|], \tag{117}$$

and

$$w_i, \quad i = 1, 2, \ldots, N, \quad \text{and} \quad z_i, \quad i = 1, 2, \ldots, N,$$

are the weights and nodes of the quadrature rule given in (107). Then,

$$\hat{u}_N(t) = u_N(t), \quad t \in [a, b].$$

The reason for which we refer to Theorem 5.3 as a *commutativity* result is that it can be summarized succinctly in the form of the commutative diagram shown in Fig. 1.

Proof. The proof essentially follows the diagram in Fig. 1. First, the initial-value-problem formulation of (116) given in Refs. 4 and 9 is reviewed. Letting

$$\hat{\gamma}_i(t) = \int_a^t \sigma_i \exp[-z_i(t-s)]\hat{u}_N(s) \, ds, \quad i = 1, 2, \ldots, N, \tag{118}$$

$$\hat{\delta}_i(t) = \int_t^b \sigma_i \exp[-z_i(s-t)]\hat{u}_N(s) \, ds, \quad i = 1, 2, \ldots, N, \tag{119}$$

where

$$\sigma_i = \sqrt{[w_i \sigma(z_i)]},$$

it was shown that $\hat{u}_N(t)$ has the representation

$$\hat{u}_N(t) = g(t) + \sum_{i=1}^{N} \sigma_i(\hat{\gamma}_i(t) + \hat{\delta}_i(t)), \tag{120}$$

where $\hat{\gamma}_i(t)$ and $\hat{\delta}_i(t)$ solve two-point boundary-value problems analogous to (76)–(77). The finite-dimensional Riccati transformation was then applied to give an initial-value formulation. Application of Casti's theorem (Ref. 30) then enabled us to show that $\hat{u}_N(t)$ could be obtained by solving initial-value problems similar in form to Eqs. (108)–(114). Denoting by $\hat{L}_i(t, z_i)$, $\hat{K}(t, z_i)$, $\hat{\rho}_i(t, z_i)$, $\hat{\delta}_i(t, z_i)$, $i = 1, 2, \ldots, N$, the functions corresponding to $L(t, z_i)$, $K(t, z_i)$, $\rho(t, z_i)$, $\delta(t, z_i)$ of this chapter, one can easily show that the following relations hold:

$$\hat{L}(t, z_i) = \sqrt{w_i} L(t, z_i), \qquad \hat{K}(t, z_i) = \sqrt{w_i} K(t, z_i), \tag{121}$$

$$\hat{\rho}(t, z_i) = \sqrt{w_i} \rho(t, z_i), \qquad \hat{\delta}(t, z_i) = \sqrt{w_i} \delta(t, z_i). \tag{122}$$

Since the proofs of (121) and (122) involve basically the same argument, we will illustrate that for $L(t, z_i)$ and $K(t, z_i)$ and leave the remaining

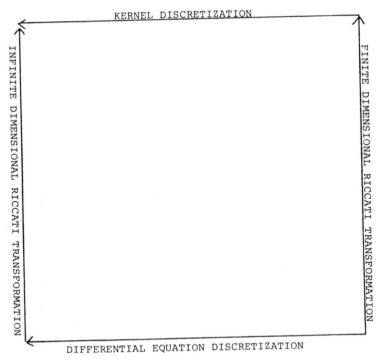

Fig. 1. Commutativity of discretization and Riccati transformations.

details to the reader. From Eq. (82) of Ref. 4, it is seen that $\hat{L}(t, z_i)$ and $\hat{K}(t, z_i)$ satisfy

$$\hat{L}_t(t, z_i) = [\sqrt{[w_i \sigma(z_i)]} - \hat{K}(t, z_i)] \sum_{j=1}^{N} \sqrt{[\sigma(z_j)w_j]}\hat{L}(t, z_j) - z_i\hat{L}(t, z_i),$$
$$\tag{123}$$

$$\hat{L}(a, z_i) = \sqrt{[w_i \sigma(z_i)]}, \quad = 1, 2, \ldots, N, \tag{124}$$

$$\hat{K}_t(t, z_i) = \left[\sum_{j=1}^{N} \sqrt{[w_j \sigma(z_j)]}\hat{L}(t, z_j)\right]\hat{L}(t, z_i), \tag{125}$$

$$\hat{K}(a, z_i) = 0, \quad i = 1, 2, \ldots, N. \tag{126}$$

Multiplying both sides of (108)–(110) by $\sqrt{w_i}$ shows that $\sqrt{w_i}L(t, z_i)$ and $\sqrt{w_i}K(t, z_i)$ satisfy (123)–(126). Since the right-hand sides of (123) and (125) are locally Lipshitz, it follows by the usual uniqueness theorem for initial-value problems that

$$\hat{K}(t, z_i) = \sqrt{w_i}K(t, z_i),$$

and

$$\hat{L}(t, z_i) = \sqrt{w_i} L(t, z_i), \qquad i = 1, 2, \ldots, N.$$

Using (121)–(122) in (120) shows that

$$\hat{u}_N(t) = u_N(t).$$

Thus, the two algorithms are equivalent, although they require slightly different internal computations. □

Theorem 5.4. Assume that

$$\lim_{N \to \infty} \left| \int_A \sigma(z) \exp[-z|t-s|] \, dz - \sum_{i=1}^{N} w_i \sigma(z_i) \exp[-z_i|t-s|] \right| = 0,$$

(127)

uniformly for t and s in $[a, b]$. Then, for N sufficiently large, the differential equations (108)–(114) have unique solutions for $t \in [a, b]$, and the function $u_N(t)$ converges uniformly to $u(t)$, the solution of (1) with kernel given by (58).

Proof. Let $C[a, b]$ be the space of continuous functions on $[a, b]$ with the uniform norm. Let K and K_N be the operators defined by the kernels $K(|t-s|)$ and $K_N(|t-s|)$, respectively. Then, a simple computation using (127) shows that K_N converges in operator norm to K. It then follows from a standard theorem (Ref. 3) that, for N sufficiently large, (116) has a unique solution $\hat{u}_N(t)$ which converges uniformly to $u(t)$, $t \in [a, b]$. From results in Ref. 4, it is known that $\hat{\gamma}(t, z_i)$ and $\hat{\delta}(t, z_i)$ satisfy a two-point boundary-value problem equivalent to (116). This is essentially Theorem 3.1 applied to this particular case. This problem was solved using finite-dimensional Riccati transformation leading to the system for $\hat{L}(t, z_i)$, $\hat{K}(t, z_i)$, $\hat{\rho}(t, z_i)$, $\hat{\delta}(t, z_i)$ in Ref. 4. From known theorems on invariant imbedding (Ref. 36), it follows that this system has a unique solution for $t \in [a, b]$. From Theorem 5.3, it now follows that Eqs. (108)–(114) have a unique solution for $t \in [a, b]$ and N sufficiently large, so that $\hat{u}_N(t)$ is well defined. Again using Theorem 5.3,

$$u_N(t) = \hat{u}_N(t), \qquad t \in [a, b],$$

and the theorem is proved. □

The above result is similar to the one established by Casti in Ref. 26 for the parameter-imbedding approach to (1) and (58).

Since Theorem 5.3 shows that, for numerical purposes, it suffices to consider semidegenerate kernels of the form (57), rather than the more

general ones given by (58), we turn our attention briefly to the consideration of more general convolution kernels which can be adequately approximated by semidegenerate ones. This topic was covered extensively in Ref.
9, so that only a short exposition is given here.

Consider the kernel

$$K(t-s) = \begin{cases} \sum_{i=1}^{N} a_i k_i(t-s) + E_N, & a \le s \le t, \\ \\ \sum_{i=1}^{M} b_i l_i(t-s) + E_M, & t < s \le b, \end{cases} \tag{128}$$

where E_N and E_M represent approximation errors. If one assumes that E_N
and E_M are small, then we solve (1) with the approximate kernel

$$\hat{K}(t-s) = \begin{cases} \sum_{i=1}^{N} a_i k_i(t-s), & a \le s \le t, \\ \\ \sum_{i=1}^{N} b_i l_i(t-s), & t < s \le b. \end{cases} \tag{129}$$

In addition, it is assumed that

$$k_i(t), \qquad i = 1, 2, \ldots, N, \qquad \text{and } l_i(t), \qquad i = 1, 2, \ldots, M,$$

solve the differential equations

$$k_i'(t) = \sum_{j=1}^{N} a_{ij} k_j(t), \qquad i = 1, 2, \ldots, N, \tag{130}$$

$$l_i'(t) = \sum_{j=1}^{M} b_{ij} l_i(t), \qquad i = 1, 2, \ldots, M. \tag{131}$$

Picking $k_i(t)$ and $l_i(t)$ to satisfy (130) and (131) is reasonable, since this
allows one to approximate $K(t-s)$ by a linear combination of polynomials,
sines, cosines, and exponentials; therefore, it contains many of the standard approximation methods in analysis (Ref. 1). In particular note that,
for (58), we have

$$k_i(t) = \exp(-z_i t), \qquad i = 1, 2, \ldots, N,$$

$$l_i(t) = \exp(z_i t), \qquad i = 1, 2, \ldots, N,$$

and it is easily verified that (129)–(131) is a semidegenerate kernel. In Ref.
9, it was shown that (1) with (129) is equivalent to a two-point boundary-
value problem with constant coefficients. Initial-value methods for solving
these were discussed, and it was shown that, under appropriate conditions,

analogous to the positivity of $\sigma(z)$, that (1) with (129) could be solved by the integration of at most $3(N+M)$ initial-value problems.

Stability. We close this section with a short discussion of numerical methods needed to solve the boundary-value problems that one encounters in our approach. In previous papers on this topic (Refs. 4 and 9), and in the above work, we have emphasized the use of initial-value methods based on the Riccati transformation. This was done in large part to compare our methods to those derived from *interval-length imbedding*, for which there is a sizeable literature (Refs. 3, 8, 27). In particular, our principal motivation has been to see if one could reduce the number of equations that have to be solved in order to approximate $u(t)$. Equation counts given in Refs. 4 and 9 indicate that the algorithms discussion in Theorems 5.1, 5.2, and 5.3 typically require 50–60% fewer equations than the interval imbedding approaches.

However, a small number of equations to be solved is not in itself a sufficient reason to consider a method as being useful. An important consideration for any numerical method is its stability. As is well known, linear two-point boundary-value problems can be very difficult to solve when they are unstable (Ref. 34), that is, when the coefficient matrices have widely separated real eigenvalues of opposite sign. Many devices have been put forth to overcome these difficulties. Among these are the orthonormalization method of Gudonov (Ref. 37) as implemented by Scott and Watts (Ref. 38), finite-difference methods (Ref. 39), the Riccati transformation (Refs. 29, 34) and interval-length imbedding (Refs. 28, 36). It is from the point of view of selecting stable algorithms that the choice of the Riccati transformation can be particularly appropriate. To elaborate, consider the following simple equation:

$$u(t) = g(t) + \sigma^2 \int_a^b \exp[-z|t-s|]u(s)\,ds, \qquad z > 0. \qquad (132)$$

Equation (132) can be considered as a special case of (1) and (58), where

$$\sigma(z') = \sigma^2 \delta(z' - z)$$

and $\delta(z)$ is the Dirac delta function. Letting

$$\alpha(t) = \sigma \int_a^t \exp[-z(t-s)]u(s)\,ds, \qquad \beta(t) = \sigma \int_t^b \exp[-z(s-t)]u(s)\,ds,$$

it is seen that

$$u(t) = g(t) + \alpha(t) + \beta(t),$$

where

$$\alpha'(t) = (\sigma^2 - z)\alpha(t) + \sigma^2\beta(t) + \sigma g(t), \qquad \alpha(a) = 0, \qquad (133)$$

$$\beta'(t) = -\sigma^2\alpha(t) - (\sigma^2 - z)\beta(t) - \sigma g(t), \qquad \beta(b) = 0. \qquad (134)$$

Since (133)–(134) have constant coefficients, the nature of the solution will be determined by the eigenvalues of the matrix

$$A = \begin{bmatrix} \sigma^2 - z & \sigma^2 \\ -\sigma^2 & -(\sigma^2 - z) \end{bmatrix},$$

which are solutions to

$$\lambda^2 = (\sigma^2 - z)^2 - \sigma^4.$$

If

$$(\sigma^2 - z)^2 - \sigma^4 < 0,$$

then the boundary-value problem is stable, and most boundary-value solvers should have little difficulty in solving it. However, if

$$z \geq 2\sigma^2,$$

then

$$\lambda^2 \geq 0$$

and the homogeneous equations

$$\alpha'(t) = (\sigma^2 - z)\alpha(t) + \sigma^2\beta(t), \qquad (135)$$

$$\beta'(t) = -\sigma^2\alpha(t) - (\sigma^2 - z)\beta(t),$$

will have the solutions of the form

$$(\alpha(t), \beta(t)) = \exp(\lambda t)\xi + \exp(-\lambda t)\eta, \qquad (\xi, \eta) \in R^2. \qquad (136)$$

If z is large, for example equal to $k\sigma^2$, then

$$\lambda = \pm\{\sqrt{[k(k-2)]}\}\sigma^2,$$

and the problem becomes unstable. In this case, simple methods like superposition may fail because the factor $\exp(\lambda t)$ will magnify any integration errors (Ref. 34). However, the Riccati equation associated with (133)–(134) has the solution

$$R(t) = [\mu_+ - \mu_-\phi(t)]/[1 - (\mu_+/\mu_-)\phi(t)], \qquad (137)$$

where

$$\phi(t) = \exp(\sigma^2 t/A) \quad \text{and} \quad A = 1/\mu_+ - \mu_-,$$

and is thus bounded for all $t \geq 0$. Since the remaining equations needed to determine $u(t)$ are all of the first order no *spurious roots* will be encountered to cause instability. From this standpoint, then, the Riccati transformation stabilizes the potentially unstable boundary-value problem (133)–(134). Although the above is far from a complete analysis of the stability of the algorithms developed above, it should offer some insight into the nature of the numerical behavior of our procedures. We expect that a more complete discussion of this problem can be based on the work of Boland and Nelson (Ref. 40).

6. Extensions and Generalizations

We conclude this chapter with a brief discussion of some possible extensions of the results in Sections 3–5 to several other classes of problems. First, consider generalizations to nonlinear equations. Since the calculations are similar to those given above, we merely quote our results and leave the details of the proofs to the reader.

Theorem 6.1. Let $u(t)$ solve

$$u(t) = g(t) + \lambda \int_a^b K(t, s) f(s, u(s)) \, ds, \tag{138}$$

where $g(t)$, $K(t, s)$, λ, and $[a, b]$ satisfy the conditions of Theorem 3.1. Assume that $f(t, u)$ is continuous for $t \in [a, b]$ and

$$|u| \leq M.$$

In addition, we assume that (138) has a unique continuous solution. Then, $u(t)$ has the representation

$$u(t) = g(t) + \lambda \left[\sum_{i=1}^N a_i(t) \alpha_i(t) + \sum_{i=1}^M c_i(t) \beta_i(t) \right], \tag{139}$$

where

$$\alpha_i(t), \quad i = 1, 2, \ldots, N, \quad \text{and} \quad \beta_i(t), \quad i = 1, 2, \ldots, M,$$

are given by

$$\alpha_i(t) = \int_a^t b_i(s) f(s, u(s)) \, ds, \tag{140}$$

$$\beta_i(t) = \int_t^b d_i(s) f(s, u(s)) \, ds, \tag{141}$$

and solve the two-point boundary value problem

$$\alpha_i'(t) = b_i(t)f\left[t, g(t) + \lambda\left[\sum_{j=1}^{N} a_j(t)\alpha_j(t) + \sum_{j=1}^{M} c_j(t)\beta_j(t)\right]\right], \quad (142)$$

$$\alpha_i(a) = 0, \qquad i = 1, 2, \ldots, N, \quad (143)$$

$$\beta_i'(t) = -d_i(t)f\left[t, g(t) + \lambda\left[\sum_{j=1}^{N} a_j(t)\alpha_j(t) + \sum_{j=1}^{M} c_j(t)\beta(t)\right]\right], \quad (144)$$

$$\beta_i(b) = 0, \qquad i = 1, 2, \ldots, M. \quad (145)$$

Conversely, if $\alpha_i(t)$, $\beta_i(t)$ solve (142)–(145), then $u(t)$ given by (139) is a solution of (138). $\qquad\square$

Theorem 6.2. Let $u(t)$ solve the equation

$$u(t) = g(t) + \lambda \int_a^b K(t-s)f(s, u(s)) \, ds, \quad (146)$$

where $g(t)$ is continuous, λ is complex, $[a, b]$ is compact, and $K(t-s)$ has the representation given by Eqs. (129)–(131). $u(t)$ and $f(u)$ satisfy the assumptions of Theorem 6.1. Then, $u(t)$ has the representation

$$u(t) = g(t) + \lambda\left[\sum_{i=1}^{N} a_i\alpha_i(t) + \sum_{i=1}^{M} b_i\beta_i(t)\right], \quad (147)$$

where

$$\alpha_i(t) = \int_a^t k_i(t-s)f(s, u(s)) \, ds, \qquad i = 1, 2, \ldots, N, \quad (148)$$

$$\beta_i(t) = \int_t^b l_i(t-s)f(s, u(s)) \, ds, \qquad i = 1, 2, \ldots, M, \quad (149)$$

and $\alpha_i(t)$, $\beta_i(t)$ solve the two-point boundary-value problem

$$\alpha_i'(t) = k_i(0) + f\left[t, g(t) + \lambda\left[\sum_{j=1}^{N} a_j\alpha_j(t) + \sum_{j=1}^{M} b_j\beta_j(t)\right]\right]$$

$$+ \sum_{j=1}^{N} a_{ij}\alpha_i(t), \qquad \alpha_i(a) = 0, \qquad i = 1, 2, \ldots, N, \quad (150)$$

$$\beta_i'(t) = -l_i(0)f\left[t, g(t) + \sum_{j=1}^{N} a_j\alpha_j(t) + \sum_{j=1}^{M} b_j\beta_j(t)\right]$$

$$+ \sum_{j=1}^{M} b_{ij}\beta_j(t), \qquad \beta_i(b) = 0, \qquad i = 1, 2, \ldots, M. \quad (151)$$

$\qquad\square$

One can now apply a variety of techniques for the solution of nonlinear boundary-value problems to solve (142)–(145) and (150)–(151), and thus obtain $u(t)$. Such techniques include finite-difference methods (Ref. 39), shooting methods (Ref. 16), quasilinearization methods (Ref. 34) and invariant-imbedding methods (Refs. 22, 29, 36). For the particular case of Volterra equations, calculations based on Theorem 6.1 are reported by Bownds and on Theorem 6.2 by us in this issue (Refs. 13, 15). The same comments concerning the use of Theorems 6.1 and 6.2 to solve problems with general kernels, made in Section 5, apply in this case as well.

6.1. Integrodifferential Equations.
In Section 1, it was stated that the methods of this chapter derive essentially from the work of Bownds and Wood on Volterra equations. In turn, their work was motivated in part by the desire to find an efficient procedure for solving the predator–prey equations of Volterra with interactions that depended on the past histories of the species (Ref. 41). Mathematically, this gives rise to the problem of solving an initial-value problem for an integrodifferential equation. Their approach was to convert this equation to an equivalent Volterra one, which could then be solved by the methods outlined in their papers (Refs. 15, 42). We now demonstrate that, by a slight modification of the ideas presented above, such equations may be converted directly to initial-value problems for a system of ordinary differential equations, without requiring a preliminary conversion to an integral equation.

Let L be a linear differential operator of the form

$$Lu(t) = \sum_{k=0}^{p} l_k(t) u^{(k)}(t), \tag{152}$$

where $\{l_k(t)\}$ are continuous functions on $[a, b]$. We consider solving integrodifferential equations of the form

$$Lu(t) = \int_{a(t)}^{t} K(t, s) f(s, u(s), u^{(1)}(s), \ldots, u^{(p-1)}(s)) \, ds. \tag{153}$$

Although our methods can handle fairly general $a(t)$'s, we restrict ourselves to the particular cases

$$a(t) = a$$

and

$$a(t) = t - a,$$

with $a > 0$. These situations are the ones that seem to occur most often in practice (Refs. 42–43). Since the two situations give rise to distinctly different types of Cauchy problems, our results are stated separately for each.

Theorem 6.3. Let $u(t)$ solve (153) with

$$a(t) = a$$

and satisfy the initial conditions

$$u^{(j)}(a) = u_j, \qquad j = 0, 1, 2, \ldots, p-1.$$

Let $K(t, s)$ be semidegenerate and $f(t, u_0, u_1, u_2, \ldots, u_{p-1})$ be continuous for $a \leq t \leq b$ and

$$\sum_{i=0}^{p-1} |u_i| \leq M.$$

Define $\alpha_k(t)$, $k = 1, 2, \ldots, N$, by

$$\alpha_k(t) = \int_a^t b_k(s) f(s, u(s), u^{(1)}(s), \ldots, u^{(p-1)}(s)) \, ds. \qquad (154)$$

Then, $u(t)$ and $\alpha_k(t)$, $k = 1, 2, \ldots, N$, solve the initial-value problem

$$Lu(t) = \sum_{k=1}^N a_k(t)\alpha_k(t), \qquad u^{(j)}(a) = u_j, \qquad j = 1, 2, \ldots, p-1,$$

$$(155)$$

$$\alpha'_k(t) = b_k(t) f(t, u(t), \ldots, u^{(p-1)}(t)), \qquad \alpha_k(a) = 0, \qquad k = 1, 2, \ldots, N.$$

$$(156)$$

Proof. Differentiate (154). □

Theorem 6.4. Let $u(t)$ satisfy (153), with

$$a(t) = t - a.$$

Assume that $K(t, s)$ and $f(t, u_0, u_1, \ldots, u_{p-1})$ satisfy the same conditions as in Theorem 6.3. Define $\alpha_k(t)$, $k = 1, 2, \ldots, N$, by

$$\alpha_k(t) = \int_{t-a}^t b_k(s) f(s, u(s), \ldots, u^{(p-1)}(s)) \, ds. \qquad (157)$$

Then, $u(t)$, $\alpha_k(t)$, $k = 1, 2, \ldots, N$, satisfy the initial-value problem

$$Lu(t) = \sum_{k=1}^N a_k(t)\alpha_k(t), \qquad (158)$$

$$u^{(j)}(t) = u_j(t), \qquad 0 \le t \le a, \qquad j = 0, 1, 2, \ldots, p-1, \tag{159}$$

$$\alpha'_k(t) = b_k(t) f(t, u(t) u^{(1)}(t), \ldots, u^{(p-1)}(t))$$
$$\qquad - b_k(t-a) f(t, u(t-a), u^{(1)}(t-a), \ldots, u^{(p-1)}(t-a)), \tag{160}$$

$$\alpha_k(a) = \int_0^a b_k(s) f(t, u_0(s), \ldots, u_{p-1}(s)) \, ds, \qquad k = 1, 2, \ldots, N. \tag{161}$$

Proof. Differentiate (157) to get (160), and evaluate at $t = a$ using (159) to give (161). □

Note that the system (158)–(161) is well posed and can be solved for $t > a$ using the method of steps (Ref. 45).

A similar treatment may be given for integrodifferential equations of Fredholm type:

$$Lu(t) = \int_b^a K(t, s) f(s, u(s), u^{(1)}(s), \ldots, u^{(p-1)}(s)) \, ds$$

with boundary conditions given at

$$t = a \quad \text{and} \quad t = b,$$

and $K(t, s)$ semidegenerate. As in the case of pure integral equations, general kernels may be treated by degenerate or semidegenerate approximations and utilizations of the above theorems.

6.2. Wiener–Hopf Equations. As we stated in Section 3, some of the conditions of Theorems 3.1 and 5.1 may be relaxed and our methods still apply. For example, integral equations of the form

$$u(t) = g(t) + \int_0^\infty K(|t-s|) u(s) \, ds$$

can be solved by looking for steady-state solutions of the associated Riccati equations. This may be accomplished in a variety of ways: for example, by integrating them from 0 to large t and then applying the algorithms of Section 5; or, more directly, it is possible to introduce the analogues of Chandrasekhar's H functions (Refs. 3, 7, 45) and proceed along lines well known in transport theory.

6.3. Eigenvalue Problems. The last class of problems that we comment on is that of finding the eigenvalues and eigenfunctions of $K(t, s)$. Here, the results of Section 4 are immediately applicable, since the eigenvectors and eigenvalues may be found by solving the homogeneous boundary-value problem 23: shooting methods, finite-difference methods, or imbedding methods may be used to locate them. Here again, we expect to give more complete coverage of this topic in later work.

7. Conclusions

We have shown how one can develop the theory of Fredholm integral equations with semidegenerate and generalized semidegenerate kernels by converting them to equivalent boundary-value or initial-value problems, or both, for ordinary differential equations. For numerical purposes, these may be solved by existing efficient codes (Ref. 38). We have also given a theoretical comparison between our approach and those of Rall and other authors based on imbedding methods. As in the work of Bownds and ourself on the special case of Volterra equations, we believe the chief advantage is to be found in the reduction of the amount of work needed to evaluate the kernel (Ref. 13). Much more work remains to be done. In particular, detailed comparisons need to be made between our methods, imbedding methods, and the more traditional techniques based on solving linear algebraic equations. We are currently developing this program.

References

1. ATKINSON, K. E., *A Survey of Numerical Methods for the Solution of Fredholm Integral Equations of the Second Kind*, SIAM, Philadelphia, Pennsylvania, 1976.
2. ANSELONE, P. M., *Collectively Compact Operator Approximation Theory and Applications to Integral Equations*, Prentice-Hall, Englewood Cliffs, New Jersey, 1971.
3. KAGIWADA, H., and KALABA, R. E., *Integral Equations via Imbedding Methods*, Addison-Wesley Publishing Company, Reading, Massachusetts, 1974.
4. GOLBERG, M. A., *The Conversion of Fredholm Integral Equations to Equivalent Cauchy Problems*, Applied Mathematics and Computation, Vol. 2, pp. 1–18, 1976.
5. AALTO, S. K., *Reduction of Fredholm Integral Equations with Green's Function Kernels to Volterra Equations*, Oregon State University, MS Thesis, 1966.
6. RALL, L. B., *Resolvent Kernels of Green's Function, Kernels and Other Finite Rank Modifications of Volterra and Fredholm Kernels*, Chapter 11, this volume.
7. CHANDRASEKHAR, S., *Radiative Transfer*, Clarendon Press, Oxford, England, 1950.

8. WING, G. M., *On Certain Fredholm Integral Equations Reducible to Initial Value Problems*, SIAM Review, Vol. 9, pp. 655–670, 1967.
9. GOLBERG, M. A., *The Conversion of Fredholm Integral Equations to Equivalent Cauchy Problems II—Computation of Resolvents*, Applied Mathematics and Computation, Vol. 3, No. 1, 1977.
10. GOURSAT, E., *Determination de la Resolvante d'un d'Equations*, Bulletin des Sciences Mathematiques, Vol. 57, pp. 144–150, 1933.
11. BOWNDS, J. M., and CUSHING, J. M., *A Representation Formula for Linear Volterra Integral Equations*, Bulletin of the American Mathematical Society, Vol. 79, pp. 532–536, 1973.
12. CASTI, J., and KALABA, R. E., *Imbedding Methods in Applied Mathematics*, Addison-Wesley Publishing Company, Reading, Massachusetts, 1973.
13. BOWNDS, J. M., *On An Initial Value Method for Quickly Solving Volterra Integral Equations*, Chapter 9, this volume.
14. BOWNDS, J. M., *A Modified Galerkin Approximation Method for Volterra Equations with Smooth Kernels* (to appear).
15. GOLBERG, M. A., *On a Method of Bownds for Solving Volterra Integral Equations*, Chapter 10, this volume.
16. ROBERTS, S. M., and SHIPMAN, S. S., *Two-Point Boundary Value Problems: Shooting Methods*, American Elsevier Publishing Company, New York, 1972.
17. KELLER, H. B., *Numerical Methods for Two-Point Boundary Value Problems*, Blaisdell Publishing Company, Waltham, Massachusetts, 1968.
18. HILLE, E., *Lectures on Ordinary Differential Equations*, Addison-Wesley Publishing Company, Reading, Massachusetts, 1969.
19. LOVITT, W. V., *Linear Integral Equations*, Dover Publications, New York, New York, 1950.
20. NASHED, M. Z., Editor, *Generalized Inverses and Their Applications*, Academic Press, New York, New York, 1976.
21. STRANG, G., *Linear Algebra and its Applications*, Academic Press, New York, New York, 1976.
22. MILLER, R. K., *Nonlinear Volterra Integral Equations*, W. A. Benjamin, Menlo Park, California, 1971.
23. GOLBERG, M. A., *A Comparison of Several Numerical Methods for Solving Fredholm Integral Equations* (to appear).
24. BOWNDS, J. M., and CUSHING, J. M., *Some Stability Criteria for Linear Systems of Volterra Integral Equations*, Funkcialaj Ekvacioj, Vol. 15, pp. 101–117, 1972.
25. BELLMAN, R. E., *A New Approach to the Numerical Solution of a Class of Linear and Nonlinear Integral Equations of Fredholm Type*, Proceedings of the National Academy of Sciences, USA, Vol. 54, pp. 1501–1503, 1965.
26. CASTI, J., *Invariant Imbedding and the Solution of Fredholm Integral Equations of Displacement Type*, The Rand Corporation, Report No. R-639-PR, 1971.
27. SCOTT, M. R., *A Bibliography on Invariant Imbedding and Related Topics*, Sandia Laboratories, Report No. SLA-74-0284, 1974.

28. GOLBERG, M. A., *Invariant Imbedding and Riccati Transformations*, Applied Mathematics and Computation, Vol. 1, No. 1, pp. 1–24, 1975.
29. MEYER, G. H., *Initial-Value Methods for Boundary-Value Problems*, Academic Press, New York, New York, 1973.
30. CASTI, J., *Matrix Riccati Equations, Dimensional Reduction, and Generalized X and Y Functions*, Utilitas Mathematica, Vol. 6, pp. 95–110, 1974.
31. KAILATH, T., *Some New Algorithms for Recursive Estimation in Constant Linear Systems*, IEEE Transactions on Information Theory, Vol. 19, pp. 750–760, 1973.
32. GOLBERG, M. A., *The Equivalence of Several Initial Value Methods for Solving Fredholm Integral Equations*, Journal of Computer and System Sciences, Vol. 6, pp. 291–297, 1972.
33. GOLBERG, M. A., *On the Relation Between a Class of Convolution Integral Equations and Differential Equations in a Hilbert Space* (to appear).
34. SCOTT, M. R., *Invariant Imbedding and its Application to Ordinary Differential Equations*, Addison-Wesley Publishing Company, Reading, Massachusetts, 1973.
35. SHUMITZKY, A., *On the Equivalence Between Matrix Riccati Equations and Fredholm Resolvents*, Journal of Computer and System Science, Vol. 3, pp. 76–87, 1968.
36. BELLMAN, R. E., and WING, G. M., *Introduction to Invariant Imbedding*, John Wiley and Sons, New York, New York, 1975.
37. GUDONOV, S. K., *On the Numerical Solution of Boundary Value Problems for Systems of Linear Ordinary Differential Equations*, Uspekhi Matematika Nauk, Vol. 16, pp. 171–174, 1961.
38. SCOTT, M. R., and WATTS, H., *SUPORT—A Computer Code for Two-Point Boundary Value Problems Via Orthonormalization*, Sandia Laboratories, Report No. SLA-75-0198, 1975.
39. AZIZ, A. K., *Numerical Solutions of Boundary-Value Problems for Ordinary Differential Equations*, Academic Press, New York, New York, 1975.
40. BOLAND, R., and NELSON, P., *Critical Lengths by Numerical Integration of the Associated Riccati Equations to Singularity*, Applied Mathematical and Computation, Vol. 1, pp. 67–82, 1975.
41. BOWNDS, J. M., Private Communication, 1977.
42. BOWNDS, J. M., and WOOD, B., *On Numerically Solving Nonlinear Volterra Integral Equations with Fewer Computations*, SIAM Journal on Numerical Analysis, Vol. 13, pp. 705–719, 1976.
43. NOBLE, B., *A Bibliography on the Numerical Solution of Integral Equations*, Mathematics Research Center, Madison, Wisconsin, 1971.
44. BELLMAN, R. E., and COOKE, K., *Differential Difference Equations*, Academic Press, New York, New York, 1963.
45. CASTI, J., *Optimal Linear Filtering and Neutron Transport: Isomorphic Theories*, IIASA Research Memorandum, No. RM-75-25, 1975.

Index